Groundwater and Ecosystems

Selected papers on hydrogeology

18

Series Editor: Dr. Nick S. Robins
Editor-in-Chief IAH Book Series, British Geological Survey, Wallingford, UK

INTERNATIONAL ASSOCIATION OF HYDROGEOLOGISTS

Groundwater and Ecosystems

Editors:

Luís Ribeiro & Tibor Y. Stigter
Instituto Superior Técnico, Lisbon, Portugal

António Chambel
Universidade de Évora, Évora, Portugal

M. Teresa Condesso de Melo
Instituto Superior Técnico, Lisbon, Portugal

José Paulo Monteiro
Algarve University, Faro, Portugal

Albino Medeiros
Grandewater – Hidrogeologia Aplicada, Azeitão, Portugal

CRC Press
Taylor & Francis Group
Boca Raton London New York

CRC Press is an imprint of the
Taylor & Francis Group, an **informa** business
A BALKEMA BOOK

Cover illustration:

Groundwater from mountain bog systems with turf and water acidified by organic acids, Chaloupská Sla, Šumava Mountains, Czech Republic. In relatively flat mountainous areas, and due to the less transmissive media of the underlying crystalline rocks, the water accumulates, forming swamps where bog mountain systems form under the right conditions of recharge (Photo: António Chambel)

CRC Press
Taylor & Francis Group
6000 Broken Sound Parkway NW, Suite 300
Boca Raton, FL 33487-2742

First issued in paperback 2019

ISBN-13: 978-1-138-00033-9 (hbk)
ISBN-13: 978-0-367-37991-9 (pbk)

Typeset by MPS Limited, Chennai, India

Library of Congress Cataloging-in-Publication Data

Groundwater and ecosystems / editors: Luís Ribeiro . . . [et al.]. — 1st ed.
p. cm. — (Selected papers on hydrogeology ; 18)
Includes bibliographical references and index.
ISBN 978-1-138-00033-9 (hardback : alk. paper) 1. Groundwater ecology. I. Ribeiro, Luís, 1955– II. Series: Hydrogeology (International Association of Hydrogeologists) ; v. 18.
QH541.5.G76G73 2013
577.6–dc23

2013000440

Published by: CRC Press/Balkema
P.O. Box 11320, 2301 EH, Leiden, The Netherlands
e-mail: Pub.NL@taylorandfrancis.com
www.crcpress.com – www.taylorandfrancis.com

Visit the Taylor & Francis Web site at
http://www.taylorandfrancis.com

and the CRC Press Web site at
http://www.crcpress.com

Table of contents

Preface ix
Foreword xi
J.T. Gurrieri
About the editors xiii
List of contributors xv

1 A toolbox for assessing the ecological water requirements of groundwater dependent ecosystems in Australia 1
R.S. Evans, P.G. Cook, P. Howe, C.A. Clifton & E. Irvine

2 Water table dynamics of a severely eroded wetland system, prior to rehabilitation, Sand River Catchment, South Africa 9
E.S. Riddell, S.A. Lorentz, W.N. Ellery, D. Kotze, J.J. Pretorius & S.N. Ngetar

3 Small-scale water- and nutrient-exchange between lowland River Spree (Germany) and adjacent groundwater 23
J. Lewandowski & G. Nützmann

4 Artificial maintenance of groundwater levels to protect carbonate cave fauna, Yanchep, Western Australia 33
C. Yesertener

5 Spatial and temporal heterogeneity in the flux of organic carbon in caves 47
K.S. Simon, T. Pipan & D.C. Culver

6 The influence of groundwater/surface water exchange on stable water isotopic signatures along the Darling River, NSW, Australia 57
K. Meredith, S. Hollins, C. Hughes, D. Cendón & D. Stone

7 A geochemical approach to determining the hydrological regime of wetlands in a volcanic plain, south–eastern Australia 69
A.B. Barton, A.L. Herczeg, P.G. Dahlhaus & J.W. Cox

8 Mapping surface water-groundwater interactions and associated geological faults using temperature profiling 81
M. Bonte, J. Geris, V.E.A. Post, V. Bense, H.J.A.A. van Dijk & H. Kooi

9 Typology of groundwater-surface water interaction (GSI typology) – with new developments and case study supporting implementation of the EU Water Framework and Groundwater Directives 95
M. Dahl & K. Hinsby

10 Conservation of trial dewatering discharge through re-injection in the Pilbara region, Western Australia 113
L.R. Evans & J. Youngs

11 Nitrogen cycle in gravel bed rivers: The effect of the hyporheic zone 125
A. Marzadri & A. Bellin

12 Groundwater recharge quantification for the sustainability of ecosystems in plains of Argentina 137
M.P. D'Elia, O.C. Tujchneider, M. del C. Paris & M.A. Perez

13 Nutrient sources for green macroalgae in the Ria Formosa lagoon – assessing the role of groundwater 153
T.Y. Stigter, A. Carvalho Dill, E.-j. Malta & R. Santos

14 Relationships between wetlands and the Doñana coastal aquifer (SW Spain) 169
M. Manzano, E. Custodio, E. Lozano & H. Higueras

15 Groundwater dependent ecosystems associated with basalt aquifers of the Alstonville Plateau, New South Wales, Australia 183
R.T. Green, R.S. Brodie & R.M. Williams

16 A shift in the ecohydrological state of groundwater dependent vegetation due to climate change and groundwater drawdown on the Swan Coastal Plain of Western Australia 197
R. Froend, M. Davies & M. Martin

17 Response of wetland vegetation to climate change and groundwater decline on the Swan Coastal Plain, Western Australia: Implications for management 207
R. Loomes, R. Froend & B. Sommer

18 Hydrogeochemical processes in the Pateira de Fermentelos lagoon (Portugal) and their impact on water quality 221
C. Sena & M.T. Condesso de Melo

19 Relationship between dry and wet beach ecosystems and *E. coli* levels in groundwater below beaches of the Great Lakes, Canada 237
A.S. Crowe & J. Milne

20 Surface water, groundwater and ecological interactions along the River Murray. A pilot study of management initiatives at the Bookpurnong Floodplain, South Australia 253
V. Berens, M.G. White, N.J. Souter, K.L. Holland, I.D. Jolly, M.A. Hatch, A.D. Fitzpatrick, T.J. Munday & K.L. McEwan

21 Hydrodynamic interaction between gravity-driven and over-pressured groundwater flow and its consequences on soil and wetland salinisation 267
J. Mádl-Szőnyi, S. Simon & J. Tóth

22 Relationship between certain phreatophytic plants and regional groundwater circulation in hard rocks of the Spanish Central System 281
M. Martín-Loeches & J.G. Yélamos

23 Surface/groundwater interactions: Identifying spatial controls on water quality and quantity in a lowland UK Chalk catchment 295
N.J.K. Howden, H.S. Wheater, D.W. Peach & A.P. Butler

24 Modelling stream-groundwater interactions in the Querença-Silves Aquifer System 307
J.P. Monteiro, L. Ribeiro, E. Reis, J. Martins, J. Matos Silva & N. Salvador

Author index 327
Subject index 329
SERIES IAH-Selected Papers 339

Preface

Recent studies of the Millennium Ecosystem Assessment concluded that over the past five decades the Earth's ecosystems have been rapidly and extensively altered by human activities. This has resulted in a substantial and often irreversible degradation of many of these ecosystems and the essential services they provide. Those depending on groundwater form no exception. Aquifers are facing increasing pressure from water consuming and contaminating activities in various socioeconomic sectors, from industry and agriculture to public supply and recreational activities. This frequently impacts environmental flows that determine the health of groundwater dependent ecosystems (GDEs). Climate change is expected to further contribute to the decrease in groundwater availability, especially in some sensitive regions such as arid and semi-arid areas and coastal and estuarine zones.

This book provides a diverse overview of important studies from across the world, on groundwater-surface water-ecosystem relationships, as well as consequences from human intervention and possible solutions for water resource and environmental management. Among others, the various chapters provide a toolbox for assessing the ecological water requirements for GDEs, relevant case studies on groundwater/surface water ecotones, interactions with vegetation and fauna, as well as the quantitative and qualitative impacts from human activities. The contributions, from Australia (nine studies), Europe (12 studies from nine countries), Argentina, Canada and South Africa, were presented originally at the 35th IAH Congress on Groundwater & Ecosystems, held in Lisbon between 17 and 21 September 2007. They have all been updated by incorporating new results obtained since then. We believe that this book is of interest to anyone dealing with groundwater and its relationship with ecosystems, whether researcher, manager or decision-maker in the field of water and environment, entrepreneur, teacher or student. It provides up-to-date information on crucial factors and parameters that need to be considered when studying groundwater-ecosystem relationships in different environments worldwide. We thank all contributing authors and are truly pleased that together we have managed to bring this challenging endeavour to a successful conclusion.

Luís Ribeiro
Tibor Y. Stigter
António Chambel
M. Teresa Condesso de Melo
José Paulo Monteiro
Albino Medeiros

Preface

Foreword

Groundwater provides a vital source of water and creates a critical habitat for a broad range of species. Groundwater dependent ecosystems (GDEs) comprise a complex and often biodiverse subset of the world's ecosystems and can be found in marine, coastal, lotic (river), lentic (lake), terrestrial, cave, and aquifer environments. These are habitats that must have access to groundwater to maintain both their ecological structure and function and are critical components in the conservation of the earth's aquatic biodiversity. GDEs support a disproportionately large number of plants and animals relative to the area they occupy and also offer multiple ecosystem services to humans, such as clean water, fish and wildlife habitat, storm-water control, ethnobotanical uses, cultural values, and sequestration of carbon. Increasingly, the water needs of communities are in direct conflict with the water needs of natural systems. Human activities have the potential to alter the fluxes, levels and quality of groundwater, which, in turn, can diminish groundwater supported biodiversity that has evolved over millennia.

In the human psyche, GDEs are inconvenient places, obstacles to travel and development, friend to neither man nor horse or pick-up truck. They have been saddled with derogatory names like swamp, quagmire, morass, and waterlogged land. They have historically been places to avoid – mosquito infested, full of malaria and deadly vapours. As such, half of the world's wetlands have been obliterated. Many have been ditched and drained for farming; most of the springs have been developed or trampled by livestock; and mangroves are disappearing to coastal development at an alarming rate.

For this reason, a collection of writings about GDEs is a peculiarity. We have been so focused on surface water flows and fisheries, and groundwater hydraulics and contamination, that the interface with GDEs has been a no-man's land for researchers. Sadly, trends in the scientific community have traditionally not been on a broad interdisciplinary view of the world, rather toward ever more specialisation on narrower topics. After all, which biologist is interested in the physics of water flow, and which hydrologist spends time musing over biological questions? Only recently have we acknowledged that surface water and groundwater are linked components of a hydrologic continuum. Fortunately, times are changing and the inner workings of these rich groundwater supported ecotones are being probed by a host of scientific interdisciplinarians, revealing the secrets of these ecologically productive, hydrogeologically unique, and biologically diverse settings.

Despite the importance of groundwater to biodiversity, the precise relationships remain unclear. The challenge for the scientific community is to improve this understanding and help develop effective approaches for protecting groundwater for biodiversity conservation. Convincing water managers that GDEs are valuable ecosystems is one thing. It is still another to answer the inevitable question, 'How much water can we take from the ecosystem before it ceases to function?' The managers' challenge is to balance the intrinsic water needs of nature with the real human need for water so that we all, Man and nature, benefit. The researchers challenge is to define and quantify the key groundwater-ecology relationships and determine ecological thresholds beyond which GDEs of all kinds may experience irreversible degradation.

Read on and take some time to listen to the lessons these natural systems have to tell us. These authors are pioneers and I thank them for their insight, careful observation, and inspiration.

Joseph T. Gurrieri

Director, IAH Groundwater & Ecosystem Network

About the editors

Luís Ribeiro graduated in Mining Engineering at the *Instituto Superior Técnico* (IST) of the Technical University of Lisbon (UTL) in Lisbon (Portugal) in 1978 and completed his PhD in Mining Engineering (Hydrogeology) at IST-UTL in 1992. He is currently Associate Professor at IST, Director of the Geo-Systems Centre/CVRM, and President of the Portuguese Chapter of the International Association of Hydrogeologists (IAH). He was also President of the Scientific Committee of XXXV IAH Congress "Groundwater and Ecosystems" held in Lisbon in 2007.

Tibor Y. Stigter completed his MSc in Geographical Hydrology in 1997 at the *Vrije Universiteit Amsterdam* (The Netherlands) and his PhD in Engineering Sciences (Hydrogeology) in 2005 at the *Instituto Superior Técnico* (IST), Lisbon (Portugal). He has worked and taught classes at the University of the Algarve (UALG) in Faro (Portugal). He currently holds a Researcher position at IST, working on groundwater, contamination, climate change and sustainability. As a member of the Geo-Systems Centre/CVRM, he works both at IST and UALG, co-ordinating research projects and supervising MSc and PhD students.

Born in 1961 in Sardoal, Portugal, **António Chambel** obtained his PhD in Geology, specialising in Hydrogeology, at the University of Évora, Portugal, in 1999. He is currently teacher in the University of Évora and ERASMUS teacher in the Universities of Prague, Czech Republic, and Huelva, Spain, as well as researcher of the Geophysics Centre of Évora. He is Vice-President of Programme and Science Coordination for the International Association of Hydrogeologists and Technical Director of Hydrogeologists Without Borders. He specialises in groundwater prospecting, hydrogeology of hard rocks, relations between groundwater and ecosystems, hydrogeological mapping and aquifer contamination. He has also focussed on legislation and water and environmental management, including the river basin management plans for Portugal.

M. Teresa Condesso de Melo graduated in Geology from the University of Coimbra (Portugal) in 1992, completed her MSc in Groundwater Hydrology at the *Universitat Politècnica de Catalunya* (UPC, Barcelona, Spain) in 1996 and her PhD in Geosciences (Hydrogeology) with a sandwich PhD project at the Hydrogeology Group of the British Geological Survey (BGS, Wallingford, UK) and the University of Aveiro (UA, Aveiro, Portugal) in 2002. In 2006 she became a Guest Professor of hydrogeology at the UA, and she now holds a Researcher position at the *Instituto Superior Técnico* (IST, Lisbon, Portugal).

José Paulo Monteiro currently holds a position as Professor at the University of the Algarve. He obtained his PhD in hydrogeology in 2001 at the Faculty of Sciences of the University of Neuchâtel (Switzerland) and his MSc in hydrogeology in 1993, at the Faculty of Sciences of the University of Lisbon (Portugal). His main research topics are: modelling and monitoring porous, fractured and karstic coastal aquifers, aquifer-ocean and river-aquifer interfaces; hydraulics of water wells; groundwater dependent ecosystems and integrated water resources management.

Albino Medeiros graduated in Applied Economic Geology at the Faculty of Sciences of the University of Lisbon (Portugal) in 1993 and completed his MSc in Engineering Geology, Faculty of Science and Technology, Universidade Nova de Lisboa – UNL (Portugal) in 1999. In 2005 he became a Guest Professor of Groundwater Modelling at the Earth Sciences Department of UNL. Currently he is a hydrogeologist at *Grandewater – Hidrogeologia Aplicada, Lda*, with development work in Portugal and Africa (Angola and Mozambique).

List of Contributors

BARTON Annette B.	Bureau of Meteorology, Kent Town, Australia
BELLIN Alberto	Department of Civil, Environmental and Mechanical Engineering, University of Trento, Trento, Italy
BENSE Victor	University of East Anglia, School of Environmental Sciences, Norwich, England
BERENS Volmer	South Australian Department for Water
BONTE Matthijs	KWR Watercycle Research Institute, Nieuwegein, The Netherlands
BRODIE Ross S.	Geoscience Australia (GA), Canberra ACT, Australia
BUTLER Adrian P.	EWRE, Department of Civil and Environmental Engineering, Imperial College London, London, UK
CARVALHO DILL Amélia	Geo-Systems Centre/CVRM – University of the Algarve, Faro, Portugal
CENDÓN Dioni	Australian Nuclear Science and Technology Organisation, Institute for Environmental Research, Kirrawee, Australia
CLIFTON Craig A.	Sinclair Knight Merz, Bendigo, Australia
CONDESSO DE MELO M. Teresa	Geo-Systems Centre/CVRM, Instituto Superior Técnico, Lisbon, Portugal
COOK Peter G.	National Centre for Groundwater Research and Training, Flinders University, Adelaide SA, Australia
COX James W.	The University of Adelaide, PMB1 Glen Osmond, Australia
CROWE Allan S.	National Water Research Institute, Environment Canada, Burlington, Ontario, Canada
CULVER David C.	Department of Biology, American University, Washington, D.C., USA
CUSTODIO Emilio	Technical University of Catalonia and International Centre for Groundwater Hydrology, Barcelona, Spain
D'ELIA Mónica P.	Facultad de Ingeniería y Ciencias Hídricas – Universidad Nacional del Litoral. Ciudad Universitaria, Santa Fe, Argentina
DAHL Mette	Sejeroe, Denmark
DAHLHAUS Peter G.	University of Ballarat, Mt Helen, Australia
DAVIES Muriel	Centre for Ecosystem Management, Edith Cowan University, Australia

ELLERY William N.	School of Life and Environmental Sciences, University of KwaZulu-Natal, Durban, South Africa
EVANS Lee R.	Pilbara Iron, Tom Price, Australia
EVANS Richard S.	Sinclair Knight Merz, Melbourne, Australia
FITZPATRICK Andrew D.	CSIRO Exploration and Mining, Western Australia
FROEND Ray	Centre for Ecosystem Management, Edith Cowan University, Joondalup, Western Australia
GERIS Josie	Newcastle University, School of Civil Engineering and Geosciences, Newcastle upon Tyne, United Kingdom
GREEN Richard T.	NSW Office of Water, Department of Primary Industries, Grafton NSW, Australia
HATCH Michael A.	University of Adelaide, South Australia
HERCZEG Andrew L.	CSIRO Land and Water and CRC LEME, PMB 2 Glen Osmond, Australia
HIGUERAS Horacio	Technical University of Cartagena, Cartagena, Spain
HINSBY Klaus	Geological Survey of Denmark and Greenland, Copenhagen, Denmark
HOLLAND Kate L.	CSIRO Land and Water, South Australia
HOLLINS Suzanne	Australian Nuclear Science and Technology Organisation, Institute for Environmental Research, Kirrawee, Australia
HOWDEN Nicholas J.K.	Queen's School of Engineering, University of Bristol, University Walk, Bristol, UK
HOWE Paul	Sinclair Knight Merz, Adelaide, SA, Australia
HUGHES Cath	Australian Nuclear Science and Technology Organisation, Institute for Environmental Research, Kirrawee, Australia
IRVINE Elizabeth	Sinclair Knight Merz, Adelaide, SA, Australia
JOLLY Ian D.	CSIRO Land and Water, South Australia
KOOI Henk	VU University, Faculty of Earth and Life Sciences, Amsterdam, The Netherlands
KOTZE Donovan	Centre for Environment and Development, University of KwaZulu-Natal, Scottsville, South Africa
LEWANDOWSKI Jörg	Leibniz-Institute of Freshwater Ecology and Inland Fisheries, Berlin, Germany
LOOMES Robyn	Department of Water, Government of Western Australia, Perth, Western Australia
LORENTZ Simon A.	School of Bioresources Engineering and Environmental Hydrology, University of KwaZulu-Natal, Scottsville, South Africa
LOZANO Edurne	GEOAQUA S.C., Zaragoza, Spain
MÁDL-SZŐNYI Judit	Department of Physical and Applied Geology, Eötvös Loránd University, Hungary
MALTA Erik-jan	ALGAE – Marine Plant Ecology Research Group, CCMAR, University of the Algarve, Faro, Portugal
MANZANO Marisol	Technical University of Cartagena, Cartagena, Spain
MARTIN Michael	Water Corporation, Western Australia

MARTÍN-LOECHES Miguel	Geology Department, Science Faculty, Alcala University, Madrid, Spain
MARTINS João	EDZ – Environmental Consulting, Lisbon, Portugal
MARZADRI Alessandra	Center for Ecohydraulics Research, University of Idaho, Boise, ID, United States
McEWAN Kerryn L.	CSIRO Land and Water, South Australia
MEREDITH Karina	Australian Nuclear Science and Technology Organisation, Institute for Environmental Research, Kirrawee, Australia
MILNE Jacqui	National Water Research Institute, Environment Canada, Burlington, Ontario, Canada
MONTEIRO José Paulo	Geo-Systems Centre/CVRM, CTA, University of the Algarve, Faro, Portugal
MUNDAY Tim J.	CSIRO Exploration and Mining, Western Australia
NGETAR Silus N.	School of Life and Environmental Sciences, University of KwaZulu-Natal, Durban, South Africa
NÜTZMANN Gunnar	Leibniz-Institute of Freshwater Ecology and Inland Fisheries, Berlin, Germany
PARIS Marta del C.	Facultad de Ingeniería y Ciencias Hídricas – Universidad Nacional del Litoral. Ciudad Universitaria, Santa Fe, Argentina
PEACH Denis W.	British Geological Survey, Environmental Science Centre Keyworth, Nottingham, UK
PEREZ Marcela A.	Facultad de Ingeniería y Ciencias Hídricas – Universidad Nacional del Litoral. Ciudad Universitaria, Santa Fe, Argentina
PIPAN Tanja	Karst Research Institute ZRC SAZU, Postojna, Slovenia
POST Vincent E.A.	Flinders University, Faculty of Science and Engineering, School of the Environment, Adelaide, Australia
PRETORIUS Jacobus J.	School of Bioresources Engineering and Environmental Hydrology, University of KwaZulu-Natal, Scottsville, South Africa
REIS Edite	Algarve River Basin Administration, Faro, Portugal
RIBEIRO Luís	Geo-Systems Centre/CVRM, Instituto Superior Técnico, Lisbon, Portugal
RIDDELL Edward S.	School of Bioresources Engineering and Environmental Hydrology, University of KwaZulu-Natal, Scottsville, South Africa
SALVADOR Núria	Geo-Systems Centre/CVRM, CTA, University of the Algarve, Faro, Portugal
SANTOS Rui	ALGAE – Marine Plant Ecology Research Group, CCMAR, University of the Algarve
SENA Clara	CESAM – Centre for Environmental and Marine Studies, Geosciences Department, University of Aveiro, Aveiro, Portugal
SILVA José Matos	Catholic University of Portugal, School of Engineering, Lisbon, Portugal

SIMON Kevin S.	Department of Environmental Sciences, University of Auckland, Auckland, New Zealand
SIMON Szilvia	Department of Physical and Applied Geology, Eötvös Loránd University, Hungary
SOMMER Bea	Centre for Ecosystem Management, Edith Cowan University, Joondalup, Western Australia
SOUTER Nicholas J.	South Australian Department for Water
STIGTER Tibor Y.	Geo-Systems Centre/CVRM – Instituto Superior Técnico, Lisbon, Portugal
STONE David	Australian Nuclear Science and Technology Organisation, Institute for Environmental Research, Kirrawee, Australia
TÓTH József	Department of Earth and Atmospheric Sciences, University of Alberta, Alberta, Canada
TUJCHNEIDER Ofelia C.	Facultad de Ingeniería y Ciencias Hídricas – Universidad Nacional del Litoral. Ciudad Universitaria, Santa Fe, Argentina
VAN DIJK Harold J.A.A.	Waterboard Aa en Maas, 's-Hertogenbosch, The Netherlands
WHEATER Howard S.	Global Institute for Water Security, University of Saskatchewan, National Hydrology Research Centre, Saskatoon, Canada
WHITE Melissa G.	South Australian Department for Water
WILLIAMS R. Michael	NSW Office of Water, Department of Primary Industries, Sydney NSW, Australia
YÉLAMOS Javier G.	Geology & Geochemistry Department, Autonomous University of Madrid, Madrid, Spain
YESERTENER Cahit	Department of Water, Perth, Australia
YOUNGS Jed	MWH, Perth, Australia

Chapter 1

A toolbox for assessing the ecological water requirements of groundwater dependent ecosystems in Australia

Richard S. Evans[1], *Peter G. Cook*[2], *Paul Howe*[3], *Craig A. Clifton*[4] & *Elizabeth Irvine*[5]

[1]*Sinclair Knight Merz, Melbourne, Australia*
[2]*National Centre for Groundwater Research and Training, Flinders University, Adelaide SA, Australia*
[3]*Sinclair Knight Merz, Adelaide, SA, Australia*
[4]*Sinclair Knight Merz, Bendigo, Australia*
[5]*Sinclair Knight Merz, Adelaide, SA, Australia*

ABSTRACT

A technically focussed 'toolbox' has been developed to provide Australian water and environmental managers with a range of techniques to assist in determining the ecological water requirements (EWR) of groundwater dependent ecosystems (GDE). In addition to providing details of the various techniques for identifying GDEs and estimating their EWR, the tools also indicate data requirements, the level of effort and expense required and the level of confidence in outputs and outcomes. The toolbox has been developed to assist in making informed decisions on water allocation that affect the health and function of groundwater dependent ecosystems. The tools themselves represent groupings of methods that can be used in some part of the GDE assessment process. A total of 14 tools have been included in the toolbox.

1.1 INTRODUCTION

The primary purpose of water allocation planning in Australia is to achieve an equitable way in which to allocate and manage a region's water resources that is consistent with the Australian Government's *Water reform framework* (COAG, 1994), and to protect the characteristics of ecosystems that are considered important to their function (services provided) and conservation value. To do this, the needs of all water users have to be considered in water planning policy, including those of the communities and ecosystems.

Typically, the needs of ecosystems dependent on surface water systems are routinely addressed within the various national approaches to water allocation planning. However, this is not often the case for groundwater dependent ecosystems (GDEs). Their ecological water requirement (EWR) is defined as the water regime needed to sustain the ecological values of water-dependent ecosystems at a low level of risk. Often, the natural water regime of GDEs will comprise a combination of one or more of groundwater, surface water and soil water.

An existing conceptual framework for GDE management devised for Australia in 2001 (Clifton and Evans, 2001) comprises of four steps: (i) identify potential GDEs; (ii) establish the natural water regime of GDEs and their level of dependence on groundwater; (iii) assess the EWRs of GDEs; and (iv) devise the necessary water provisions for GDEs. However, work undertaken to date in Australia to provide water for GDEs has generally stalled at the identification stage and has not progressed through the three remaining steps of the 2001 conceptual framework. EWRs have generally been subjectively derived and are based on 'best estimates' of ecosystem interactions with groundwater. EWRs that have been determined in this way provide only limited confidence that any later environmental water provision (EWP) will actually sustain ecosystem function, and consequently there was an identified need to develop a national framework for assessing the EWRs of GDEs so as to provide a platform from which to take advantage of current and future information.

In 2007, Land & Water Australia published '*A framework for assessing the Environmental Water Requirements of Groundwater Dependent Ecosystems*' (LWA, 2009), which included a toolbox of methods and approaches that can be applied to developing an understanding of water requirements necessary to sustain GDEs, which has since been updated by the Australian National Water Commission (Richardson *et al.*, 2011a and 2011b).

The Australian GDE Toolbox ('the toolbox') presents a suite of practical and technically robust tools and approaches that allows water resource, catchment and ecosystem managers, to identify GDEs, assess the reliance of those ecosystems on groundwater, and assess possible changes to ecosystem state or function due to changes in the groundwater environment in response to anthropological use, climate change etc. The tools are based on a review of methods reported in the literature. In addition to providing details of the various techniques for identifying GDEs and estimating their ecological water requirements, the toolbox includes for each tool: (i) data requirements; (ii) level of effort and expense required in use and in results analysis; and (iii) level of confidence that can be had in outputs and outcomes arising from use.

This paper presents in summary the framework and recent updates to the suite of 'tools' presented in the *Australian GDE Toolbox*. For full context the reader is referred to the National Water Commission website (http://nwc.gov.au/publications/waterlines/australian-groundwater-dependent-ecosystems-toolbox).

1.2 DEVELOPMENT AND APPLICATION OF GDE TOOLS

1.2.1 Overview

The toolbox has been developed to assist in making informed decisions on water allocation that affect the health and function of groundwater dependent ecosystems. The information provided is structured in a way that allows users to commission appropriate specialist studies. It does not provide such detailed descriptions of the assessment tools that users could directly undertake such studies. However, each of the tool descriptions includes references to case studies of their application.

1.2.2 Development of assessment tools

Australian and international literature was reviewed to determine what methods have been used in GDE assessments. The review identified a wide range of approaches that can be adapted to different types of ecosystems, many of which were actively being researched but few of which were being applied. They range in sophistication, technical rigor and the level of insight provided into the nature of groundwater dependency and the EWR.

The tools themselves represent groupings of methods that can be used in some part of the GDE assessment process. For example, the "analysis of water chemistry" tool comprises various approaches for assessing the chemistry of interactions between groundwater and surface water, including isotopic analysis, use of environmental tracers and analysis of conservative anions and cations.

A total of 14 tools have been included in the toolbox. Each tool contains a brief description of how it works and indicates its applicability to GDE assessments and the level of certainty expected from the results. The scales at which the tools may be applied have also been identified. The principles behind each tool's application to GDE/EWR assessments are described to provide the user with an understanding of the approach to be applied. A short discussion has been provided on how the tool is applied, outlining the tools components and how they are used in GDE assessments. Methods used to interpret the data used by the tools, such as presentation options or specialist software, have been included in the tools to ensure that the user obtains the greatest benefits. The major limitations of the tools in GDE assessments are listed together with a representative range of costs to apply each main component of the tool. The main types of specialist skills and resources required to apply the tool have been listed along with the data requirements.

Each tool identifies potential links with other tools within the toolbox and indicates whether it may provide inputs, or require outputs from the related tool. This can assist in GDE management by providing opportunities to support findings and build the users knowledge of the ecosystem. A list of key references and Australian case studies that may be referred to for further information is also provided for each tool.

A summary of the toolbox and brief descriptions of the tools are given in Table 1.1. The reader is referred to the online version of the toolbox for key references for each technique and references to relevant Australian case studies. A description of the typical template for each of the tools is presented in Table 1.2.

1.2.3 Toolbox application

The Australian GDE Toolbox provides an intuitive reference document, presented in two parts, *Part 1 Assessment Framework* and *Part 2 Assessment Tools* (Richardson et al., 2011a and 2011b, respectively). The assessment framework outlines a structure for deriving EWRs for GDEs that is presented within a broader context of water management planning (Fig. 1.1) that involves three assessment stages: (i) GDE location, classification and basic conceptualization; (ii) characterisation of groundwater reliance; and (iii) characterisation of ecological response to change.

Each of the three stages of the framework (see Fig. 1.1 and Table 1.3) is framed around answering 'key questions' as they apply to GDE conceptualization, effective monitoring and adaptive management, dealing with challenges presented by change and, importantly, GDE typology:

- Type 1 – Aquifer and cave ecosystem
- Type 2 – Ecosystems dependent on the surface expression of groundwater
- Type 3 – Ecosystems dependent on the subsurface presence of groundwater.

Table 1.1 Summary of tools comprising the Toolbox (summarised from Richardson *et al.*, 2011b).

#	*Tool*	*Brief description*
T1	Landscape mapping	Locating and identifying ecosystems that are potentially groundwater dependent based on a number of biophysical parameters such as depth to water table, soils and vegetation type. Assessing primary productivity, water relations and/or condition of vegetation communities using remotely sensed images to infer use of groundwater.
T2	Conceptual modelling	Documentation of a conceptual understanding of the location of GDEs and interaction between ecosystems and groundwater.
T3	Pre-dawn leaf water potential	Identification of groundwater uptake by components of vegetation on the basis of pre-dawn measurements of leaf water potential.
T4	Plant water stable isotopes	Use of naturally occurring stable isotopes to identify sources of water used for plant transpiration.
T5	Plant water use modelling	Identification of sources and volumes of water used for plant transpiration, by using mathematical simulations of plant function.
T6	Plant rooting depth and morphology	Comparison of the depth and morphology of plant root systems with measured or estimated depth to the water table, in order to assess the potential for groundwater uptake.
T7	Plant groundwater use determination	Measures of Leaf Area Index and climatic data are used to estimate groundwater discharge from terrestrial ecosystems that have access to groundwater.
T8	Water balance – vegetation	Use of water-balance measurements and/or calculations to determine whether and to what extent plant water use is dependent on groundwater uptake.
T9	Stygofauna sampling	Techniques available to observe, monitor and measure ecosystem interactions and biological activity within the groundwater system.
T10	Evaluation of groundwater – surface water interactions	Analysis of the hydraulics of groundwater-surface water interactions. The processes by which groundwater discharges into a surface water system provides insight into the nature of groundwater dependency in wetlands and baseflow river ecosystems.
T11	Environmental tracers	Environmental tracers are a physical or chemical property of water, or any substance dissolved in water, that can be used to trace origin. Analysis and interpretation of these properties of surface water and groundwater can be used to identify groundwater contribution to dependent ecosystems.
T12	Introduced tracers	Analysis of deliberately introduced hydrochemical tracers to identify water sources and groundwater-surface water mixing relationships.
T13	Long-term observation of system response to change	Long-term observations of GDEs to establish ecosystem responses to changes in groundwater regime due to climate or anthropogenic influences.
T14	Numerical groundwater modelling	Construction of mathematical models to simulate of groundwater flow systems.

The key questions to be addressed for each stage are as follows:

- Stage 1 *GDE location, classification and basic conceptualisation*
  - Where are ecosystems that might use groundwater located?
  - What is the broad type of GDE and functional grouping?

Table 1.2 Tool Description Template (adapted from Richardson *et al.*, 2011b).

Tool number and title		
Description	A brief description of how the tool works and the information and insights it may provide	
Application to components of EWR studies	Applicability of the tool to the stages of GDE assessment as described in *Part 1: Assessment Framework* For applicability: ✓ low applicability; ✓✓ moderate applicability; ✓✓✓ high applicability; blank – not applicable. For level of certainty: ✓ low applicability; ✓✓ moderate applicability; ✓✓✓ high certainty; blank – not applicable	
Scale of measurement	Usefulness of the tool at common scales of application ✓ poor suitability; ✓✓ moderate suitability; ✓✓✓ high suitability; blank – not suitable	
Principles	Principles behind the application to GDE assessments	
How the tool is applied in GDE assessments	Outlines tool components and how they are used	
Analysis approach	An outline of the methodological approach	
Limitations	Any limitations associated with the tool	
Advantages	Specific strengths of the tool	
Disadvantages	Any weaknesses of the tool	
Costs	Indicative costs for application of the tool	
Specialist skills and resources	Outline of specialist skills and resources required to apply the tool	
Main data types required	*Data* Data requirements	*Likely source(s)* Indication of likely sources of existing data or process to obtain new data
Complementary tools	Other tools for which the tool may provide data or require data ✓ indicates relevance	
Key references and Australian case studies	Summary of Australian or international case studies where the tool has been applied in order of relevance	
Further Information	Links to resources that contain further information on tool application	

- Stage 2 *Characterisation of groundwater reliance*
  - Is groundwater used by the ecosystem?
  - How reliant is the ecosystem on groundwater?
- Stage 3 *Characterisation of ecological response to change*
  - What are the threats posed to the groundwater system / ecosystem?
  - How might the ecosystem change if groundwater conditions change?
  - What is the predicted long-term ecosystem state in response to change?

1.3 CONCLUSIONS

It is intended that the presentation of the various GDE assessment methods in the form of the 14 tools will assist the systematic investigation of the EWR of GDEs. Through

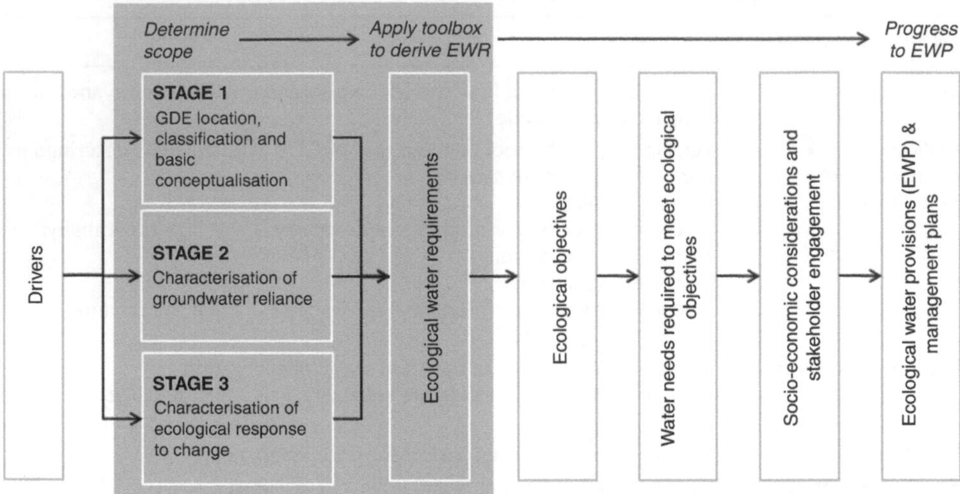

Figure 1.1 Three stage assessment framework for determination of EWRs (shaded area represents the scope of content of the GDE Toolbox; adapted from Richardson *et al.*, 2011a).

Table 1.3 Links between the GDE toolbox assessment framework, GDE type and tools useful to each assessment stage and GDE type (adapted from Richardson *et al.*, 2011b).

Type 1	*Type 2*	*Type 3*
Stage 1: GDE location, classification and basic conceptualization		
←	T1 Landscape mapping T2 Conceptualization	→
Stage 2: Characterisation of groundwater reliance		
T9 Stygofauna sampling T10 Evaluation of groundwater-surface water interactions T11 Environmental tracers T12 Analysis of introduced tracers	T10 Evaluation of groundwater-surface water interactions T11 Environmental tracers T12 Analysis of introduced tracers	T3 Leaf water potentials T4 Plant water-stable isotopes T5 Plant water use modeling T6 Root depth and morphology T7 Plant groundwater use estimation T8 Water balance – vegetation
Stage 3: Characterisation of ecological response to change		
←	T3–T12 as appropriate T13 Long-term observation of system response to change T14 Numerical groundwater modelling	→

Notes:

Type 1 – aquifer and cave ecosystems

Type 2 – ecosystems dependent on the surface expression of groundwater

Type 3 – ecosystems dependent on the sub-surface presence of groundwater use of the toolbox and framework described in this paper, GDE studies can be undertaken with scientific rigor and should give rise to a greater understanding of whether ecosystems are interacting with groundwater and, if so, to what degree. Through this understanding it will be possible to identify what levels or types of management strategies can be developed to mitigate threats to ecosystem function as a result of planned development that has the potential to cause change to groundwater conditions.

ACKNOWLEDGMENTS

The information presented in this paper, and the ideas on which the discussion is based, result from the experience of many people across Australia. The authors wish to thank all who contributed to this work. In particular, the authors wish to acknowledge the efforts of a number of people. Firstly those involved in the early development of the LWA framework, including Jodie Pritchard, Tony Fas, Brendan Cossens, Chris McAuley, Marcus Cooling and Andrew Boulton. Secondly, those involved in more recent work undertaken for the National Water Commission in revising and updating the framework, including Stuart Richardson, Stephanie Barber and Bonnie Bonneville (Sinclair Knight Merz), Ray Froend (Edith Cowan University), Paul Boon (Dodo Environmental), Ian Overton and Sébastien Lamontange (CSIRO Land and Water), Anthony O'Grady (CSIRO Ecosystem Services), Lindsay Hutley (Charles Darwin University) and Des Yin Foo (Northern Territory Department of Natural Resources, Environment, the Arts and Sport). Lastly, the efforts and feedback received from the reference group set up for the framework update. Finally, the enthusiastic support of Land and Water Australia and the National Water Commission is gratefully acknowledged.

REFERENCES

COAG 1994. Water reform framework. Extracts from Council of Australian Governments: Hobart, 25 February 1994 communique.

Clifton C & Evans R 2001. Environmental Water Requirements of Groundwater Dependent Ecosystems Environmental Flows Initiative Technical Report No. 2. Commonwealth of Australia.

LWA 2009. A framework to provide for the assessment of environmental water requirements of groundwater dependent ecosystems. [Online] (Updated July 30th, 2009) Available at: http://lwa.gov.au/node/3225

Richardson S, Irvine E, Froend R, Boon P, Barber S, Bonneville B 2011a. The Australian GDE Toolbox Part 1 Assessment Framework, Waterlines report 69, National Water Commission Canberra, available at http://nwc.gov.au/publications/waterlines/australian-groundwater-dependent-ecosystems-toolbox

Richardson S, Irvine E, Froend R, Boon P, Barber S, Bonneville B 2011b. The Australian GDE Toolbox Part 2 Assessment Tools, Waterlines report 69, National Water Commission Canberra, available at http://nwc.gov.au/publications/waterlines/australian-groundwater-dependent-ecosystems-toolbox

Chapter 2

Water table dynamics of a severely eroded wetland system, prior to rehabilitation, Sand River catchment, South Africa

Edward S. Riddell[1], *Simon A. Lorentz*[1], *William N. Ellery*[2], *Donovan Kotze*[3], *Jacobus J. Pretorius*[1] & *Silus N. Ngetar*[4]

[1] *Centre for Water Resources Research, University of KwaZulu-Natal, Scottsville, South Africa*
[2] *Department of Environmental Science, Rhodes University, Grahamstown, South Africa*
[3] *Centre for Environment and Development, University of KwaZulu-Natal, Scottsville, South Africa*
[4] *School of Life and Environmental Sciences, University of KwaZulu-Natal, Durban, South Africa*

ABSTRACT

The severe degradation of the headwater wetlands of the Sand River, South Africa, by large erosion gullies has been the focus of a hydrological monitoring programme in order to determine the hydrodynamic response of these wetlands to rehabilitation interventions. This paper presents the findings relating to the behaviour of a headwater wetland's groundwater phreatic surface prior to rehabilitation. The findings of hydrometric observations include the delineation of a stratified water table system. The behaviour of this suggests the occurrence of hydrodynamically distinct regions within the wetland, and loss of groundwater through head-cut erosion. Two Dimensional Electrical Resistivity Tomography (ERT) surveys identified a zone of finer sediment which is thought to act as a sub-surface flow buffer within this otherwise sandy wetland substrate. These findings suggest that these impacted wetlands need to be 'plugged' with a hydraulic stopper structure in order to restore their hydrological regime.

2.1 INTRODUCTION

'Hydrology is probably the single most important determinant of the establishment and maintenance of specific types of wetlands and wetland processes' (Mitsch & Gosselink, 1993). Without the development of an understanding of how wetlands form from a hydrological and geomorphological perspective, the predicted function and proper management of such systems would be difficult to achieve (Ellery *et al.*, 2008). It is generally accepted that wetlands at the headwaters of river systems act as regulators of flow by sustaining baseflow and attenuating peak flow. Although this is the subject of debate (e.g. Bullock & Acreman, 2003), it is only recently that this hypothesis has begun to be tested in the southern African context (e.g. McCartney, 2000). Meanwhile these wetland hydrogeomorphic types do serve a purpose in the region in terms of food security through both wet and dry season subsistence cultivation.

South Africa in particular experiences the continued conversion of these wetland systems to agriculture, with little or no control (Kotze & Silima, 2003).

The Sand River is the main tributary of the Sabie River, the last remaining perennial river to flow through the Kruger National Park and into Mozambique, although the Sand River itself was once considered a perennial river it is now considered as being severely degraded. The degradation of the Sand River is thought to be due to two main principle factors; inappropriate forestry and commercial agriculture within the contributing catchment; and the degradation of the extensive wetlands at the headwaters of the river system through their conversion to agriculture for subsistence cultivation (Pollard *et al.*, 2005), a situation that has arisen largely as a result of enforced settlement under a previous political regime.

These wetland systems exist in an area subject to tectonic uplift at the granitic foothills of the Klein Drakensberg escarpment, subject to weathering processes and prone to large and intense rainfall events. These two factors contribute to the natural landscape scale erosional processes evident in the foothill zone of the Klein Drakensberg Escarpment. Meanwhile the recent population pressures have probably led to the degradation of the wetland environment and their contributing micro-catchments through a reduction in vegetation cover and mechanical alteration of the catchment soils. It is postulated that excessive sediment delivery to the wetland substrate, resulting from these effects, have led to the over-steepening of the wetland surface topography and as a consequence made them extremely vulnerable to perturbation by large and intense precipitation events. This would conform to the concept of crossing a geomorphic threshold of slope stability-instability (Ritter *et al.*, 2002; Grenfell *et al.*, 2009). As a result the initiation and rapid head-ward retreat of large erosion gullies now characterise these wetland systems leading to a significant loss of wetland sediment and severe desiccation of the wetland environment. Meanwhile, recent literature (e.g. Dewandel *et al.*, 2006) has suggested that hydrodynamic properties of composite aquifers in hard rock areas, such as granite, is significantly influenced by the various porosities expressed in weathered profiles in the landscape. These porosities vary between mutli-layered clay-rich saprolites formed by in-situ decomposition of the parent material overlaying laminated coarse-grained regions at greater depth.

The aim of this study is to determine the behaviour of the shallow subsurface water table, hereafter referred to as the phreatic surface, in one of these wetlands prior to technical rehabilitation. The aim of the rehabilitation is to mimic zones of finer sediments or 'clay-plugs' within the sub-surface wetland aquifer that are thought to have previously buffered the flows of sub-surface water in these very sandy (highly conductive) wetland soils (Pollard *et al.*, 2005). This paper presents initial interpretations from the first year of a three year study that will comprehensively examine the overall impacts of rehabilitation on the wetland's hydrodynamics.

2.2 METHOD

The research site is located within the Manalana sub-catchment at the headwaters of the Sand River, approximately 300 km east-north-east of Tshwane (Pretoria) (Fig. 2.1), adjacent to the village of Craigieburn. This is one of many such headwater catchments containing riparian wetlands situated within a dry sub-humid belt on the periphery of

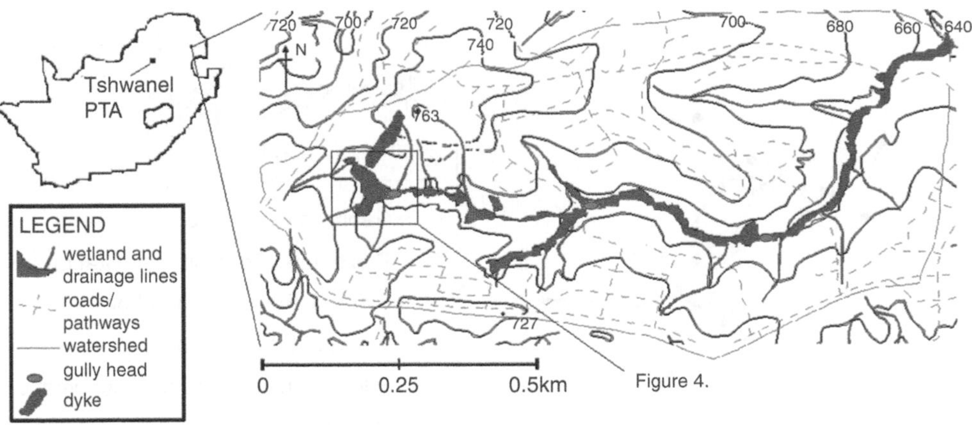

Figure 2.1 1:15,000 relief map of the Manalana sub-catchment and its location within South Africa.

the South African semi-arid lowveld savanna at the foothills of the Klein Drakensberg Escarpment. The lowveld is an area of low to moderate relief averaging 300 m above sea level on the Mozambique coastal plain or Eastern Plateau Slope (Kruger, 1983). The Sand River is unusual in that its entire catchment is situated within the lowveld complex, while most other north-eastern South African river systems have their origins on the high altitude grasslands (highveld >1500 m).

The Manalana is a long and narrow sub-catchment whose relief is relatively steep with an average 13% interfluve slope, whilst its altitude ranges from 744 m above sea level (m.asl) at the highest point along the watershed to 654 m.asl at its confluence with Motlamogasana stream. The wetlands upstream of the erosion gullies within the Manalana are typically un-channelled valley bottom wetlands. The Manalana sub-catchment lies within a granitic geological zone (medium- to coarse-grained porphyritic biotite granite) while a medium grained diabase (dolerite) dyke intersects this on its northern flank. Due to the proximity of the Klein Drakensberg Escarpment the study area bears imprints of tectonic uplift with remnant erosion terraces apparent at certain reaches within the catchment, where outcrops of both granite parent material and grus occur. Hydrolysis during hot, humid summers is the leading cause of in-situ deep weathering of both granite and dolerite in this catchment, and moreover cross-section surveys within the catchment have revealed a stratified geological structure where dolerite sills alternate with granite (Ngetar, 2012).

Soils within the catchment are dominated by coarse- to medium-grained sand and are classified as being sandy-clays with a high plasticity and generally free draining. Significantly there appears to be higher clay content on the slopes, thus sand concentrations increase towards the valley bottom such that the wetland substrate itself is predominantly sand. Detailed hydrogeomorphic description of this catchment may be found in Riddell *et al.* (2010) whilst a schematic of the hydropedological formations on two hillslopes are given in Figure 2.2. Gully erosion has been identified as being due in part to the exposure of the lower dispersive silt-clay horizon (>0.8 m deep) beneath a more stable sandy-clay A-horizon. The Manalana catchment is 2.61 km^2

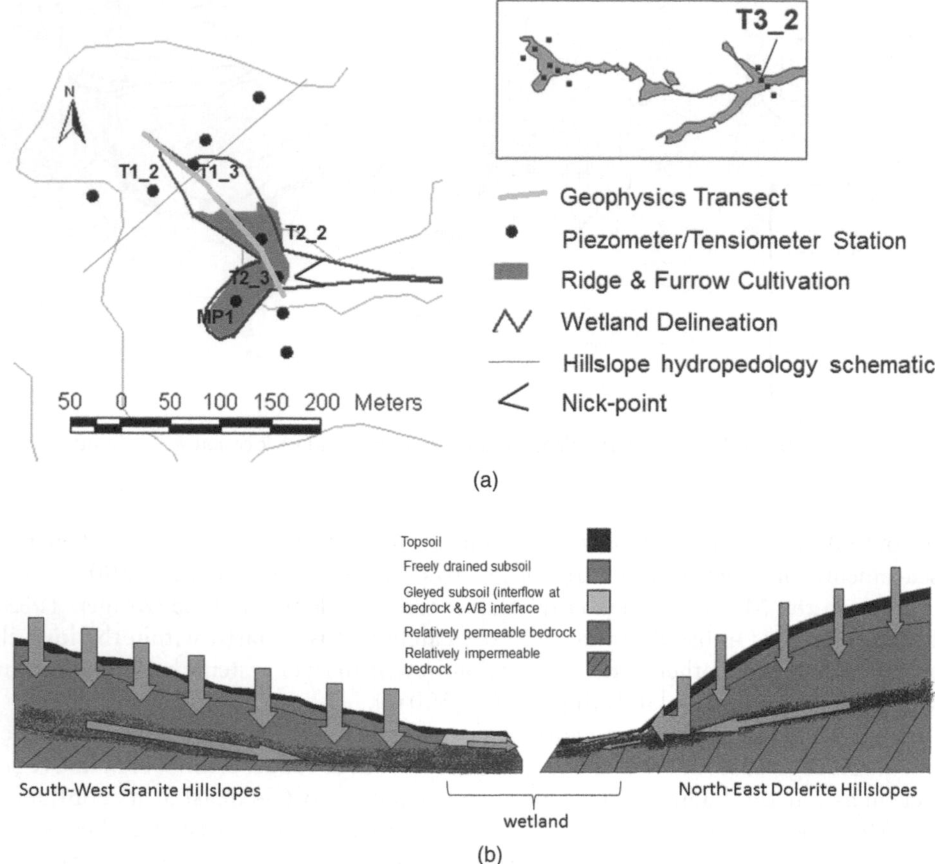

Figure 2.2 a – The most head-ward region of the Manalana wetland with hydrometric observation stations relevant to this manuscript, and site of active gully erosion (nick-point); b – hydropedological schematic of Manalana hillslopes (adapted from Le Roux *et al.*, 2010).

of which 2.5 km^2 and 0.11 km^2 (or 95.6% and 4.4%) make up the area of interfluve and wetland respectively. Rainfall is strongly seasonal in the Sand River catchment occurring during the summer months October to March. A mean annual precipitation of 1075 mm a^{-1} (1904–2000) for this area has been derived from the nearest long-term dataset at the Wales rain gauge, approximately 2.3 km away, although the Manalana probably receives somewhat less than this since it is not as close to the Klein Drakensberg Escarpment.

Vegetation within the Manalana catchment is characterised by grassy shrubland communities on the interfluves to communities more suited to permanently flooded conditions within the wetlands (e.g. *Phragmites mauritianus*). Land use within the catchment comprises densely populated rural housing with dry crop smallholdings (e.g. maize), there is also a dense network of roads and pathways as well as heavy grazing by cattle and goats. Within the wetland itself there is a high density of plots

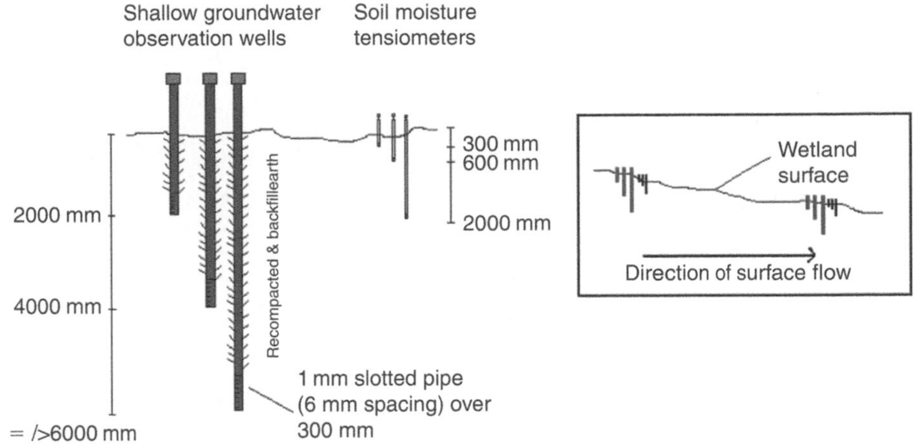

Figure 2.3 Schematic of groundwater PVC piezometer and tensiometer instrumentation (not to scale).

with raised bed and furrow systems, cultivated predominantly with another regional staple, madumbe (*Colocasia esculenta*).

Instrumentation was installed during the latter half of the 2005 dry season (August–October) in order to determine the hydrological processes within the wetland catchment at the commencement of the first rains October–November. Hydrological monitoring stations with automated soil moisture tensiometers and shallow groundwater observation 63 mm diameter PVC piezometers were installed along three transects that ran perpendicular to the longitudinal orientation of the catchment. In both cases these were installed via hand-augered holes to the required depth, with compaction and backfilling of the original substrate in order to line the instruments. These manual groundwater observation piezometers were replicated at different depths at the wetland monitoring stations in order to account for possible phreatic surface stratification (see Fig. 2.3). Further manual groundwater observation piezometers were installed at other relevant positions within the wetland.

Automated monitoring stations recorded soil moisture status and groundwater levels on a 12 minute time-step using a HOBO® 4-channel logger and University of KwaZulu-Natal timing board system. Additional manual groundwater measurements were made using a dip meter at least weekly and successively after rainfall events.

The longitudinal topography of the wetland was recorded using a builder's auto level on 15 September 2005, and the depth to the wetland phreatic surface was recorded via hand-augered holes.

A dry season Two-Dimensional Electrical Resistivity Tomography (ERT) survey was conducted during October 2006 when the phreatic surfaces and soil moisture values were deemed to be at their lowest, this was necessary to delineate sub-surface materials in the absence of groundwater which encompasses the same resistivity range as clay materials (Loke, 1999). A Trimble™ Pro_XRS Differential GPS was used to determine altitudes for this geophysics survey and the locations of piezometer and tensiometer instrumentation.

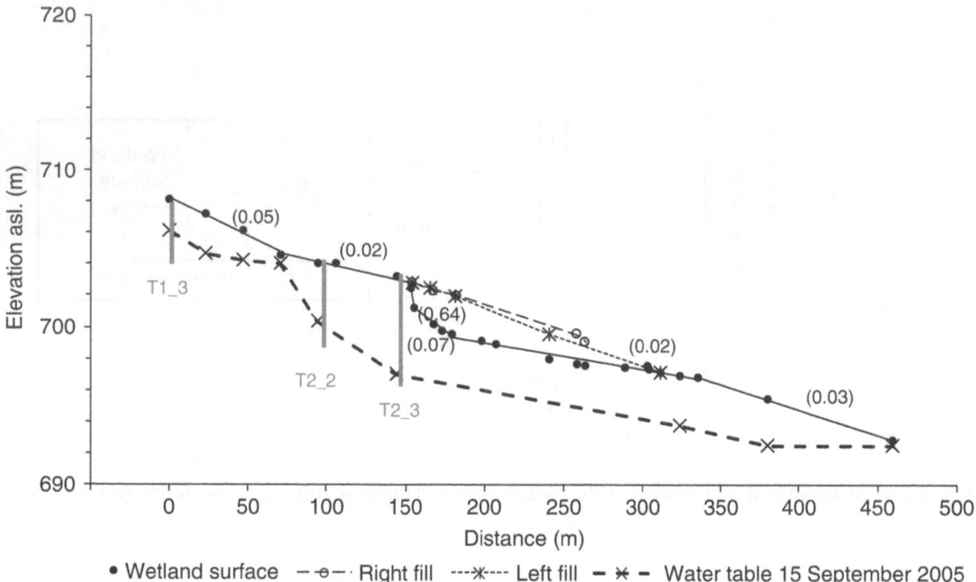

Figure 2.4 Longitudinal profile of the Manalana wetland, numbers in parenthesis represent the mean topographic slope (Survey 15 September 2005).

2.3 RESULTS

Figure 2.4 displays the longitudinal profile of the Manalana wetland from its most head-ward position to just over 450 m downstream. This illustrates the relatively steep gradient of the wetland in these headwater settings, with slopes ranging between 2.08% and 4.99%. This slope increases at sites of down-cutting erosion to 63.5%. Figure 2.3 also displays the longitudinal gradient of the wetland prior to this down-cutting erosion through the mapping of the left and right walls (fills) of the gully, suggesting that the slope in this region could have been 1.67% in the past. Moreover, the phreatic surface encountered during this survey suggests that the relatively shallow phreatic surface close to the start of the survey at 0 m may be fed from a spring arising from the contributing hillside. Whilst at approximately 70 m some hydrogeological effect is causing the phreatic surface to appear close to the wetland topographic surface, beyond which there is a sharp increase in the depth of the wetland phreatic surface particularly adjacent to the eroded region of the system.

Figure 2.5 suggests that the phreatic surface of the wetland responds very rapidly to the rapid wetting of deep soils at the toe of the interfluves, since these hill slope toe positions experience a rapid wetting at ≥2.0 m soil depths following substantial rainfalls during the early part of the rainy season. This can be seen from Figure 2.5 where the 2.04 m tensiometer at T1_2 (hill slope toe position) experiences a sharp fall of capillary pressure head moving into the positive pressure head range, hence a rapid phreatic surface elevation appears relatively quickly, as seen on 8 of January 2006 suggesting rapid vertical infiltration from above and/or a significant lateral transfer of

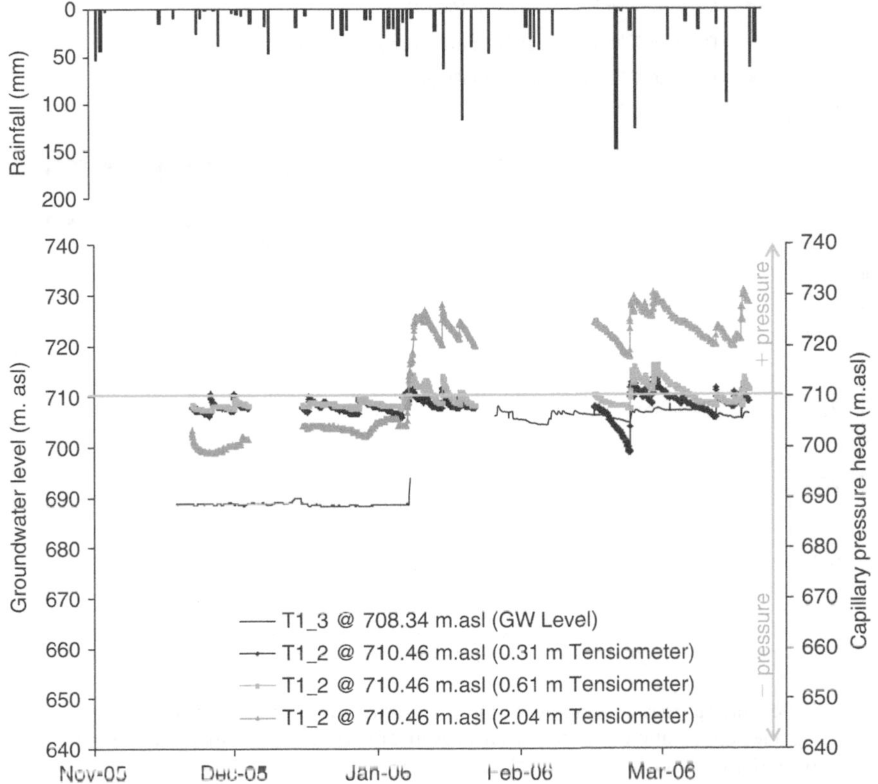

Figure 2.5 Soil moisture time series (hourly mean) at hill slope toe position T1_2, and phreatic surface elevation at T1_3. Rainfall data from the nearest accurate record, Hebron Forestry Station approximately 1.6 km away.

water from the hill slope to the wetland. The gaps in the data are due to sorting for unreliable readings and exceeding of transducer pressure range.

This rapid phreatic surface elevation was also observed at observation station T2_2, on 8 January 2006 as shown by the 4.0 m automated groundwater reading in Figure 2.6. The delineation of a stratified phreatic surface system is also displayed in Figure 2.6 whereby a seasonally perched phreatic surface observed in a 2.0 m piezometer first appears quite far into the rainy season at 703.8 m.asl (0.88 m below ground surface) on 14 January 2006 rising to 0.41 m below ground surface on 10 February 2006. Prior to this no phreatic surface was observed in the 6.0 m observation piezometer suggesting that the operative phreatic surface in this vicinity is within a 2.0–4.0 m depth. However only following the substantial rains early (>368 mm 26/10/05–14/01/06) in the season does a deeper phreatic surface first appear at 701.1 m.asl (3.61 m below ground level) on 14 January 2006 rising up to 703.6 m.asl (1.15 m below ground surface) by 23 March 2006. Since there is no stream within this wetland it means that recharge into the wetland aquifer does not occur via transmission losses, but more likely that a broader catchment hill slope recharge process is

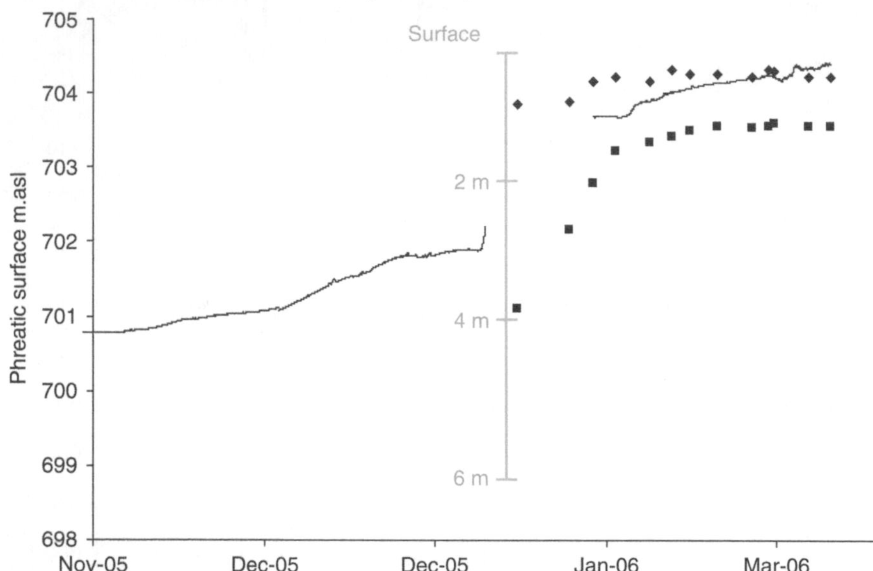

Figure 2.6 Differential phreatic surface levels at hydrometric observation station T2_2 (grey vertical line signifies appearance of phreatic surfaces in 2 m and 6 m piezometers, cross-lines represent total depth of each piezometer).

occurring to maintain and replenish the phreatic surface observed in the 4.0 m piezometer throughout the year. Meanwhile the deeper and shallower regions of the wetland substrate only become connected to a rainfall input during the latter part of the hydrological season, perhaps when certain antecedent conditions are met for hydrological connectivity in the sub-surface.

Figure 2.7 displays the longevity of the existence of perched phreatic surfaces as observed in 2.0 m piezometers at four locations longitudinally downstream along the wetland, where piezometer T3_2 is approximately 750 m downstream. The permanency of the perched phreatic surface increases downstream as one would expect since at a lower topographic position in the catchment the wetland is likely to intersect groundwater flowing out from the wetland aquifer. However the piezometer at T2_3 does not satisfy this expectation, and indeed has the shortest duration. The hydrometry station T2_3 is adjacent to the site of active gully erosion (see Fig. 2.2).

Figure 2.8 shows the positions of the perched and permanent phreatic surfaces at three hydrometry stations in order to describe their longitudinal elevations. Here it is observed that throughout the rains of 2005–2006 the perched phreatic surface (encountered within 2.0 or 4.0 m deep piezometers) elevates toward the wetland surface meanwhile there is a deeper recharge to a permanent phreatic surface recorded in the ≥6.0 m piezometers (as this phreatic surface was observed during the prior dry season)

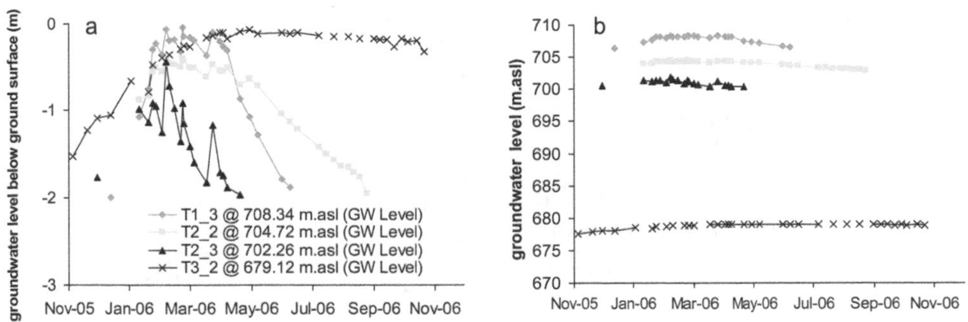

Figure 2.7 Perched phreatic surface permanency (a – to datum wetland surface elevation; b – to datum above sea level).

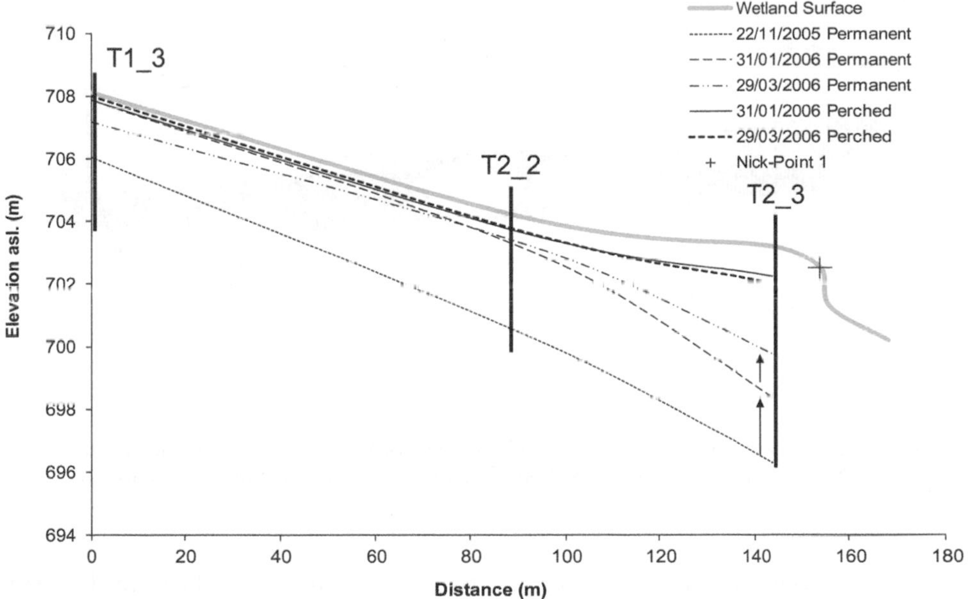

Figure 2.8 Longitudinal phreatic surface stratification during the 2005–2006 rains.

that also elevates during the rainy period. The significance of this display is the loss of groundwater elevation in the vicinity of active gully erosion for both the perched and deeper permanent phreatic surface systems at the site of erosion (nick-point).

Figure 2.9 documents the temporal fluctuation of the perched phreatic surface system at the three stations shown in Figure 2.8, but also at site MP1 (a stand-alone 6.0 m observation piezometer, refer to Fig. 2.2). Here it is noted that all sites seem to experience phreatic surface fluctuations at the same time, obviously due to subsurface water inputs to the wetland. However the amplitude of these fluctuations suggests that

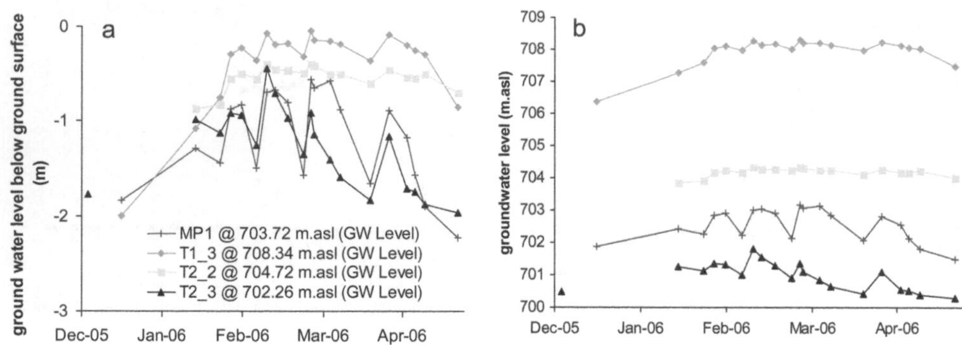

Figure 2.9 Perched phreatic surface behaviour during the 2005–2006 rains at each monitoring station. (a – to datum wetland surface elevation; b – to datum above sea level).

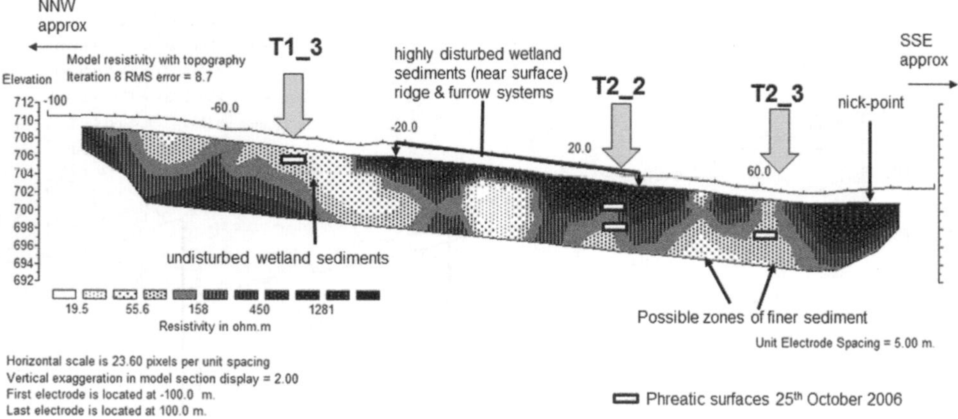

Figure 2.10 Dry season Two-Dimensional ERT survey longitudinally through headward zone of Manalana wetland (conducted 24th October 2006 using Wenner-α short array).

different hydrodynamic conditions exist between two zones in this headward wetland region. Notably stations T1_3 and T2_2 exist in a larger distinct region were these fluctuations are dampened, cause unknown, while MP1 and T2_3 exist in a smaller and narrower wetland zone with a much more pronounced phreatic surface fluctuation.

Figure 2.10 displays the sub-surface ERT survey conducted at the end of the dry season 2006. Here the zones of high resistivity indicated areas of substrate disturbance, such as altered wetland topography through subsistence cultivation practices (ridge and furrow systems, see Fig. 2.2) against areas of undisturbed substrate further upstream. The notable finding from this image is what would appear to be two vertical bodies (narrow bands) of low resistance substrate, which are highly likely to be zones of finer re-deposited sediment such as colluvium or illuvium, delineated here in the absence of groundwater (>4.0 m) at the end of the dry season, since clays have a resistivity which range between 1–100 Ωm.

2.4 DISCUSSION

Since the findings presented here were collated from the first year in a three year monitoring study their interpretation will be aided by successive hydrological sequences. Nevertheless some important interpretations can be made from them.

Firstly it was observed that the head-ward region of the Manalana wetland responds very quickly to the rapid saturation of deep substrate (2.0 m) at the toes of the contributing catchment. It remains to be determined whether this input is caused by lateral throughflow resulting from excessive ponding of water at the soil bedrock interface, which is a distinct possibility as calcrete was observed within 0.5 m of the soil surface at these hill slope toe positions whilst conducting soil characterisation during the dry season of 2006.

The phreatic surface stratification observed is highly likely due to the stratification of different soil textural classes in the wetland substrate, with zones of more conductive coarser sandy materials overlying much less, near impermeable silt-clay bands. Initial soil characterisation in the wetland during the dry season of 2006 at site T1_3 revealed very thin layers of silt-clay, sometimes as narrow as 0.10 m within the predominantly sandy substrate and these were repeated successively up to a depth of 2.0 m. This was the final depth of analysis and these could exist at still greater depths. These clay bands have probably been created through leaching and capillary forces/vegetation hydraulic lift during the wet-dry cycles, acting to sort the finer sediments from the coarser sedimentst, it could also be an event driven phenomena. However this multiple stratification of soil textural classes suggests that multiple perched phreatic surfaces (temporally variable), exceeding those observed, may exist in the wetland subsurface. Furthermore, areas of substrate disturbance such as in the cultivated parts of the wetland may have reduced this stratification phenomenon to depths where the soil structure has not been artificially altered, in which case certain aspects of the wetlands hydraulic regime may have been irreversibly altered.

The appearance of a deep phreatic surface well into the rains, as highlighted by site T2_2 in Figure 2.6 also presents an intriguing observation. Since there is a shallower phreatic surface (i.e. within 4.0 m of the surface) prior to the appearance of the deeper phreatic surface recorded in the 6.0 m observation piezometer, this would suggest one of two possibilities. First, either this deep zone is linked to a separate input, such as a deeper regional phreatic surface. Second, the hydrostatic pressure in the above perched water layer increases to a critical threshold level, such that the positive pressure head of this water induces a rapid infiltration of water through an otherwise low permeability silt-clay layer.

The fact that both permanent and perched phreatic surfaces when presented longitudinally (Fig. 2.8) seem to slope downward adjacent to the site of active erosion suggests a loss of moisture from the wetland subsurface at this point. This would be expected in this highly conductive substrate. There appears to be some factor that causes a deflection in the hydraulic gradient of the phreatic surfaces between sites T2_2 and T2_3, this would be a typical seepage zone with a rapid drawdown of subsurface water at the exit (nick-point). Interestingly this deflection is observed in both the perched and permanent phreatic surfaces during the wet period, whilst at the end of the dry period (i.e. 22/11/2005) there is no deflection in the hydraulic gradient of the permanent phreatic surface. This suggests much lower hydraulic conductivity of the wetland substrate at deeper depths, which could be inferred from the deeper low resistivity soils in Figure 2.10, underlying the modified, high resistant soils nearer the wetland surface.

As a 2-Dimensional ERT survey revealed what would appear to be zones of finer sediments between the two hydrometry stations T2_2 and T2_3, the assumption is that this is a controlling factor in the behaviour of the phreatic surfaces in two regions of the wetland (Fig. 2.9), between the first larger area upstream containing sites T1_3 and T2_2 and a smaller region with sites T2_3 and MP1. It has been shown by Riddell *et al.* (2010) through coupled geophysics and hydraulic conductivity data that these zones of finer sediment have been created at zones of valley confinement. Whereupon illuviated fine sediments from the surrounding hill slopes converge in the valley bottom to create regions of the wetland system, termed 'clay-plugs' that maintain a degree of hydrological disconnectivity in the subsurface by buffering longitudinal flows through the wetland substrate. Moreover, the same study shows that the near surface hydraulically conductive sandy wetland substrate is underlain by a deep clay lens (generally >2 m) and this has similarities with other headwater wetland systems in the southern African sub-continent, for instance where it has been documented in Malawi (McFarlane, 1992) and in Zimbabwe (McCartney & Neal, 1999).

Since the phreatic surfaces in these two regions either side of the clay-plug fluctuate accordingly but with differing amplitudes, and there is a loss of phreatic surface elevation adjacent to the site of active gully erosion, it can confidently be assumed that these zones of finer sediments have a retaining influence on the sub-surface hydrodynamics in this wetland. It is now known that through rehabilitation of the system with technical structures in the form of buttress weirs, that the hydrodynamic behaviour of the wetland adjacent to erosion gully has been restored to a similar amplitude as that observed upstream of the clay-plug region observed in Figure 2.10 (Riddell *et al.*, 2012).

2.5 CONCLUSIONS

The first year of hydrometric observation and geophysical analysis of the Manalana wetland revealed some important observations regarding the hydrodynamic processes in the headwaters of the Sand River system. The fact that zones of finer sediment were found to exist supports the original hypothesis that this is how moisture in the highly conductive sub-surface is retained. This has consequences for further rehabilitation of similarly degraded headwater wetlands in this river system that seek not only to halt the mass removal of wetland sediment but also restore to more natural hydrological regimes in the wetlands. This also has consequences for crucial river system processes. The research however still requires considerable investment to determine the impact of rehabilitation on the wetland water balance, specifically inputs to this and the geomorphic controls existing in these headwater settings.

ACKNOWLEDGEMENTS

This research was funded by the South African Water Research Commission (WRC). Research and logistical support was provided by Dr Sharon Pollard and team at the

Association for Water and Rural Development (AWARD). Many thanks to Mr Ronny Maaboi for assistance with installation of site instrumentation. Many thanks also to Mr Bertram Koning and Mr Victor Kongo for assistance with ERT measurement and interpretation.

REFERENCES

Bullock, A. & Acreman, M. (2003) The role of wetlands in the hydrological cycle. *Hydrology and Earth System Sciences*, 7 (3), 358–389.

Dewandel, B., Lachassagne, P., Wyns, R., Maréchal, J.C. & Krishnamurthy, N.S. (2006) A generalised 3-D geological and hydrogeological conceptual model of granite aquifers controlled by single or multiphase weathering. *Journal of Hydrology*, 330, 260–284.

Ellery, W.N., Grenfell, M., Grenfell, S., Kotze, D., McCarthy, T., Tooth, S., Grundling, P.L., Beckedahl, H., le Maitre, D. & Ramsay, L. (2008) *WET-Origins: Controls on the distribution and dynamics of wetlands in South Africa.* Water Research Commission, Pretoria. WRC Report No TT 334/08.

Grenfell, M.C., Ellery, W.N. & Grenfell, S.E. (2009) Valley morphology and sediment cascades within a wetland system in the KwaZulu-Natal Drakensberg Foothills, Eastern South Africa. *Catena*, 78 (1), 20–35.

Kotze, D. & Silima, V. (2003) Wetland cultivation: Reconciling the conflicting needs of the rural poor and society at large through wetland wise use. *International Journal of Ecology & Environmental Sciences*, 29, 65–71.

Kruger, G.P. (1983) *Terrain morphological map of southern Africa.* Pretoria, Soil and Irrigation Research Institute, Department of Agriculture.

Le Roux, P.A.L., van Tol, J.J., Kuenene, B.T., Hensley, M., Lorentz, S.A., Everson, C.S., van Huyssteen, C.W., Kapangaziwiri, E. & Riddell, E. (2010) *Hydropedological interpretations of the soils of selected catchments with the aim of improving the efficiency of hydrological models.* Water Research Commission, Pretoria. WRC Report No. 1748/1/10.

Loke, M.H. (1999) Electrical imaging surveys for environmental and engineering studies: A practical guide to 2-D and 3-D surveys [http://www.terrajp.co.jp/lokenote.pdf].

McCartney, M. & Neal, C. (1999) Water flow pathways and the water balance within a headwater catchment containing a dambo: inferences drawn from hydrochemical investigations. *Hydrology and Earth System Sciences*, 3 (4), 581–591.

McCartney, M.P. (2000) The water budget of a headwater catchment containing a dambo. *Physics and Chemistry of the Earth (B)*, 25 (7–8), 611–616.

McFarlane, M.J. (1992) Groundwater movement and water chemistry associated with weathering profiles of the African surface in parts of Malawi. *Geological Society of London, Special Publications*, 66 (1), 101–129.

Mitsch, W.J. & Gosselink, J.G. (1993) *Wetlands.* 2nd edition. New York, John Wiley & Sons. 720 pp.

Ngetar, S.N. (2012) *Causes of wetland erosion at Craigieburn, Mpumalanga province, South Africa.* PhD Thesis. Durban, School of Environmental Sciences, University of KwaZulu-Natal. 157 pp.

Pollard, S., Kotze, D., Ellery, W., Cousins, T., Monareng, J., King, K. & Jewitt, G. (2005) *Linking water and livelihoods. The development of an integrated wetland rehabilitation plan in the communal areas of the Sand River Catchment as a test case.* South Africa, Association for Water and Rural Development. Available from: http://www.award.org.za/File_uploads/File/WL%20Phase%20I%20final%20report% 202005-2.pdf [Accessed December 2011].

Riddell, E.S., Lorentz, S.A. & Kotze, D.C. (2012) The hydrodynamic response of a semi-arid headwater wetland to technical rehabilitation interventions. *Water SA*, 38 (1), 55–66.

Riddell, E.S., Lorentz, S.A. & Kotze, D.C. (2010) A geophysical analysis of hydro-geomorphic controls within a headwater wetland in a granitic landscape, through ERI and IP. *Hydrology and Earth System Sciences*, 14 (8), 1697–1713.

Ritter, D.F., Kochel, R.C. & Miller, J.R. (2002) *Process Geomorphology*. Long Grove, IL, Waveland Press. 560 pp.

Chapter 3

Small-scale water- and nutrient-exchange between lowland River Spree (Germany) and adjacent groundwater

Jörg Lewandowski & Gunnar Nützmann
Leibniz-Institute of Freshwater Ecology and Inland Fisheries, Berlin, Germany

ABSTRACT

Water- and nutrient-exchange between surface water and groundwater affects the water quality of both water bodies and, thus, is important for their management. To find out more about the exchange processes a site surrounded by the present river bed and an old branch of the lowland River Spree was equipped with a transect of twelve observation wells. Infiltration and exfiltration of the surface water into the groundwater alternates depending on precipitation and water level fluctuations of the River Spree. Biogeochemical data revealed much more spatial variability than previously thought which might partly be caused by peat mineralisation of relict layers in the soil and by upwelling deep groundwater. Temporal variability of the biogeochemical data is usually low, there are few wells where alternation of in- and exfiltration causes considerable concentration changes.

3.1 INTRODUCTION

Traditionally rivers and groundwater have been treated as distinct entities (Brunke & Gonser, 1997). Nowadays, surface water and groundwater ecosystems are viewed as linked components of a hydrologic continuum (Sophocleous, 2002). The hyporheic zone is an active ecotone (Boulton *et al.*, 1998). The study reported here aims to identify and quantify exchange processes and exchange patterns between groundwater and surface water for a lowland river. Besides investigating the hydrology of in- and exfiltration, turnover of nutrients and other biogeochemical impacts of the water exchange are identified and analysed.

3.2 MATERIALS AND METHODS

The field site is located in north-eastern Germany near the small village of Freienbrink, close to Berlin (average rainfall about 550 mm). The site is located in the floodplain of the River Spree, which is used as pasture land. It is surrounded by the present day river bed and an oxbow of the lowland River Spree, a 6th order river (Fig. 3.1). The present river bed was built around 1960 to shorten the length of the waterway and thus, to decrease travel times. At the same time, the oxbow was closed at the upstream end. In 1992 the oxbow was partly reopened, by placing some tubes at the upstream

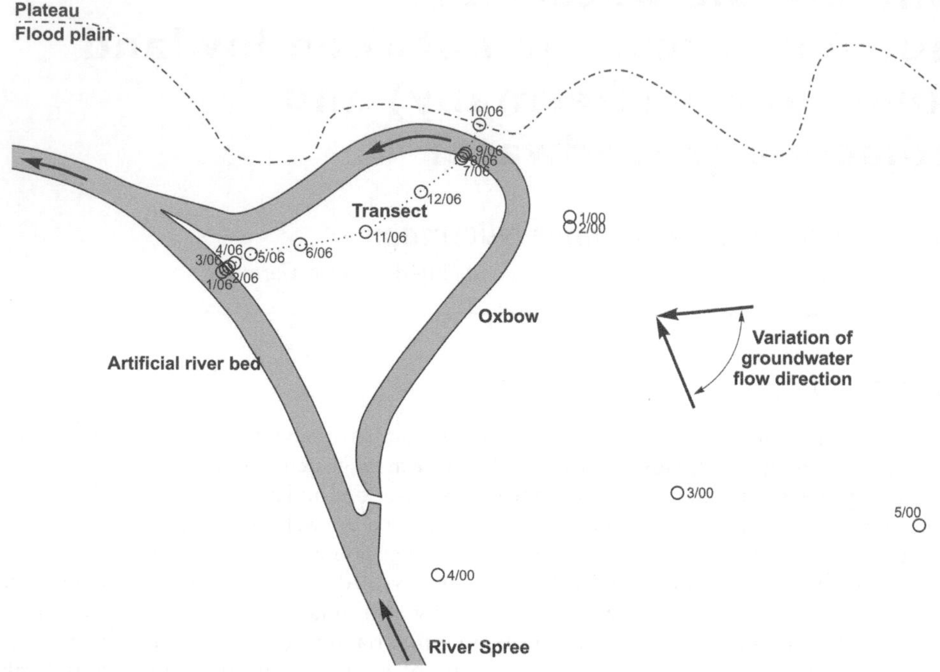

Figure 3.1 Map of the study site east of Berlin close to Freienbrink. Variation of groundwater flow direction estimated from biweekly water level measurements from January to June 2007 conducted in all observation wells shown on the map.

end. Nowadays, due to sediment deposits at the upstream end, there is only a flow through the oxbow at high water levels, whereas the downstream end of the oxbow is always connected to the River Spree. There is a slope of the water level of 2 cm along the 370 m long current river bed and the 780 m long oxbow. The water level of the River Spree is controlled by the Weir Grosse Tränke, where a part of the water is diverted into the Oder-Spree-Channel. The amplitude of water level fluctuations is thus lowered and the height of floods reduced.

The geology of the field site is characterised by medium-grained sands of the Warsaw-Berlin-glacial valley. These medium-grained sands with some finer grade material, fine gravel and few peat burials are highly permeable. The thickness of the aquifer is approximately 30 m. The groundwater flow direction ranges from parallel to the river bed and perpendicular to the river bed. To investigate groundwater flow, groundwater composition and biogeochemical turnover, 12 groundwater wells were installed on a 310 m long transect and equipped with data loggers for automatic measurement of water levels and water temperatures in one-hour-intervals (8 September 2006 to 10 May 2007). Since the groundwater level is close to the surface, all wells have one or two 1 m long filters close to the land surface. Additionally, two wells equipped with data loggers measured the temperature and water level in the River Spree and in the oxbow. Bi-weekly pH, conductivity, redox potential, oxygen, phosphate, nitrate, ammonium, sulphate, dissolved iron and chloride concentrations were measured in the wells to investigate nutrient concentrations (25 September 2006 to 19 June 2007, except iron: 25 September 2006 to 22 May 2007). In five additional wells (1/00–5/00) the water level was measured bi-weekly from January to June 2007 to determine the groundwater flow directions.

3.3 RESULTS AND DISCUSSION

Figure 3.2 shows the water level fluctuations of the River Spree, the oxbow and all observation wells in the transect. All the hydrographs are similar, i.e. they respond quickly to adjacent water level changes. The amplitude of the water level fluctuation in the investigated period was about 1 m. To analyse in more detail the response of the adjacent observation wells to water level fluctuations on the River Spree, a fast increase of the water level of the River Spree (Fig. 3.3A) and a fast decrease in the water level (Fig. 3.3B) have been investigated. Figure 3.3, like Figure 3.2, reveals a close coupling of the observation well adjacent to the River Spree. The time lag of the peak 25 m away from the River Spree (Well 5/06) is approximately 2 hours.

The study site, surrounded by the present river bed and the oxbow (Fig. 3.1), can be regarded as an island with border conditions defined by the water levels of the River Spree on the one side of the island and the water levels of the oxbow on the other side. Thus, there are three cases (Fig. 3.4A). The water level of the River Spree and the oxbow are lower than the groundwater level of the study site. In this case there is an infiltration of groundwater into the oxbow and the river Spree until a new equilibrium is reached. This situation might be a consequence of decreasing water levels in the River Spree or a consequence of increasing water levels of the study site due to groundwater recharge during hydrological winter half-year. The second case (Fig. 3.4B) shows the same water levels in the river Spree, the oxbow and the groundwater. As a consequence the flow rate is negligible. In the third case (Fig. 3.4C) the water level in the River Spree and the oxbow are higher than the groundwater level in the island. Consequently, there is an exfiltration of surface water into the aquifer. This case might be a consequence of increasing water levels in the River Spree.

There are some periods with a clear difference in the water levels in Well 10/06 and in the oxbow (Well 9A), although Well 10/06 is located at a distance of only 8 m from the oxbow shore line (see Fig. 3.2). A high resolution graph (not shown here) shows that there are also periods with clear differences in the water levels in Wells 8 and 7 (also located only 2 or 6 m from the shore line of the oxbow, respectively) compared to the water levels of the oxbow. Even the water level of Well 9, with its screen located directly below the oxbow, shows a clear difference to the water level of the oxbow, measured in Well 9A, in the same position as the previous one, but with a filter screen in the oxbow, instead on the underlying aquifer. The differences in the water levels of adjacent observation wells were the reason for calculating the mean deviation of each observation well compared to the water level of the Spree (Well 1A). These calculations were conducted separately for water levels below and above 33.3 m, because data evaluation revealed that the water level differences were eliminated at higher water levels (Fig. 3.5). The different reactions to water level fluctuations at low and high water levels are a consequence of clogging of the oxbow bed: Fine autochthonous

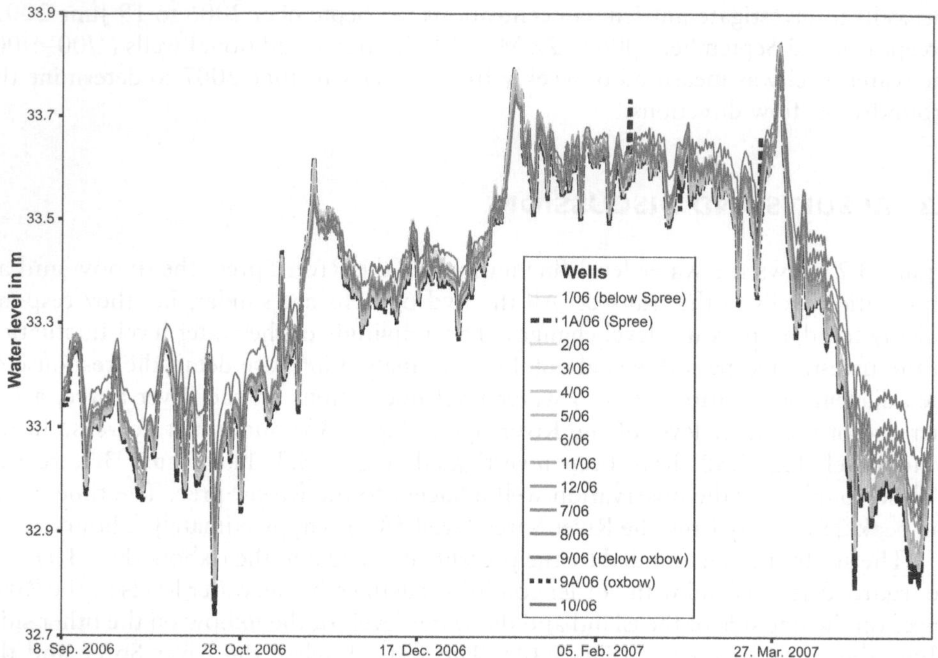

Figure 3.2 Water level fluctuations of the River Spree, the oxbow and all observation wells of the transect (8 September 2006 to 10 May 2007).

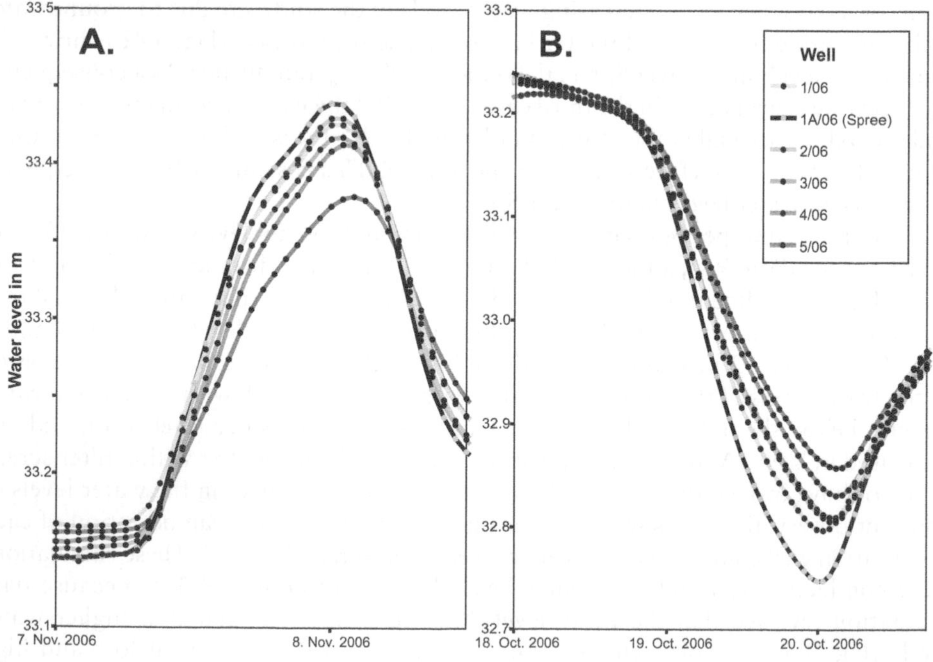

Figure 3.3 Detail of two events shown in Figure 3.2: A. Fast increase of the water level on River Spree and effects on adjacent observation wells. B. Fast decrease of the water level on River Spree and effects on adjacent observation wells.

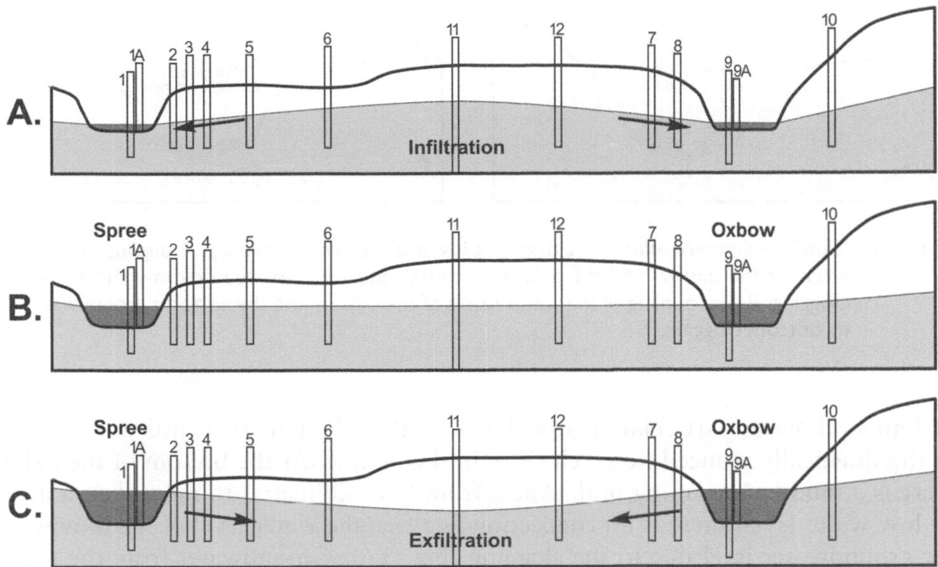

Figure 3.4 Section through the study site along the transect of the observation wells, showing the supposed coupling of water level fluctuations. Not to scale. Well names abbreviated, for complete names add /06 to each one. Cases A, B, and C described in the text.

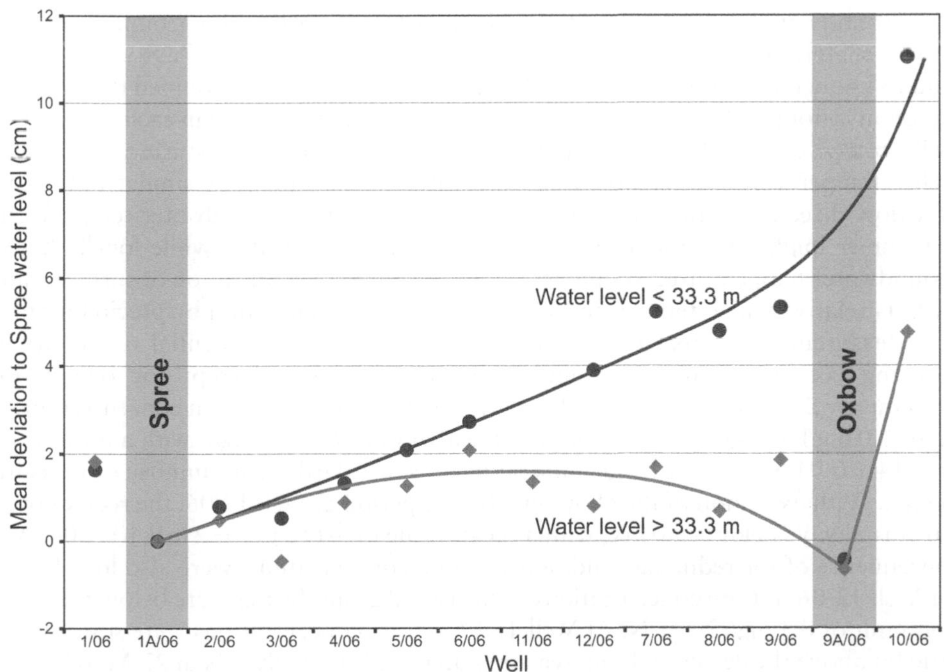

Figure 3.5 Different reactions to water level fluctuations shown as mean deviation to the River Spree water level at low (<33.3 m: 9 September 2006 to 7 November 2006 and 17 April 2007 to 10 May 2007) and high (>33.3 m: 14 November 2006 to 10 April 2007) water levels.

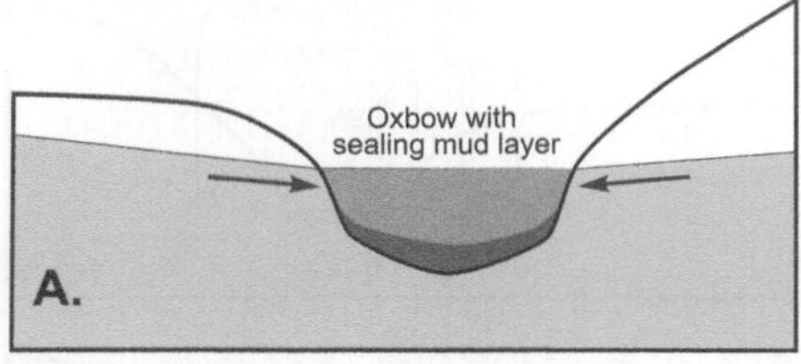

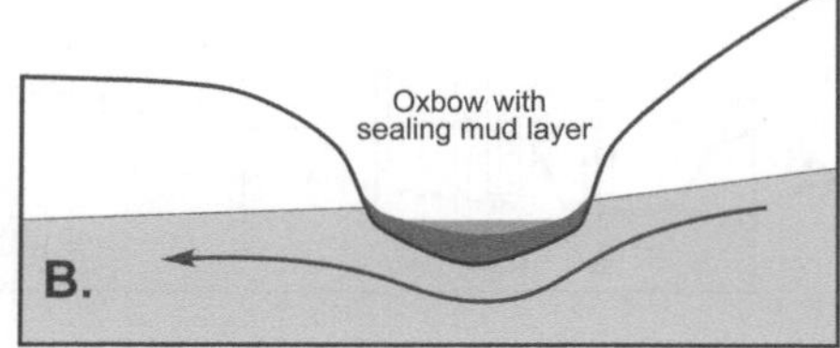

Figure 3.6 Sketch of different effects of clogging at low and high water levels. A. Coupling of the water levels of the oxbow and the adjacent groundwater, due to steep oxbow shores without clogging. B. Decoupling of the water levels of the oxbow and the adjacent groundwater due to bottom clogging.

and allochthonous particulate material settled after the new river bed was built due to the drastically reduced flow velocities in the oxbow. At the bottom of the oxbow there is a mud layer varying in thickness from few millimetres to more than a metre. At low water levels there is no connection between the water level of the oxbow and the groundwater level due to the clogging (Fig. 3.6). Groundwater from the nearby plateau flows beneath the oxbow towards the River Spree. That causes the water level differences between the oxbow and the adjacent observation wells. In the case of higher water levels (Fig. 3.6A) there is a connection of the oxbow to the underlying groundwater since the clogging seals only the bottom of the oxbow and not the steep oxbow shores.

Groundwater levels measured outside the island area reveal a groundwater flow approximately perpendicular to the observation transect. Although there is insufficient data to know if this flow direction is also true for the island, it is assumed that the flow direction is not parallel to the transect. Temperature data recorded in each observation well showed quite different groundwater temperatures on neighbouring observation wells. Temperatures of neighbouring observation wells should be similar in the case of a flow direction parallel to the transect. Locally lower groundwater temperatures in summer might be a result of upwelling deeper groundwater, while locally higher groundwater temperatures in summer might be caused by exfiltration of surface water with its relatively high temperatures or groundwater regeneration by precipitation.

Measurements of oxygen concentrations and the redox potential reveal strictly reducing conditions in the groundwater of the island. Except for one period (30 January 2007 to 27 March 2007) at Well 12/06, with an oxygen concentrations of up to 1.0 mg/l, oxygen concentrations were always below 0.5 mg/l, with a mean value of 0.14 ± 0.01 mg/l, $n = 215$ (arithmetic mean $\pm$ standard error, number of measurements). With exception of the above mentioned period at Well 12/06, the redox potential was always below -80 mV, with a mean value of -150 mV $\pm$ 2 mV, $n = 204$. As a consequence of the reducing conditions, nitrate concentrations were also low. Except for Well 12/06, nitrate concentrations of the island groundwater were below the detection limit of 0.1 mg/l NO_3-N. At Well 12/06 nitrate concentrations were sometimes a little bit above the detection limit, reaching up to 3.74 mg/l NO_3-N at 27 March 2007.

In contrast to the reducing condition of the island groundwater located in the flood plain, the groundwater of the plateau (Well 10/06) had oxygen concentrations

Table 3.1 Mean SRP and dissolved iron concentrations at the observation wells, in the River Spree and in the oxbow, measured biweekly in the period 25 September 2006 to 19 June 2007; ± standard error; n number of measurements.

Well	*SRP conc. (μg PO_4-P/l) (arithmetic mean)*	*Dissolved iron conc. (mg /l) (arithmetic mean)*
Spree	46 ± 5.3 (n = 20)	<0.1 (limit of quantification)
1/06	278 ± 6.0 (n = 20)	2.9 ± 0.21 (n = 18)
2/06	463 ± 17.0 (n = 20)	12.3 ± 1.16 (n = 18)
3/06	473 ± 11.7 (n = 20)	11.9 ± 1.28 (n = 18)
4/06	517 ± 9.9 (n = 20)	12.3 ± 0.98 (n = 18)
5/06	1192 ± 17.1 (n = 20)	36.6 ± 1.07 (n = 18)
6/06	419 ± 7.1 (n = 20)	36.6 ± 1.85 (n = 18)
11/06	420 ± 8.7 (n = 20)	7.4 ± 0.34 (n = 18)
12/06	445 ± 50.7 (n = 20)	11.6 ± 0.65 (n = 18)
7/06	353 ± 10.6 (n = 20)	3.9 ± 0.45 (n = 18)
8/06	283 ± 8.0 (n = 20)	2.6 ± 0.50 (n = 18)
9/06	151 ± 4.7 (n = 20)	0.5 ± 0.05 (n = 18)
Oxbow	84 ± 16.1 (n = 20)	<0.1 (limit of quantification)
10/06	82 ± 1.4 (n = 20)	<0.1 (limit of quantification)

of 2.3 to 3.9 mg/l, with an arithmetic mean of 3.0 ± 0.1 mg/l (n = 20). The redox potential of that groundwater was between 110 and 230 mV, with an arithmetic mean of 190 ± 10 mV (n = 19). Nitrate concentrations at Well 10/06 were between 17.1 and 36.3 mg/l NO_3-N, with one outlier of only 4.1 mg/l. The arithmetic mean of 25.1 ± 1.4 mg/l NO_3-N is surprisingly high for groundwater originating from a forested area.

At Well 12/06 a clear change throughout the study period could be observed, with highest values of approximately 700 μg PO_4-P/l in October and November 2006 and lowest values of approximately 200 μg PO_4-P/l in April and May 2007. Annual fluctuations of the measured soluble reactive phosphorus (SRP) concentrations also occurred in the surface water samples of the River Spree and the oxbow. At the other observation wells, SRP concentrations were pretty constant throughout the study period, as shown by a mean coefficient of variation (mean relative standard deviation) of 11% (Table 3.1). The SRP concentrations determined for the groundwater of the flood plain are high, which might be a consequence of cattle breeding or of the mineralisation of peat locally present in the soil. The locally higher SRP concentration at Well 5/06 is also caused by peat mineralization.

Compared to the SRP concentrations, the concentrations of dissolved iron were fluctuating much more (Table 3.1, mean coefficient of variation 38%). At the neighbouring Wells 2/06, 3/06 and 4/06 iron concentrations increased during the winter (November to April) from below 10 mg/l up to 20 mg/l. At the other wells no trend in the fluctuations could be detected. Highest iron concentrations of approximately 37 mg/l were detected at the neighbouring Wells 5/06 and 6/06.

Chloride concentrations in the Spree were very constant (coefficient of variation 10%) throughout the investigated period: 50 ± 1.1 mg/l (n = 20). Chloride concentrations in the groundwater of the plateau (Well 10/06) were lower 19 ± 1.0 mg/l (n = 20) but fluctuated a little bit more (coefficient of variation 23%) than the chloride concentrations in the surface water. In Wells 12/06 (10 ± 0.3 mg/l, n = 20), in 7/06 (15 ± 0.7 mg/L, n = 20), and partly in 9/06 (25 September 2006 to 24 April 2007: 14 ± 0.4 mg/l, n = 16) even lower chloride concentrations than in Well 10/06 were detected. Since chloride can be regarded as an ideal tracer, those lower concentrations indicate that the groundwater at that wells originates from a different groundwater resource with lower chloride concentrations than the groundwater of the plateau or originates from precipitation.

At high water levels (>33.4 m) in the oxbow chloride concentrations in the oxbow were similar to the chloride concentrations in the Spree, since a considerable part of the Spree water flows through the oxbow (mean relative deviation of chloride concentrations in the oxbow to that of the Spree 1 ± 0.2%, n = 8). At low water levels (<33.4 m) the Spree water was diluted by groundwater with its low chloride concentrations, although groundwater infiltration was small due to clogging of the oxbow bed. Dilution resulted in a mean relative deviation of chloride concentrations in the oxbow to that of the Spree of 19 ± 5.2%, n = 11. The highest deviation 41 ± 3.7%, n = 3 occurred at water levels below 33 m and was probably caused by the fact that, at such low water levels, there is little to no flow in the oxbow over long time periods, resulting in an enrichment of groundwater with its extremely low chloride concentrations in the oxbow.

Interestingly, for the observation wells closest to the River Spree, chloride concentrations higher than the ones occurring in the groundwater of the plateau or in the surface water of the River Spree were determined: Well 1/06: 60 ± 1.9 mg/L, n = 19; Well 2/06: 144 ± 15 mg/L, n = 19; Well 3/06: 147 ± 14, n = 19; Well 4/06: 174 ± 8, n = 19. Since chloride is an ideal tracer it is hard to explain chloride concentrations higher than the concentrations occurring in the surface water or in the groundwater. At first, it was assumed that mineral cattle feed given to cattle close to those observation wells caused the high chloride concentrations. However, sulphate was also high. Like for chloride, highest sulphate concentrations occurred close to the River Spree and there was much temporal fluctuation (Fig. 3.7). However, the cattle feed used in that area contains no sulphate, so there must be different reasons for the observed sulphate concentrations. Further investigations, not yet published (geoelectric measurements, groundwater dating, hydrogeochemical genesis model) revealed that the high chloride and sulphate concentrations are caused by local upwelling of deep saline groundwater. Drastic fluctuations (coefficient of variation at Well 2/06 46% and at Well 3/06 42%) of the chloride concentrations at the observation wells close to the River Spree might be caused by alternation of in- and exfiltration or changes in the flow direction.

In total, there are a couple of observation wells where the measured chemical data are quite homogeneous over time. That homogeneity of the data proves the reliability of sampling and analytic techniques. There are also some observation wells with less temporal homogeneity. Some show an irregular fluctuation caused by alternation of infiltration and exfiltration, others follow a temporal trend. However, measured concentrations differ spatially much more than they differ temporally. This heterogeneity can only be explained by different origin and travel times of the groundwater, although some higher concentrations of phosphate and other ions might be a result of small amounts of peat present in the aquifer due to former fens. Peat mineralisation

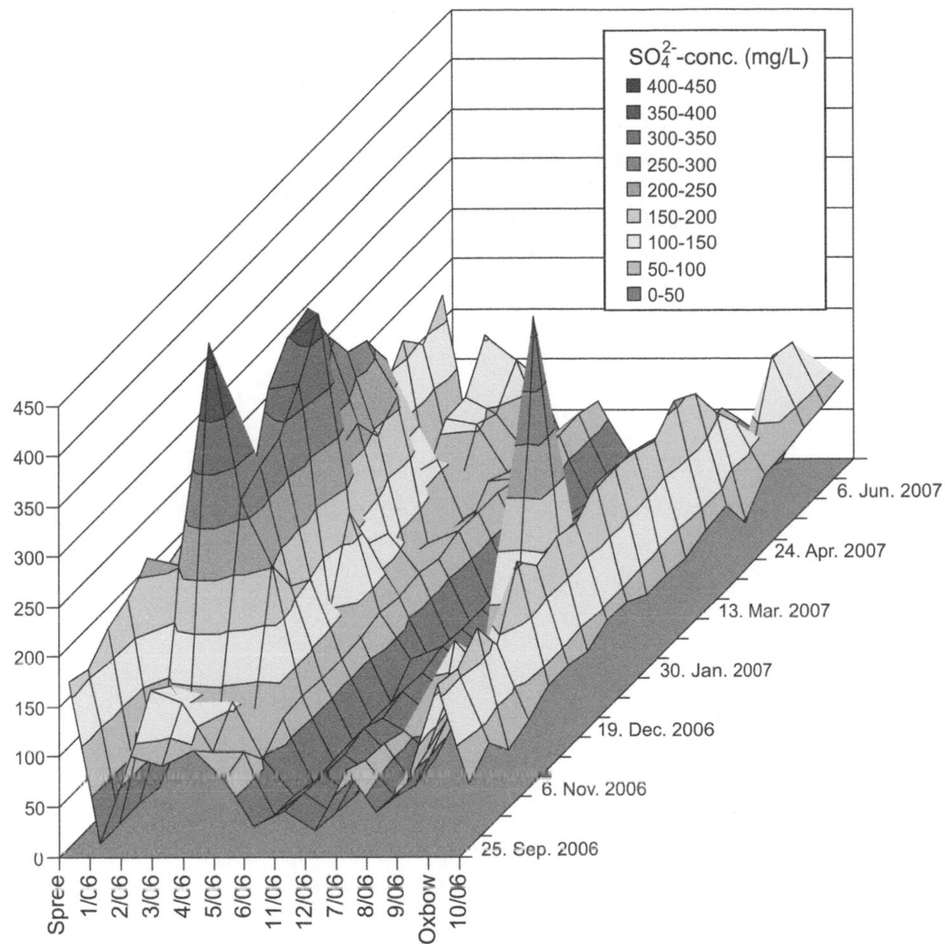

Figure 3.7 Sulphate concentration in groundwater of the study site, in the River Spree and in the oxbow, from 25 September 2006 to 19 June 2007.

might result in high phosphate, ammonium and iron concentrations (Lewandowski and Nützmann, 2010). Additionally, the extremely high concentrations of chloride detected close to the shore line of the River Spree were probably caused by upwelling of deep salty groundwater.

ACKNOWLEDGEMENTS

The authors gratefully acknowledge the help of Grit Siegert, Juliane Ackermann, Elke Zwirnmann, Sarah Schellenberg, Antje Lüder, Hans-Jürgen Exner and Christiane Herzog with sampling and chemical analyses.

REFERENCES

Boulton, A.J., Findlay, S., Marmonier, P., Stanley, E.H. & Valett, H.M. (1998) The functional significance of the hyporheic zone in streams and rivers. *Annual Review of Ecology and Systematics*, 29, 59–81.

Brunke, M. & Gonser, T. (1997) The ecological significance of exchange processes between rivers and groundwater. *Freshwater Biology*, 37, 1–33.

Lewandowski, J. & Nützmann, G. (2010) Nutrient retention and release in a floodplain's aquifer and in the hyporheic zone of a lowland river. *Ecological Engineering*, 36, 1156–1166.

Sophocleous, M. (2002) Interactions between groundwater and surface water: The state of the science. *Hydrogeology Journal*, 10, 52–67.

Chapter 4

Artificial maintenance of groundwater levels to protect carbonate cave fauna, Yanchep, Western Australia

Cahit Yesertener

Department of Water, Perth, Australia

ABSTRACT

The Yanchep Caves, Western Australia, Perth, contain a unique ecological system that is dependent on groundwater for its survival. Due to declining groundwater levels in the Gnangara Groundwater Mound in the same area since 1969, mainly from reduced rainfall, increased use of groundwater for public and private water abstraction and pine plantations, the groundwater levels under the caves have declined thus increasing the stress on the dependent cave fauna. Despite the artificial sprinkling system for protecting the fauna, the water levels have continued to decline since monitoring began in 1991. It has become obvious that a more robust and generous provision of water is needed. To evaluate the amount of groundwater required for each of the caves and to determine the effects of pumping groundwater from the superficial aquifer, a three-dimensional groundwater flow model was constructed using Visual MODFLOW Pro version 4.1 coupled with MODFLOW SURFACT. The models estimate that a total discharge rate of up to 3.6 Gl per annum would maintain water in ponds in each priority cave in both summer and winter until 2015. The effects of pumping on groundwater dependent environments have been evaluated. The cone of depression resulting from pumping of these bores stabilises in two years and indicates that there may be a 0.5 m drawdown within approximately 1 km of the boreholes with no impact on the lake system towards the east after the second year of operation.

4.1 INTRODUCTION

The Yanchep Caves contain a unique ecological system that is dependent on groundwater for their survival. The caves are located within the Yanchep National park, located about 45 km north of Perth, Western Australia (Fig. 4.1). The National Park covers an area of 26.8 km^2 and contains about 300 caves. Seven of these caves previously had permanent streams and pools supporting cave root mat communities. Due to declining groundwater levels within the Gnangara Groundwater Mound, which started in 1969 due to reduced rainfall, increased use of groundwater by public and private water abstraction and new pine plantations, the groundwater levels under these caves have also been declining (Yesertener, 2005, 2008). The decline in groundwater levels has increased the stress on the cave fauna since the mid 1990s as the fauna is dependent on groundwater for survival.

An artificial sprinkling system designed and installed to support the fauna, appeared to work for a few years. In recent years, as groundwater levels have reduced on average by 0.8 m annually since monitoring began in 1991, it has become clear

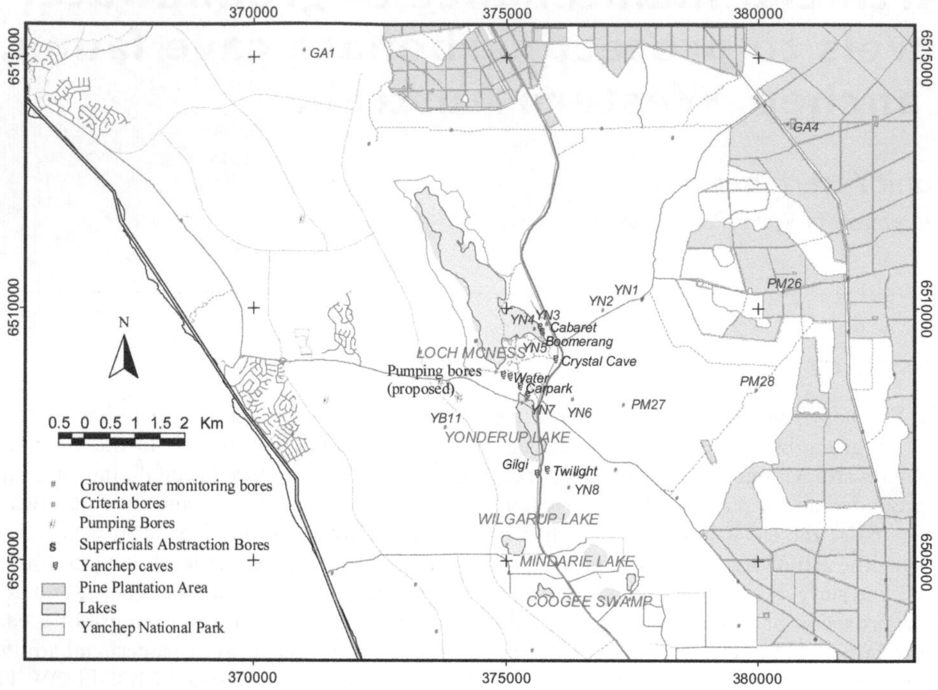

Figure 4.1 Yanchep Caves.

that a more robust and generous provision of water is needed. Over the last two years, the Department of Environment and Conservation, Water Corporation and the Department of Water (DoW) have co-operated to design and trial a new emergency re-watering system for these caves.

The short term objective of the emergency re-watering system is to test the feasibility of re-hydrating the cave system by establishing and maintaining local groundwater mounds at seven of the faunal caves. The longer-term objective is to develop a permanent artificial system to reinstate and protect the threatened ecological invertebrate communities of Stygofauna associated with Tuart tree root mats. The artificial maintenance trial, which was carried out between December 2002 and September 2003, provided valuable information on estimation of the required water to maintain the caves, and it was agreed that a permanent maintenance system could be accomplished (Calvert & Yesertener, 2005).

To evaluate the amount of groundwater required for each of the seven caves and to determine the effects of pumping groundwater from the superficial aquifer, a three-dimensional groundwater flow model was constructed. The groundwater flow model estimates the recharge augmentation required to maintain water in one or more ponds in each priority cave in summer and winter from 2006 and up to 2015.

This paper discusses a model to predict the long term effects of declining water levels and artificial recovery of levels for sustainability.

4.2 HYDROGEOLOGY

The Yanchep groundwater area is a part of the Gnangara Groundwater Mound, which is the major groundwater supply to Perth. Groundwater occurs in pervious superficial formations, which are mainly sand and limestone. Limestone, which covers most of the study area, is an extensive, karstic and highly productive unconfined aquifer (Fig. 4.2). The permeable units are mainly sand decomposed from limestone on the east of the lake system, and mainly limestone in the west. The dashed line in Figure 4.2 separates a fissured aquifer on the west from a mainly intergranular aquifer on the east. The unconfined superficial aquifer is connected with underlying confined aquifer on the east of the dotted line in Figure 4.2; elsewhere they are disconnected by impervious or semi-pervious layers. The groundwater flow direction is southwest towards the ocean. Hydraulic gradients generally range from an average of 0.005 within the sandy aquifer in the east to 0.0015 within the karstic limestone towards the west depending on hydraulic conductivity changes in the aquifers. Lake Loch McNess, Yonderup Lake, Wilgarup Lake, Pippidinny Swamp and Coogee swamp occur within the inter barrier depression with prominent karstic phenomena and are located on the eastern margin of the limestone.

Loch McNess and Pippidinny Lake are permanent lakes, which are surface expressions of the groundwater table and are also considered as groundwater throughflow lakes. There are no known pump test analyses for the study area, however, Davidson (1995) provided various hydraulic conductivity values (K) modified from Hazel (1973). According to these values, in sandy parts of the superficial aquifers K values vary between 4 to 50 m/d depending on the grain sizes. Limestone K values vary from 100–1000 m/d. Recharge to the unconfined aquifer is mainly from rainfall percolation, even though some limited recharge from direct rainfall occurs through the lake system. Rainfall recharge estimations were conducted in a number of studies since 1970 (Davidson, 1995).

Groundwater levels within the Yanchep Caves area are on decline due to declining groundwater levels in the Gnangara Groundwater Mound since 1969. Hydrograph analysis shows that the major cause of this decline within the Yanchep area is the reduction in rainfall, however there is also some impact from local groundwater abstraction. Groundwater levels within the caves area have declined over 1.0 m within the last 10 years (Fig. 4.3).

Groundwater salinity of the superficial aquifer ranges from 170 mg/l (GA4) to 240 mg/l TDS (YN2) in the intergranular aquifer of mainly calcareous sands in the east, and from 300 mg/l (YN6) to 710 mg/l TDS (YN8) in the mostly carbonate aquifer west of the dashed line in Figure 4.2.

The pH values of groundwater vary within the different superficial aquifers. Within the intergranular aquifer groundwater at the top of the aquifer is acidic, with a range of pH 5.8 to 6.6. The carbonate aquifer, however, has groundwater with pH ranges from 7.0 to 8.0.

Examination of the Schoeller diagram indicates that groundwater is NaCl type within the intergranular aquifer towards the east and gradually mixes with a $CaHCO_3$ type of groundwater while passing through the calcareous sands (Fig. 4.4). The concentrations of Ca and HCO_3 ions gradually increases within the calcareous sands and become dominant within the carbonate aquifer (YN5, GA1, and YN8).

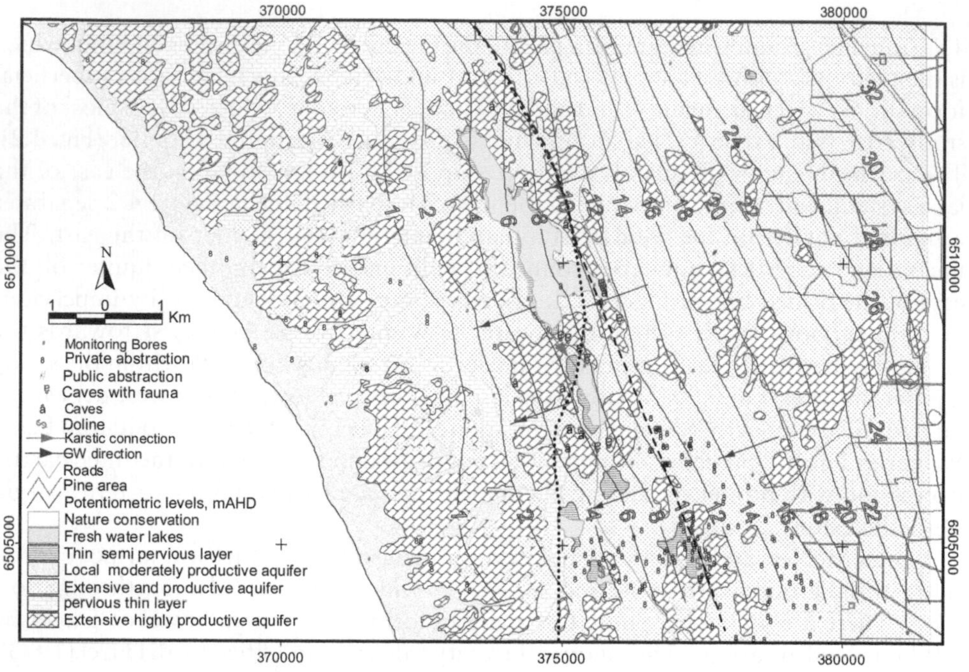

Figure 4.2 Hydrogeology map of the study area.

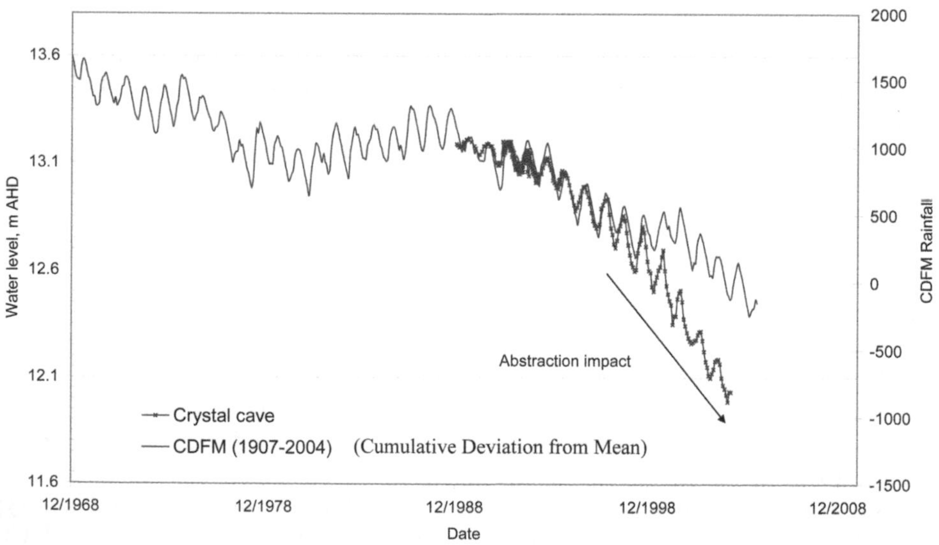

Figure 4.3 Groundwater Hydrograph, Crystal cave.

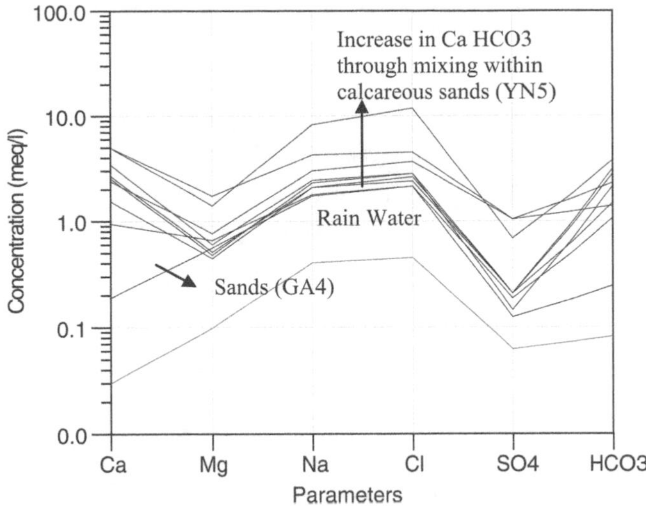

Figure 4.4 Schoeller plot showing the groundwater types.

Groundwater is generally unsaturated with respect to calcite within the carbonate aquifer except in GA1 and YN5. Barber (2003) also showed that the carbonate aquifer under and in the vicinity of the cave system is unsaturated except in some ponds in Crystal cave. The Calcite saturation index of the groundwater is −0.067, showing that groundwater is near saturation, therefore, it is not aggressive water that would dissolve more limestone within the caves.

4.3 GROUNDWATER FLOW MODELLING

Visual MODFLOW Pro version 4.1 coupled with MODFLOW SURFACT, developed by HydroGeoLogic Inc., was used for the simulation of groundwater flow in the superficial aquifer. MODFLOW SURFACT is a fully integrated groundwater flow package based on the USGS MODFLOW code (McDonald and Harbaugh, 1988) and has the capability of modelling unsaturated moisture and air movement, which reduces the unsaturated flow problems, accurately delineates the water table elevations, and captures the delayed yield response of an unconfined system to pumping and recharge.

The model domain covers an area of 16 km from east to west and 12 km north to south. It is bounded by the Indian Ocean to the west and the pine plantation in the east and covers the whole Yanchep National Park (Fig. 4.5). The northern and southern extents of the model were selected to ensure minimum boundary effects. The ocean is assigned as a constant head boundary along the coast, with inactive cells beyond it. The northeastern extent of the model is assigned as an inflow boundary and simulated as an infinite source of water (constant head). Regional decline of the groundwater level is introduced to the inflow boundary at northeast. The model domain was divided into 160 × 120 uniform cells each measuring 100 m × 100 m and with 5 layers. A total

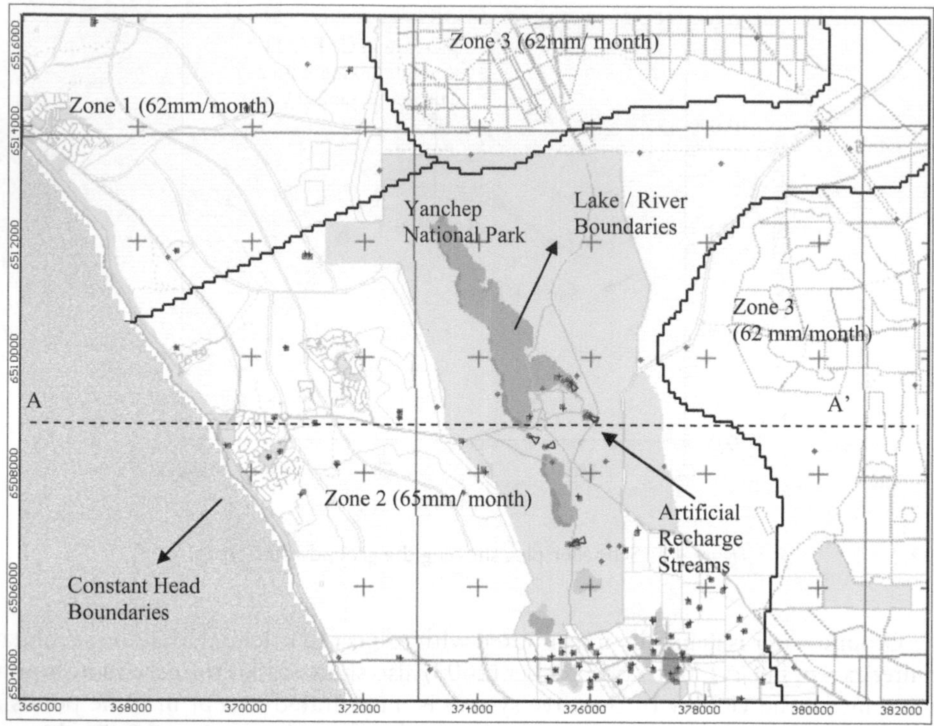

Figure 4.5 Model Domain, boundaries, observation bores (squares), abstraction bores (circle with plus sign), and recharge zones with monthly figure. Triangle signs show the proposed artificial recharge streams.

of 96 000 model cells were generated. The top of the model corresponds to the surface topography.

Based on the conceptual model, five physical model layers were constructed to represent the hydrogeological units. The first three layers represent the superficial aquifers (sand and limestone) and the remaining two represent the confined aquifer and confining or semi confining layers in between the aquifers.

The aquifer parameters assigned to each of the modelled hydrogeological units are the optimised values for each unit using PEST within the given ranges, even though they are known to vary locally (Table 4.1, Fig. 4.6).

Loch McNess, Yonderup, Wilgarup Lake, Pippidinny Swamp and Coogee Swamp (see Fig. 4.1) were simulated using the river package from MODFLOW. This allows both inflows and outflows depending on the river stage and the surrounding groundwater levels. Monitored monthly water levels of these large surface bodies were assigned as their river stage levels and the rates of the inflow or outflows were governed by the conductance of the riverbed and river stages.

The licensed public and private production bores in domain area were simulated in the model as part of the steady state and transient calibration starting from 1996. Boreholes are represented as sinks with specified discharge rates, which can vary over time.

Table 4.1 Aquifer parameters used in the model.

Layers	*K Zones*	K_x	K_y	K_z	S_y/S_s
Sand	(3)	5–10	5–10	0.5–2.0	0.15–0.25
Calcareous sand	(4)	5–10	5–10	0.5–2.1	0.15–0.25
Limestone	(2 and (8)	20–300	20–300	2–7	0.2–0.35
Confining layer (Shale)	(6)	0.001	0.001	0.0001	1E-3–1E-5
Confined aquifer (sand)	(7)	10–20	10–20	1–2	0.2–1E-5
Confined layer (clayey sand)	(5)	1–5	1–5	0.1–1	0.1–0.01

K = Hydraulic conductivity, m/day: x, y and z show the directions in Cartesian coordinate system
S_y = Specific yield, S_s = Storage coefficient

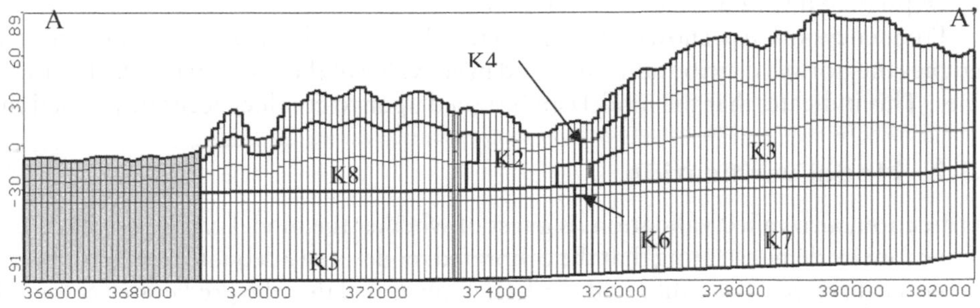

Figure 4.6 Cross-section (A-A') showing the hydraulic conductivity (Kx) zones.

Since the information on the abstraction rates for most private boreholes is lacking, annual allocated discharge rates have been used for these private sources. However, six public borehole abstraction rates were supplied by the Water Corporation, therefore, monthly discharge rates for these have been used in model calibration and 2004 discharge rates have been applied to the model until 2015 for further predictions.

Groundwater levels in a total of 32 monitoring boreholes from 1996 onward were used in steady state and transient model calibrations. The 1996 minimum water levels were selected as the initial head and the model converged to a steady state solution. The converged and predicted head water levels were then used in transient model calibration.

The main source of recharge into the unconfined aquifer comes from rainfall infiltration. The model used the monthly recharge rates and evapotranspiration rates calculated from the CDFM technique (Yesertener, 2005, 2008). The surplus monthly rainfalls above the long-term average (threshold value) are considered as net recharge to the groundwater and deficient rainfalls below the long-term average are considered as real evapotranspiration from the groundwater. The recharge zones and threshold values are given in Figure 4.5.

Recent DoW studies show that mature pines, Zone 3 in the model, are reducing the groundwater recharge by approximately 30% (Yesertener, 2005). This reduced recharge has been included in the model. The recharge is applied to the highest active cell, to simulate water entering and recharging the water table aquifer.

4.3.1 Model calibrations

Calibration of the steady state model was accepted with a correlation coefficient of 0.997 between predicted and measured water levels. The standard error of the estimate was 0.135 m. After achieving a steady state run with high correlation, the model was run for transient state from January 1996 to end of 2015. The transient data set for monitoring bores and abstraction bores is for the period 1996 to 2003. It has been assumed that the abstraction rate in 2003 will stay the same during the modelled period. There are small changes made for the hydraulic conductivities along the transition zone between the layers using PEST parameter optimization package to get the best correlation coefficient.

Best calibration is achieved with 0.016 m standard error of the estimate for all times. The correlation coefficient is 0.988, which shows that about 98% of the data can be predicted with high level of reliability.

Frequency analysis shows that residual values for all times in the transient run matches the normal distribution curve and mean value of the residual is −0.31 m for all times. This indicates that the model can be relied upon to conduct accurate predictions.

4.3.2 Model verification

Verification, also called validation, is a test of whether the model can be used as a predictive tool by demonstrating that the calibrated model is an adequate representation of the physical system. The calibrated model demonstrated that the prediction reasonably matches the observations of the reserved data set, deliberately excluded from consideration during calibration (Murray-Darling Basin Commission, 2001). The Crystal cave monitoring data set is deliberately excluded from consideration during the calibration; however, model prediction for Crystal Cave closely matches the observation of the Crystal Cave data set as seen in Figure 4.7.

4.3.3 Model simulations

The management committee suggested constructing a groundwater flow model to simulate the groundwater levels around the caves. It was designed to predict the groundwater requirement to maintain the groundwater levels under the caves in a year, and also the water requirements for a longer period, until 2015, and evaluate the environmental impact of the discharge wells to the surrounding wetlands and ecosystems.

Two scenario runs were selected for this paper as follows:

- Scenario 1: Long term water requirement (10 years)
- Scenario 2: 1 day, 7 days and 30 days failure in operation to maintain the levels under the caves

Model results for the scenario runs for water requirement for short term (1 to 3 yrs) is subjected to other paper.

Scenario 1: Long term water requirement

To predict the optimum water requirement for long-term to maintain the groundwater levels under the caves, needed to evaluate target minimum groundwater levels for each individual caves for 2015.

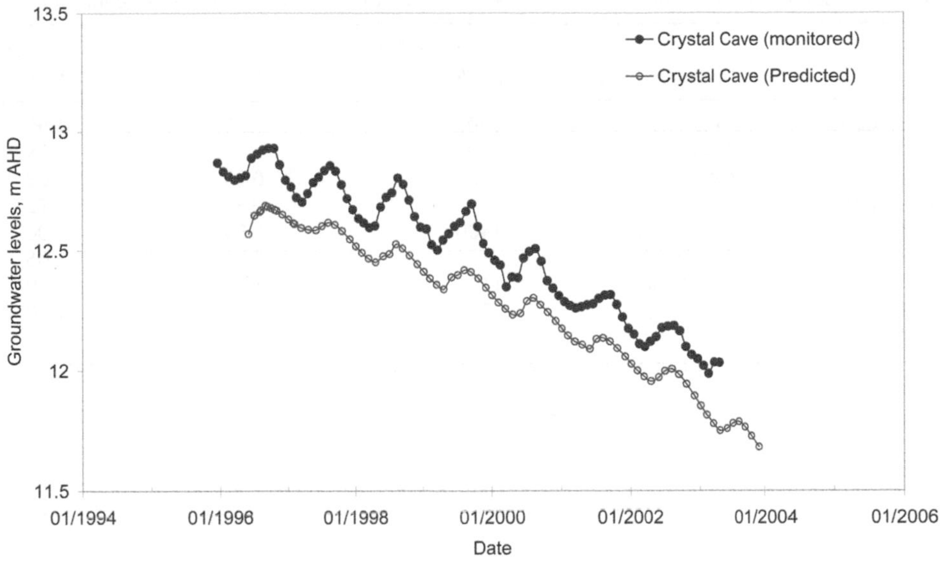

Figure 4.7 Crystal Cave groundwater level observations and model predictions in same location.

Table 4.2 Local groundwater decline levels per annum.

Cave	*Name*	*Easting*	*Northing*	*Trend[1], m/yr*	*Trend[2], m/yr*	*Target 2015[1]*	*Target 2015[2]*
Yn1	Crystal	375946	6508974	0.090	0.040	2.25	1.75
Yn5	Cabaret	375637	6509606	0.070	0.030	1.30	0.9
Yn11	Water	374999	6508634	0.030	0.020	0.90	0.8
Yn18	Carpark	375247	6508443	0.040	0.020	1.05	0.85
Yn27	Gilgi	375685	6506740	0.050	0.020	1.75	1.45
Yn99	Boomerang	375664	6509515	0.070	0.030	1.35	0.95
Yn194	Twilight	375780	6506795	0.050	0.020	1.75	1.45

[1]Target rise estimated using Hydrograph analysis; [2]Target rise estimated from the groundwater model runs
*All target levels are minimum groundwater levels rise calculated for summer period.

The local groundwater decline trend under the caves has been calculated using hydrograph analysis and estimated form do nothing scenario model run for individual caves and given in Table 4.2. Target groundwater levels, for example for Crystal Cave, have two figures, target level generated from modelling and the target level generated from hydrograph analysis using the CDFM. Due to continuous artificial recharge to the caves has positive impact in reducing the magnitude of the declining trend of the groundwater, target level for the year 2015 given at column 9 is less than the target level given in column 8 in Table 4.2. These figures are 2.25 m and 1.75 m for Crystal caves, respectively.

The model was run a number of times for achieving the target groundwater levels including 2015, using a trial and error approach using different artificial recharge rates

Table 4.3 Modelled artificial recharge estimates for seven caves for long-term water requirements

Caves	*Discharge Point*	*Discharge Rates, (m^3/d)*	*Target Level Rise, m*	*Prediction Rise, m*
Crystal Cave	Crystal 1	1800	1.75	1.99
	Crystal 2	2000		1.99
Cabaret	YN6	1200	0.90	1.03
Boomerang			0.95	1.92
Water	Water Cave	1400	0.80	1.06
Car Park	Car park cave	800	0.85	1.05
Gilgi	Gilgi Cave	1100	1.45	1.60
Twilight	Twilight Cave	1300	1.45	1.70
Total Discharge Rate:		9600 (3.6 Gl)		

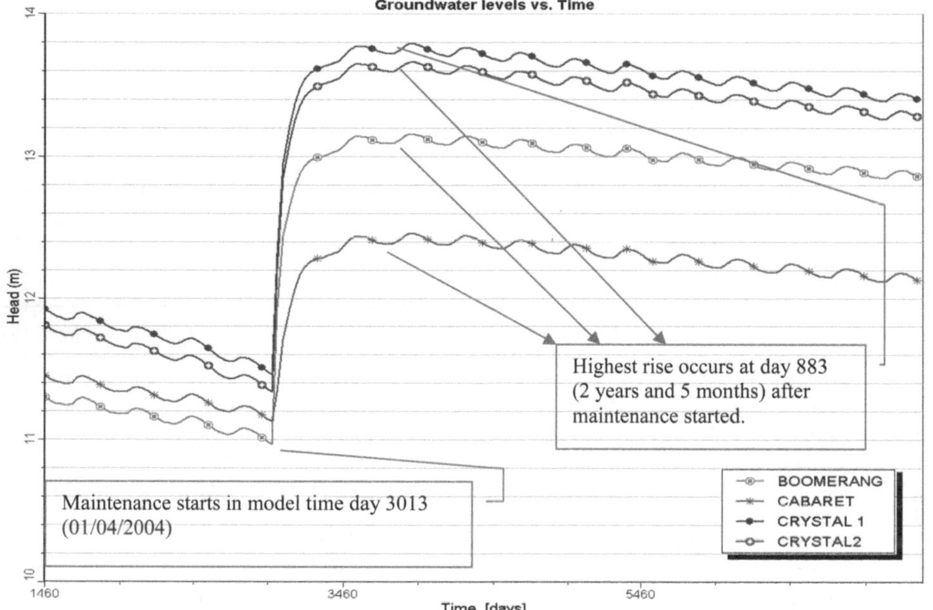

Figure 4.8 Groundwater level changes under caves resulting from the long-term artificial maintenance programme.

and recharge points for each cave. The best result was achieved with the discharge rates and recharge points given in Table 4.3. Note that there are two discharge points in Crystal Cave, which are assumed to be upstream of Jewel City and in the Pantheon Cavern.

The model estimates artificial recharge required to maintain water in one or more ponds in each priority cave in summer and winter 2006 and up to 2015. A total artificial recharge rate of up to 3.6 Gl/yr should maintain ponds in the caves to 2015 (Table 4.3).

Figure 4.8 shows the groundwater levels changes in m AHD under the individual caves as predicted from the long-term artificial maintenance programme. Model results

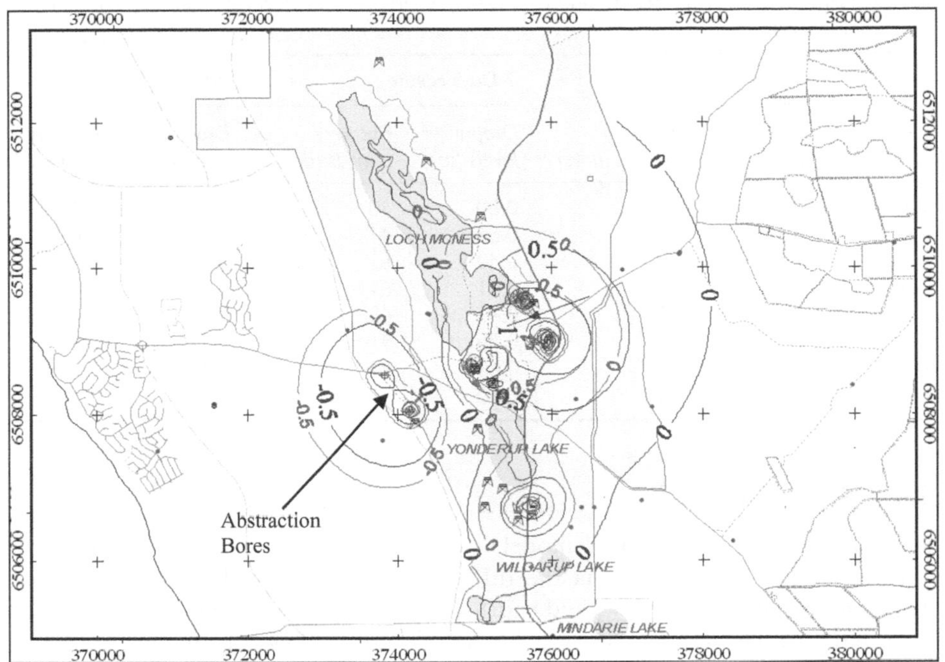

Figure 4.9 Artificial Recharge to Yanchep Caves – Maximum rising effects (883rd day – September 2006 (larger fonts) and 10th year-2015 effect (smaller fonts)).

show that groundwater levels reach their maximum levels after 883 days, which is equivalent to about 2.5 years and establishes a new equilibrium with a declining trend due to the effect of the regional groundwater level decline.

To show the environmental effects of the artificial recharge to Yanchep caves and the depression cone resulting from the supply bores, areal distribution of the rise and decline contours have been given for the maximum water level rise (883rd day ≈ September 2006), and year 2015 (Fig. 4.9).

The predicted drawdown at the two supply bores (see Fig. 4.9) is between 2–3.2 m. In respect to the environment, there are no GDEs other than Lake McNess that may be affected by pumping. The effect of pumping from the bores indicates that there may be a 0.5 m drawdown within approximately 1 km of the bores, particularly toward the west after the second and tenth year of operations, however there is no negative impact to Loch McNess (Fig. 4.9).

Scenario 2: Failures in operation for 1, 7 and 30 days

It is possible that artificial water supply into the groundwater system may fail for short periods of time due to power failure, pump failure or other factors. The results show that the decline of groundwater levels and duration required for recovery increase in proportion to the duration of failure. Individual groundwater responses under the caves and the duration required to recover the water levels are summarised in Table 4.4.

Table 4.4 Groundwater level decline and recovery periods under the caves resulting from failures in operation.

	1 Day Failure		*7 Days Failure*		*30 Days Failure*	
Caves	*Drop in levels (m)*	*Recovery periods (days)*	*Drop in levels (m)*	*Recovery periods (days)*	*Drop in levels (m)*	*Recovery periods (days)*
Crystal	0.24	32	0.80	70	1.24	131
Cabaret	0.13	10	0.51	41	0.83	106
Boomerang	0.08	15	0.40	61	0.74	115
Water	0.20	24	0.61	61	0.69	91
Car Park	0.14	30	0.47	65	0.59	111
Twilight	0.21	50	0.75	91	1.03	156
Gilgi	0.18	50	0.68	91	0.97	157
Average	0.17	30	0.60	69	0.87	124

4.3.4 Model limitations

The Yanchep groundwater flow model is a local model, which is connected to the regional groundwater system via the inflow boundary parallel to the groundwater potentiometric line on the east, and outflow boundary along the coastline. The level of regional groundwater decline over the next decade is obviously an important factor affecting recharge rates to maintain cave pools. Although the best possible estimate of the future regional groundwater trend has been calculated by a number of methods and has been considered on the inflow boundary in the model, it will depend on the rainfall trend and any changes in the land and water usage pattern across the Gnangara Groundwater Mound. Because of the uncertainty in future trends, there is a risk that a system designed to meet the predicted recharge rates for 10 years will need to be modified to achieve the objectives.

The private licensed and Water Corporation bores are presented by a well package in the Yanchep model. Although the Water Corporation provides details of their production bores and monthly water usage data, the private licensed data contains the latest licensed allocation amount. The model generated time series data for the private licensed bores for the modelling period using a scaling file, based on historical data with the assumption that most of the allocated water is used in the summer months. There are uncertainties in private borehole data including record duplications, incomplete records and misallocation of bores in the wrong aquifer. These may affect the results used for predicting the required recharge rates.

4.4 CONCLUSIONS

This study presents a brief geology, hydrogeology, hydrochemistry, and a three-dimensional groundwater flow model constructed using Visual Modflow Pro 4.1 to investigate and evaluate artificial recharge of groundwater to maintain groundwater levels within key cave systems to protect cave fauna in the Yanchep National Park.

The model considers all current information about the hydrogeology of the Yanchep Area, including private/public abstractions and nearby pine plantation areas.

The best calibration is achieved with a correlation coefficient of 0.997. The standard error of the estimate is 0.135 m in steady state and 0.016 m for all times in transient calibration.

The model is designed to use groundwater from two production bores situated in the superficial aquifer west of the Yanchep Caves about one km south west of the Loch McNess. Water will be discharged directly and simultaneously into these seven caves and will flow back in the westerly direction toward the supply bores.

The model takes into account the regional groundwater declines. Although the best possible estimate of the future regional groundwater trend has considered on the inflow boundary in the model, the result depends on rainfall trend and any changes of land and water use pattern across the Gnangara Mound. Because of the uncertainty in future trends, there is a risk that a system designed to meet the predicted recharge rates for 10 years will need to be modified to achieve the objectives.

The model has estimated that a total recharge rate of up to 3.6 Gl/yr would maintain water in ponds in each priority cave in both summer and winter up to ten years. In the area of Loch McNess and the caves, this simultaneous artificial supplementation into the caves creates an artificial groundwater mound of approximately 18 km^2 covering all targeted caves and Loch McNess and Yonderup Lake (Fig. 4.9). There are no other GDEs in the vicinity that would be affected by pumping from the bores. The cone of depression resulting from pumping of these bores stabilizes in two years indicating that there may be a 0.5 m drawdown within approximately one km west of the bores with no impact on lake system after the second year of operation.

The results of this study also indicate that groundwater within the calcareous sands is not saturated by calcite, however the Calcite saturation index of the groundwater sampled the proposed abstraction bores is −0.067, near the saturation by calcite, and less aggressive than groundwater under the caves, therefore weathering of limestone is probably less than weathering which occurs during natural discharge of groundwater within the cave systems.

Overall, it is concluded that artificial recharge using groundwater from the calcareous aquifer would have the least impact on the cave systems, in terms of additional weathering of limestone and nutrient status. The developed groundwater flow model is very effective in predicting the recharge rates for each cave to reach the target groundwater levels and also effective in predicting the time required for recovery in case of 1, 7 or 30 days failures in operation.

REFERENCES

Calvert, T. & Yesertener, C. (2005) *Yanchep Caves – Artificial Maintenance Scheme Operation Strategy*. Department of Environment, January 2005.

Davidson, A. (1995) *Hydrogeology and Groundwater Resources of the Perth Region, Western Australia, Bulletin 12*, Geological Survey of Western Australia, Department of Minerals and Energy. 257p.

Hazel, C.P. (1973) *Lecture notes on groundwater hydraulics*. Adelaide, Australia, Australian Water Resources Council, Groundwater School.

McDonald, M.G. & Harbaugh, A.W. (1988) *A Modular Three Dimensional Finite Difference Groundwater Flow Model*. Reston, VA, US Department of the Interior, USGS, National Centre.

Murray-Darling Basin Commission (2001) *Groundwater–Groundwater Flow Modelling Guideline*, Canberra, Australia. 112p.

Yesertener, C. (2005) Impacts of climate, land and water use on declining groundwater levels in the Gnangara Groundwater Mound, Perth, Australia. *Australian Journal of Water Resources*, 8 (2), 143–152.

Yesertener, C. (2008) *Assessment of the declining groundwater levels in the Gnangara Groundwater Mound, Hydrogeological Record Series HG14*, Department of Water, Western Australia.

Chapter 5

Spatial and temporal heterogeneity in the flux of organic carbon in caves

Kevin S. Simon[1], Tanja Pipan[2] & David C. Culver[3]

[1]*Department of Environmental Sciences, University of Auckland, Auckland, New Zealand*
[2]*Karst Research Institute ZRC SAZU, Postojna, Slovenia*
[3]*Department of Biology, American University, Washington, D.C., U.S.A.*

ABSTRACT

A conceptual model for the movement of organic carbon through a karst basin requires three main inputs: (1) localised flow of particulate organic carbon (POC) and dissolved organic carbon (DOC) through sinks and shafts, (2) diffuse flow of POC and DOC from soils and epikarst, and (3) deep groundwater inputs. Inputs of the localised and diffuse components were measured in the Postojna-Planina Cave System in Slovenia. Most DOC enters through streams entering the cave but diffuse flows from soils and epikarst is more generally distributed and appears to play an important role in subterranean food webs. Considerable spatial variation in DOC among epikarst drips was detected, and these differences were consistent through time. DOC concentration was not a good predictor of the abundance of copepods in dripping water.

5.1 INTRODUCTION

Simon *et al.* (2007) recently presented a conceptual model of the flow and distribution of organic carbon in caves. They point out that there have been two significant steps forward in ecosystem thinking in cave environment. One was the extensive work of Rouch and his colleagues on the Baget Basin, small karst drainage in France. He made an important conceptual advance of using an entire drainage basin rather than only a cave (Rouch, 1977) as the appropriate unit of analysis in karst. Gibert (1986), in what is the first true ecosystem study in karst, quantified the flux of organic carbon from springs draining the epikarst and the saturated zones of the Dorvan-Cleyzieu basin in France. She found that dissolved organic carbon (DOC) represented a larger flux than particulate organic carbon (POC), organic carbon availability was both spatially and temporally variable, and that microbes were likely to be key players in mediating energy transfer of organic carbon in karst. Simon and colleagues applied the methods and paradigms of surface stream ecology to the study of organic carbon cave streams. Among their findings were that most coarse particulate organic matter moved relatively short distances (tens of meters) before it was broken down or consumed (Simon & Benfield, 2001), that cave streams were more likely carbon rather than nutrient limited (Simon & Benfield, 2002), and that microbial films fueled by DOC are an important food in cave streams (Simon *et al.*, 2003).

The goals in this paper are to: 1) summarise the conceptual model of energy flow through karst; 2) investigate the spatial variation in DOC entering from epikarst

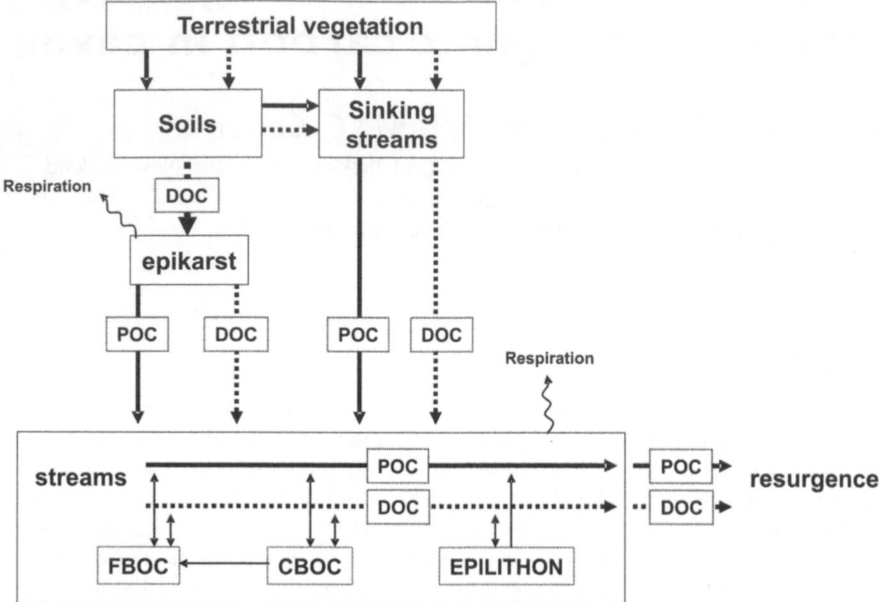

Figure 5.1 A conceptual model of energy flow and distribution (as organic carbon) in a karst basin. Standing stocks are particulate (POC) and dissolved (DOC) organic carbon in the water column and fine (FBOC) and coarse (CBOC) benthic organic carbon and microbial films on rocks (epilithon). Solid and dashed arrows represent fluxes. Within the cave stream double headed arrows connecting DOC to BOC and epilithon represent leaching and microbial uptake. Double headed arrows connecting POC and BOC represent deposition and suspension. The arrow connecting CBOC and FBOC represents breakdown. Adapted from Simon *et al.* (2007).

3) explore the relationship between DOC and invertebrate abundance, measured by copepod abundance.

5.1.1 A conceptual model of energy flux in karst

The most appropriate scale for an ecosystem approach to studying energy flow in karst is one that includes the relevant energy sources to caves and one for which input-output budgeting may be used. The karst basin used by Rouch satisfies both requirements. Energy inputs to karst basins can include internal production and the import of DOC and POC ultimately derived from surface vegetation. Input of DOC and POC may arrive via two different pathways (Fig. 5.1). Localised openings such as sinking streams and shafts permit entry of DOC or POC, such as leaves, wood, and fine detritus from streams and soils. Water percolating through soils and the epikarst, the zone of contact between soils and bedrock lying above caves, also carries with it DOC, but POC is effectively filtered by soils (Gibert, 1986). Deep groundwater may be another source of DOC, but considering the long residence time and distance from organic matter sources, deep groundwater inputs likely contribute little organic carbon within karst basins.

After input into the karst basin, POC and DOC are used or processed to different forms before eventually being exported through resurgences (Fig. 5.1). The major standing stocks of organic carbon within the basin include DOC and POC in the epikarst, epikarst drip pools, sinking streams, and cave streams, and POC (either fine (FPOC), <1 mm, or coarse (CPOC) >1 mm) and microbial films on rocks (epilithon) in epikarst and cave streams. The major fluxes of organic carbon include transport of DOC and POC in drips and streams (sinking and within caves), suspension and deposition of CPOC and FPOC in streams, breakdown of CPOC to FPOC, leaching of DOC from POC and epilithon, uptake of DOC by epilithon and microbes associated with benthic POC, losses from respiration along the flowpath of water through the basin, and export from springs (Fig. 5.1). While ideally the standing stocks and fluxes within the epikarst should be included within the karst basin, operationally this is virtually impossible given the inaccessibility of the epikarst. Therefore, only the fluxes out of the epikarst are likely to be directly measurable.

5.2 METHODS AND MATERIALS

The study site was Postojna-Planina Cave System in Slovenia. The Postojna-Planina Cave System contains approximately 23 km of surveyed passage (17 in Postojna and 6 in Planina connected by 2 km of flooded passage), and is arguably the most extensively studied cave system in the world (e.g., Pipan & Brancelj, 2004; Sket, 2004). There are two main streams in the Postojna-Planina Cave System which join and exit at Planinska jama entrance. One stream is formed by the Pivka River, a moderate-sized river draining approximately 230 km^2 of carbonate and flysch which sinks near the Postojnska jama entrance. The other stream (Rak) is a somewhat smaller stream draining approximately 27 km^2 of carbonate and flysch. The land over the Postojna-Planina Cave System, which is developed in Upper Cretaceous carbonate rocks, is forested, and the Pivka River drains land with a variety of uses, including forest and agriculture, as well as several small towns. The area over the cave system itself is approximately 20 km^2. Samples were taken at epikarst drips in Postojnska jama (see Pipan [2005] for locations), the two cave streams, and the Pivka River where it enters the cave, and one resurgence—the Unica River.

Data about inputs of organic carbon through epikarst drips and sinking streams were colleted as well as DOC in cave streams and the resurgences of the Postojna-Planina Cave System. Details of the sampling and measurement procedures for organic carbon can be found in Simon *et al.* (2007). For drips, water was collected in acid-washed 50 mL HDPE sample bottles over the course of at most one hour, depending on drip rate, which ranged between 75 and 1500 ml/h. Water was then placed in a 60 cc syringe and passed through a 0.45 μm glass fiber filter (Gelman GF/F) into a second bottle and preserved to pH < 2 with a drop of concentrated HCl . The samples were analyzed for DOC concentration using the persulfate digestion method on an OI Analytical Total Organic Carbon Analyser Model 1010.

In August and September of 2006, water samples were collected from 25 drips in three different sections of the Postojna-Planina Cave System: Postojnska jama (12 drips), Črna jama (7 drips), and Pivka jama (6 drips). The spatial pattern of

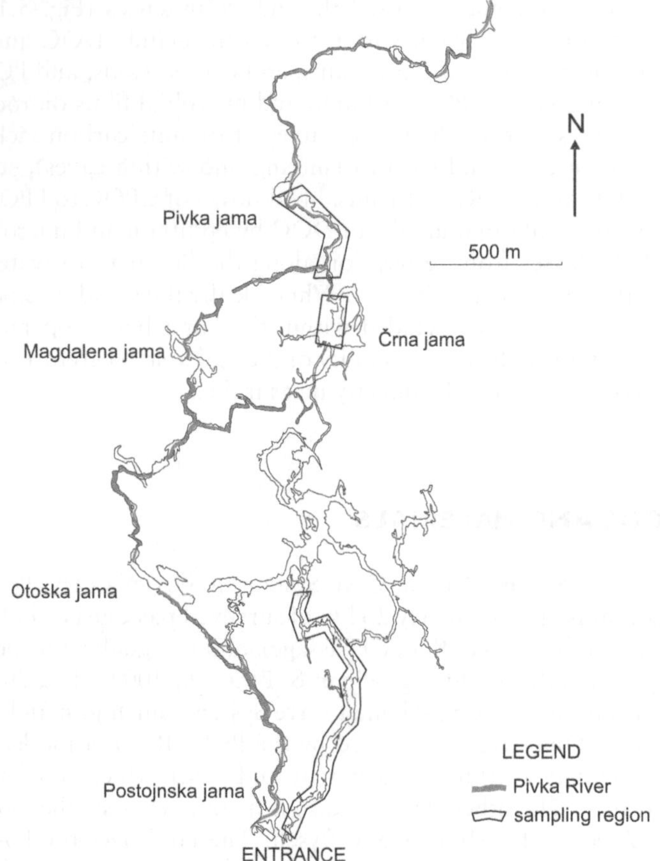

Figure 5.2 Location of sampling areas in Postojna Planina Cave System.

DOC in the three sections of the cave was plotted on detailed cave maps (see Pipan, 2003, 2005). The location of the three sampling areas is shown in Figure 5.2.

The values for DOC obtained from these epikarst drips were used to investigate whether they had any power to predict the richness and abundance of copepods continuously collected from many of these same sites by Pipan (2003, 2005) between March 2000 and February 2001. Copepods and other epikarst invertebrates were collected by continuously funneling dripping water through a bottle with 60 μm mesh sides in a manner that kept the copepods in water (see Pipan 2005 for details).

A simple linear regression model was used to examine the connection between DOC and copepod richness and abundance.

5.3 RESULTS

DOC concentration in the sinking streams was 4 times higher than that in the epikarst drips. DOC concentration in the cave stream was similar to that in the sinking streams,

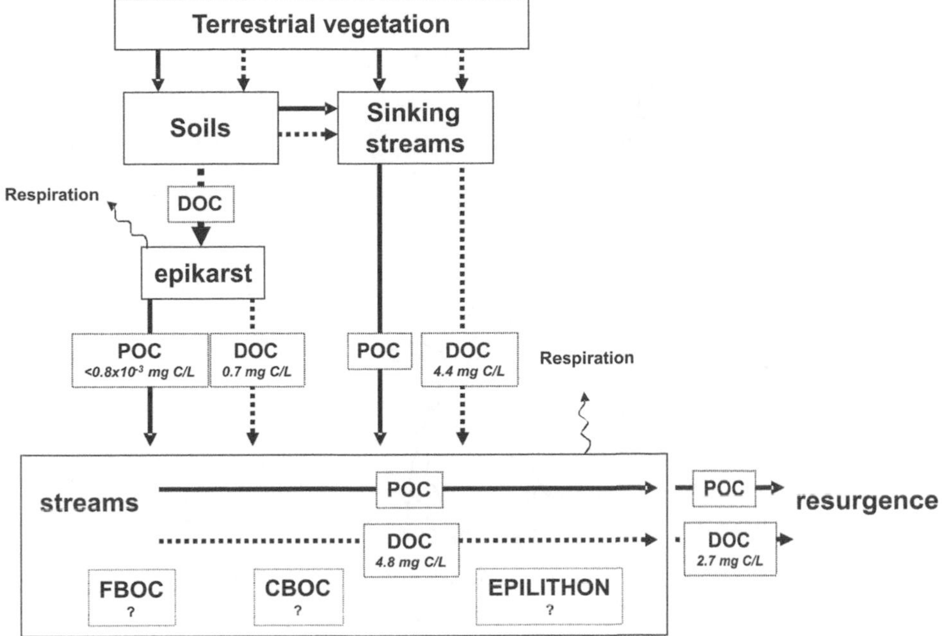

Figure 5.3 Schematic diagram of organic carbon flux in Postojna Planina Cave System (PPCS). Data are standing stocks of carbon. POC, DOC, FBOC and CBOC are particulate, dissolved, fine benthic, and coarse benthic organic carbon, respectively. From Simon *et al.* (2007).

but at the resurgence of the Postojna-Planina Cave System, DOC concentration was intermediate between the streams and epikarst drips (Simon *et al.*, 2007, Fig. 5.3).

5.3.1 Postojnska jama

With one exception, there were relatively minor differences between August and September values, and only the means are shown in Figure 5.4. In August mean DOC in drips was 0.39 ± 0.03 mg/l and in September it was 0.43 ± 0.05 mg/l. The exception was the second most northerly drip, which had a high value in September (0.90 mg/l). Overall, values tended to be low relative to Pivka jama (Fig. 5.5) and Črna jama (Fig. 5.6). In the central part of Postojnska jama, DOC values were particularly low. In general, drips within 100 m of each other had very similar DOC values.

5.3.2 Pivka jama

Overall, DOC values (Fig. 5.7) tend to be higher than in Postojnska jama, and except for the two most northerly points near the entrance, the mean was about 0.75 mg/l. The mean of the northernmost points, those closest to the entrance, was approximately 1.50 mg/l. Taking all points into account, the August mean was 0.97 ± 0.28 mg/l and the September mean was 0.75 ± 0.10 mg/l, but the difference was not statistically significant. As was the case with Postojnska Jama, there seemed to be little spatial variation within a range of about 100 m.

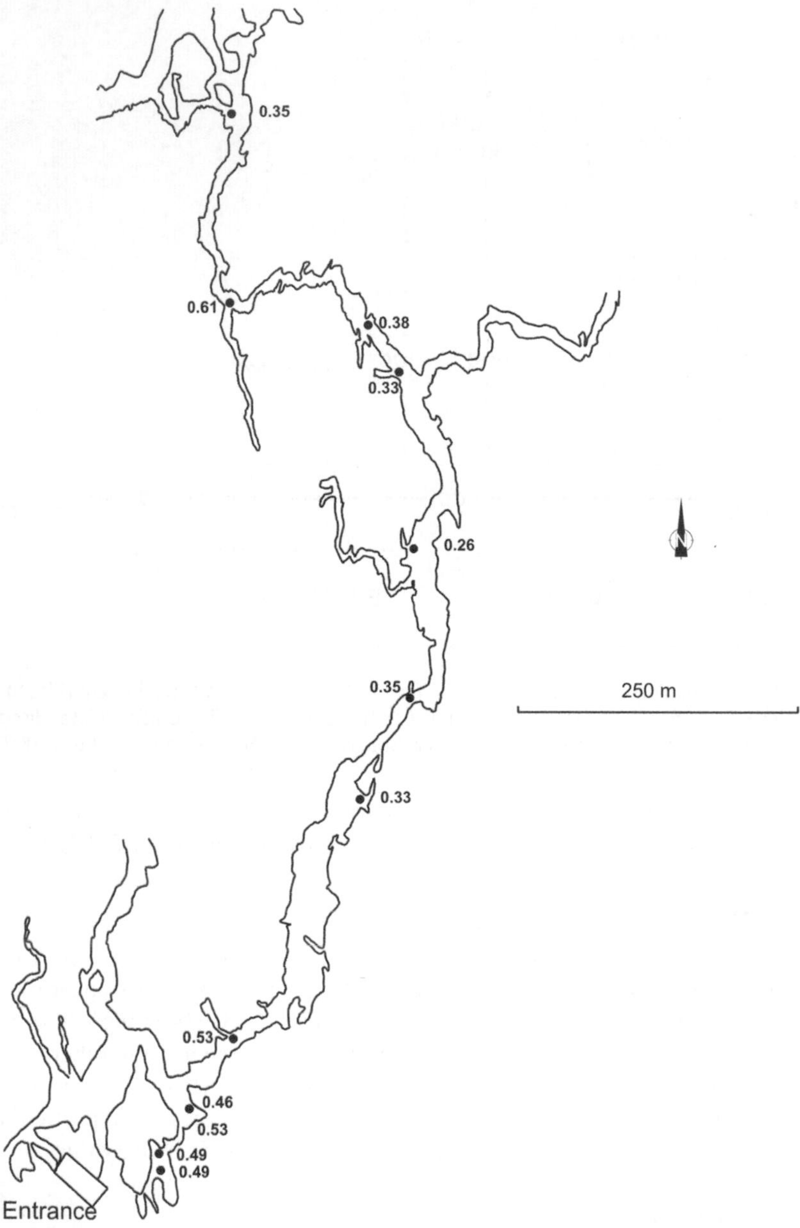

Figure 5.4 Map of Postojnska jama, showing epikarst drip sampling sites with average values of DOC in mg/L.

5.3.3 Črna jama

In the case of Črna jama, DOC values were generally intermediate between Postojnska jama and Pivka jama. The average in August was 0.66 ± 0.15 mg/l and in September it was 0.43 ± 0.06 mg/l, not a significant difference. The pattern was one of generally low

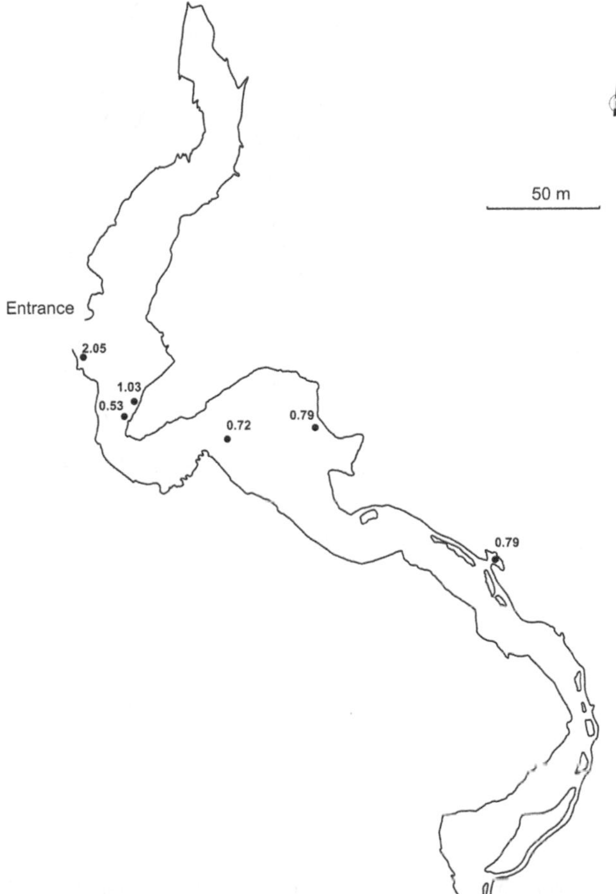

Figure 5.5 Map of Pivka jama, showing epikarst sampling sites and average DOC values (mg/L).

DOC values, similar to those of Postojnska jama, with two isolated spots of high DOC, one in the north part of the area and one in the south part of the area. Nevertheless, the pattern in Črna jama is not as clear as the other two areas, and nearby drips did not always have similar DOC values.

5.3.4 Relationship between DOC and copepod abundance

In 2000/2001 both copepod richness (0–10 species) and abundance (0–265 animals/l) varied considerably across drip sites. However, neither copepod richness ($p = 0.30$) nor abundance ($p = 0.11$) were related to mean DOC concentration in the drip water.

5.4 DISCUSSION

In the context of the conceptual model of Simon *et al.* (2007), there is considerable spatial variation in dissolved organic carbon entering a cave from epikarst. In general,

Figure 5.6 Map of Črna jama, showing epikarst sampling sites and average DOC values (mg/l).

this variation is at the scale of 100 m or more, but there were a couple of cases where closely adjoining drips had different DOC values. This is in accord with the observation that water chemistry often varies among nearby drips (*e.g.*, Tooth & Fairchild, 2003). They suggest a variety of scenarios within epikarst to explain these differences. One explanation for the differences in DOC is proximity to an entrance as several drips near an entrance had high DOC values (Figs. 5.4 and 5.5).

Most invertebrates in epikarst are copepods (Pipan, 2005), and the number of copepods could be a good predictor of total invertebrate abundance. There was no relationship between copepod diversity or abundance and the mean amount of DOC in the drip water. There are several important caveats to this pattern. Estimates of copepods and DOC were not made at the same time. The degree of temporal variation in both of these parameters is not well known, but it is possible both vary considerably. The total DOC abundance is comprised of a complex mix of compounds that vary considerably in quality. It may be the case that DOC quality (see Simon *et al.*, 2010) is more important than bulk quantity in influencing food webs. Finally, the exit point of the epikarst was sampled and this may not accurately represents the flux of

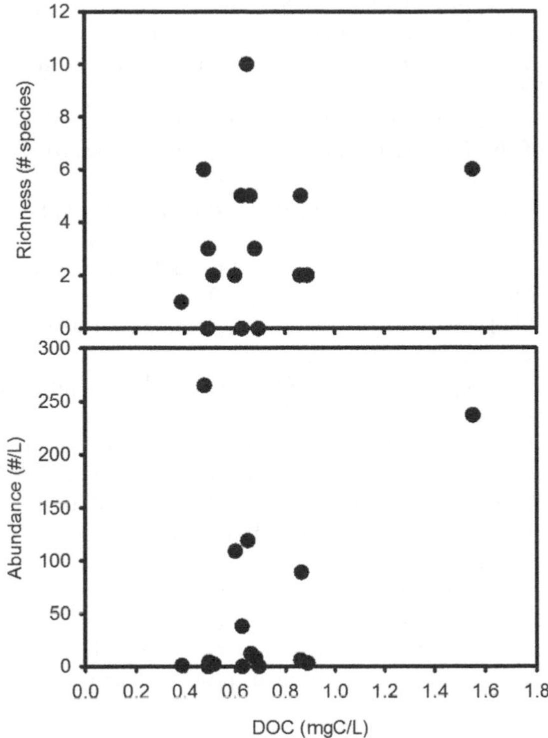

Figure 5.7 Relationship between copepod species richness or abundance and dissolved organic carbon (DOC) concentration in drip water across 20 drips in PPCS.

DOC moving into epikarst zones where it is likely consumed and altered before exiting through dips. For example, Pipan (2005) demonstrated that copepod abundance was negatively correlated with ceiling thickness, suggesting a potential role of water flow-path through the epikarst is a key feature. How DOC enters and is processed through the epikarst zone needs detailed study.

ACKNOWLEDGMENTS

Research was supported by the Karst Waters Institute, the College of Arts and Sciences of American University, and the Ministry of Higher Education, Science and Technology of the Republic of Slovenia, and the Slovenian Research Agency.

REFERENCES

Gibert, J. (1986). Ecologie d'un systeme karstique jurassien. Hydrogéologie, dérive animale, transits de matières, dynamique de la population de *Niphargus* (Crustacé Amphipode). *Memoires de Biospeologie*, 13, 1–379.

Pipan, T. (2003) *Ekologija ceponožnih rakov (Crustacea: Copepoda) v prenikajoèi vodi izbranih kraških jam*. PhD, Univerza v Ljubljani [in Slovene].

Pipan, T. (2005) *Epikarst – A Promising Habitat. Copepod Fauna, Its Diversity and Ecology: A Case Study from Slovenia (Europe)*. Ljubljana, Slovenia, ZRC Publishing, Karst Research Institute at ZRC SAZU.

Pipan, T. & Brancelj, A. (2004) *Distribution patterns of copepods (Crustacea: Copepoda) in percolation water of the Postojnska Jama Cave System (Slovenia)*. *Zoological Studies*, 43, 206–210.

Rouch, R. (1977) Considerations sur l'ecosystem karstique. *Compte Rendus Academie des Sciences de Paris Serie D*, 284, 1101–1103.

Simon, K.S. & Benfield, E.F. (2001) Leaf and wood breakdown in cave streams. *Journal of the North American Benthological Society*, 482, 31–39.

Simon, K.S. & Benfield, E.F. (2002) Ammomium retention and whole stream metabolism in cave streams: *Hydrobiologia*, 482, 31–39.

Simon, K.S., Benfield, E.F. & Macko, S.A. (2003) Food web structure and the role of epilithic films in cave streams. *Ecology*, 84, 2395–2406.

Simon, K.S., Pipan, T. & Culver, D.C. (2007) A conceptual model of the flow and distribution of organic carbon in caves. *Journal of Cave and Kart Studies*, 69, 279–284.

Simon, K.S., Pipan, T., Ohno, T. & Culver, D.C. (2010) Spatial and temporal patterns in abundance and character of dissolved organic matter in two karst aquifers. *Fundamental and Applied Limnology*, 177, 81–92.

Sket, B. (2004) Postojna–Planina cave system: Biospeleology. In: Gunn, J. (eds.) *Encyclopedia of Caves and Karst Science*. New York, NY, Fitzroy Dearborn. pp. 601–603.

Tooth, A.F. & Fairchild I.J. (2003) Soil and karst aquifer hydrological controls on the geochemical evolution of speleothem-forming drip water. *Journal of Hydrology*, 273, 51–68.

Chapter 6

The influence of groundwater/surface water exchange on stable water isotopic signatures along the Darling River, NSW, Australia

Karina Meredith, Suzanne Hollins, Cath Hughes, Dioni Cendón & David Stone

Australian Nuclear Science and Technology Organisation, Institute for Environmental Research, Kirrawee, Australia

ABSTRACT

Stable isotopes have been analysed in river waters collected from the Barwon-Darling River over the years (2002 to 2007). Run-of-river sampling results were compared with temporal data from three gauging stations located along the river. Darling River surface water samples are generally enriched in heavy isotopes due to evaporation. Partitioning of distinctly labelled isotopic waters such as enriched surface water and depleted groundwater allowed for the identification of groundwater/surface water exchange. Results showed that large flood events recharge the shallow aquifer with fresh-enriched waters and during low flow conditions, saline-depleted groundwaters rebound towards the river. Consequently, during drought periods saline groundwaters discharge into the river system. The flux of saline groundwaters into the surface water system was found to not only increase the salinity of scarce fresh water supplies in the region but also create environmental conditions ideal for cyanobacteria blooms to develop.

6.1 INTRODUCTION

Stable isotopes were analysed for surface waters collected from the Darling River, over a five year period (2002–2007) in the northern part of the Murray-Darling Basin (MDB). In Australia, particularly in the MDB, extended periods of drought during this time have occurred placing enormous pressures on surface water resources in the region. Ensuing drought conditions during this time were experienced in much of Australia, providing optimum conditions for the increase in natural surface water losses from evaporation. The pressure for water resource development in dryland regions in recent years has been high because of both the overall scarcity of water and the insecurity of supply (Young *et al.*, 2006).

The Barwon-Darling River, at 2735 km in length, is the longest river within Australia. Its headwaters are located in the highland regions of south-eastern Queensland and it flows in a westerly direction towards the semi-arid interior of Australia. Consequently, the water that flows inland through this system is exposed to intense evaporation. Groundwater/surface water exchange between fresh river waters and saline groundwaters during low flow conditions also places this system at risk of environmental degradation. Burrell and Ribbons (2006) carried out a stream flow analysis

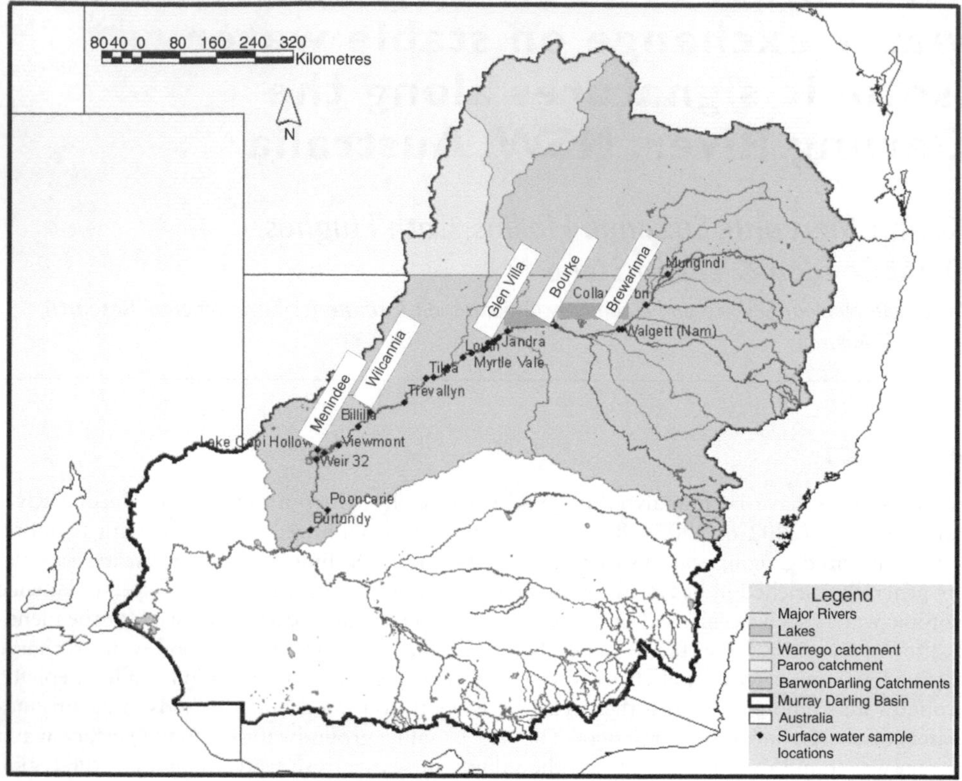

Figure 6.1 Location of Murray-Darling Basin with river water sample locations along the Darling River.

on the Darling River at Wilcannia (Fig. 6.1) and found below average flows due to a drought sequence which began around 2001. They also found that this drought was one of the most extreme on record for the Darling River.

In 1991, the most extensive cyanobacteria bloom ever reported worldwide occurred in the Barwon-Darling River system (Bowling & Baker, 1996). It was found that low river flow was the major factor triggering the blooms which were exacerbated by drought conditions that existed over much of western New South Wales (NSW) during this time. Davis & Koop (2006) concluded that during periods of low flow in the Darling River, an influx of sulphate-rich, saline groundwaters occurred. They found that the presence of these groundwaters was another major environmental trigger for the 1991/92 bloom. In this work they mention that there is very little understanding of the role that these saline sulphate-rich groundwaters play in triggering and sustaining phytoplankton blooms in shallow water bodies.

The Darling River forms one of the only freshwater supplies in western NSW and suffers environmental pressures from over-abstraction, unpredictable flow regimes, evaporative losses, drought and saline groundwater exchange. Stable isotopes are extremely useful in elucidating hydrological components of a river system, such as rainwater and groundwater inputs. A database of monthly results from a five year period (2002–2007) together with run-of-river sampling events are compared with flow and rainfall data from various locations to evaluate how various hydrological components influence the resultant stable isotope signature of the Barwon-Darling River.

6.2 THE DARLING RIVER

The MDB is a large surface water catchment located in the eastern part of Australia which incorporates the Murray Geological Basin and the Darling River Drainage Basin (DRDB) (Ife & Skelt, 2004). Within the DRDB, the Barwon-Darling River drains the north-westerly portion of the MDB in south-eastern Australia and has a catchment size of 650 000 km^2 (Thoms & Sheldon, 2000) (Fig. 6.1). Historical and present day drainage patterns of the Darling River are largely influenced by the underlying geological units and structures (White, 2000). Run-of-river sample events of the Barwon-Darling River have focused on towns from Mungindi in the north to Burtundy near the confluence of the Murray River in the south-west (Fig. 6.1). The main aquifer system in the area is the Narrabri Subsystem which is made up of shallow alluvial fan deposits that form discontinuous minor aquifers and reach up to 200 m in thickness (Ife & Skelt, 2004). Hydrogeological investigations conducted in the Narrabri Formation show that there is a good degree of interconnectivity between surface and groundwater systems (Barnett *et al.*, 2004). The geology and hydrogeology of the area have been described in Meredith *et al.* (2009).

6.2.1 Anthropogenic influence on the hydrology of the Darling River

The Barwon-Darling River is a highly regulated system that contains nine major headwater dams with a combined storage capacity of over $4.4 \times 10^9\, m^3$, with 15 main channel weirs and numerous small weirs on tributaries and anabranch channels (Thoms & Sheldon, 2000). An increased demand on water supplies from the Barwon-Darling River has been demonstrated by a tenfold increase in licensed water abstractors between Mungindi and Menindee over the past 35 years. For the years 1997 to 1998, diversions from the river were approximately 87% of the long-term mean annual flow (Thoms & Sheldon, 2000). Researchers in the area have found that major water extractions from the Darling River have led to bank slumping, sedimentation and poor health of the riparian zone (Thoms & Sheldon, 2000; Thoms *et al.*, 2005; Sheldon & Thoms, 2006). Major infrastructure emplaced along the river has also led to environmental issues associated with varying the ideal temperature of the river and increased sediment load in the river (DeRose *et al.*, 2003).

6.2.2 Stable isotope studies in the Barwon-Darling River

The stable isotopes of oxygen (^{16}O, ^{18}O) and hydrogen (^{1}H, ^{2}H) occur naturally in the water molecule and are often used as hydrologic tracers in various surface water environments (Kendall & Coplen, 2001; Gibson *et al.*, 2002a; Gibson *et al.*, 2005).

In regulated river system such as the Darling River, SWI signatures can be used to delineate various hydrological components of a river system. In semi-arid environments, surface waters are continuously exposed to evaporation, where they accumulate an evaporative signature (Gat & Airey, 2006). In a drought influenced river system, a progressive evaporative isotopic signature would be expected to dominate the stable isotope results.

Simpson & Herczeg (1991a and b) used stable isotopes together with major ion chemistry to estimate salt loads, evaporation and the hydrochemical evolution of surface waters in the Murray River, which is part of the larger Murray-Darling Basin. For the Murray River system, they estimated that the cumulative evaporation losses along the length of the river were approximately 35% of the initial volume of surface water in the system. The most comprehensive study to date in the Barwon- Darling River used stable isotopes combined with runoff discharge data and baseline monthly data from Global Network for Isotopes in Precipitation (GNIP) (Gibson *et al.*, 2008). They used this data to initially evaluate the ungauged gains and losses along the Darling River. They found that by evaluating the physical monitoring data alone, the Darling River appeared to be a slow draining river with little exchange. However, the inclusion of isotopic data in the evaluation revealed that the system is much more complex. It was surmised that several reaches of the river were volumetrically gaining, including the Bourke to Louth and Louth to Wilcannia reaches (see Fig. 6.1). SWIs were used to evaluate the amount of evaporation occurring along the river and groundwater baseflow into the system was found to be a very important parameter in the water balance.

6.3 METHODOLOGY

Monthly samples were collected from nine stream gauging stations operated by the Department of Water and Energy (DWE) located along the river from Mungindi in the north-east to Burtundy in the south-west during the period from June 2002 to October 2005. This sampling programme contributes to the Global Network for Isotopes in Rivers (GNIR) (Gibson *et al.*, 2002b) data set which, in Australia was focused along the Darling River (Stone *et al.*, 2003). River discharge volumes (Ml d^{-1}) were extracted from a database of archived hydrological data for rivers in Australia (DNR NSW, 2005).

River water samples were collected in HDPE (High Density Poly-Ethylene) bottles by DWE. These bottles were sub-sampled and the remaining water samples were frozen and sent to ANSTO (Australia Nuclear Science and Technology Organisation) for further analysis. The samples were thawed and sub-samples were analysed for δ^{18}O and δ^2H. The δ^{18}O and δ^2H values were measured by Isotope Ratio Mass Spectrometry (IRMS) and are reported as per mil (‰) deviations from the international standard V-SMOW (Vienna Standard Mean Ocean Water). The δ^{18}O and δ^2H measurements were reproducible to ± 0.15 and ± 1.0‰, respectively. Extended methods can be found in Meredith *et al.* (2009).

In addition, higher spatial frequency water samples were collected by ANSTO during run-of-river sampling campaigns in January and May 2007. All river water samples collected by ANSTO were from an in-line, 0.45 μm; high volume filter attached to a submersible pump located approximately 0.2 m below the water surface and 10 m from the bank of the river. River waters were tested for EC, in-stream. Samples for SWI were collected in 30 ml HDPE bottles, with no further treatment. Nineteen nested groundwater monitoring wells were sampled in August 2007 from various depths ranging from 20–130 m bgs. Groundwater samples were collected from an in-line, 0.45 μm, high volume filter that was attached to a submersible GRUNDFOS pump. Water samples were collected for various analytes but only SWI and EC values are discussed herein.

6.4 RESULTS AND DISCUSSION

6.4.1 Stream flow

For this study, three gauging stations of the nine were chosen to demonstrate the broadly different hydrological responses of the Barwon-Darling River system with distance from the mouth of the Darling River. Stream flow versus rainfall data for Brewarrina (Barwon River), Bourke (Darling River) and Menindee (Darling River) is presented in Figure 6.2. Surface water flows at Brewarrina and Bourke show a degree of seasonality with higher flows experienced in the summer months (December to April) and coincide with larger rainfall events. The river appears to be well connected at Brewarrina and Bourke, however, further downstream at Menindee, the river only responds to very large flow events, such as the one observed in February 2004 to early March 2004. In the upper reaches of the river, discharge is more closely linked to local catchment rainfall and in the lower reaches low flow conditions prevail.

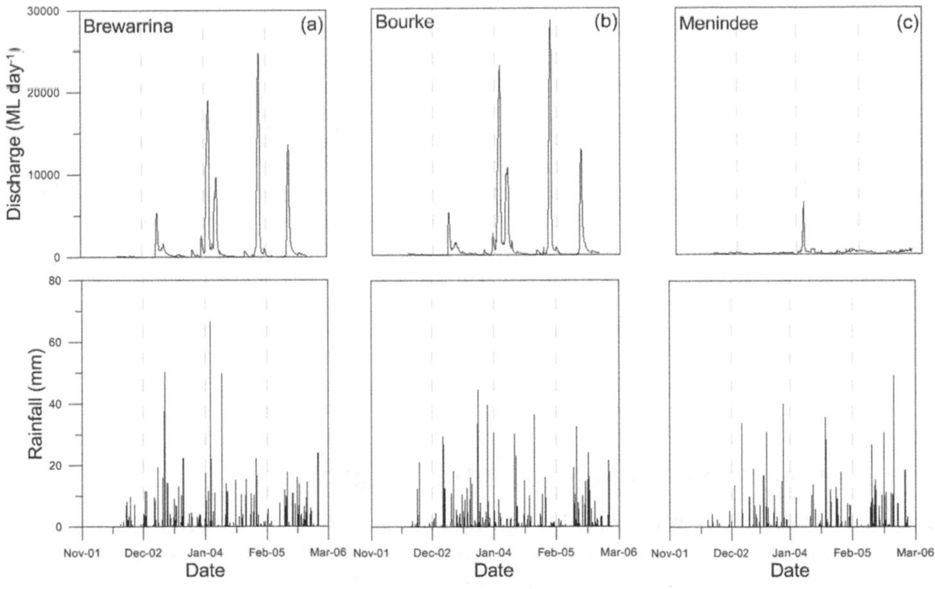

Figure 6.2 Rainfall data compared with stream discharge for three gauging stations (a) Brewarrina, (b) Bourke and (c) Menindee along the Darling River (July 2003 to January 2006).

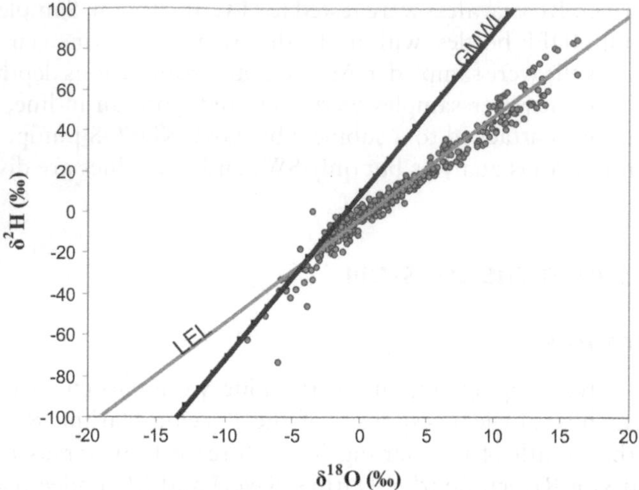

Figure 6.3 Bivariate plot of δ^2H vs. $\delta^{18}O$ for all river water samples from the Barwon-Darling River (July 2003 to January 2006) plotted against the GMWL ($\delta^2H = 8\delta^{18}O + 10$) and LEL ($\delta^2H = 5.00\delta^{18}O - 4.99$) for the region.

6.4.2 SWI results for the Barwon-Darling River

Over 400 samples have been collected and analysed for stable isotopes from various locations along the Barwon-Darling River system (Fig. 6.1). The $\delta^{18}O$ and δ^2H values ranged from −8.32‰ ($\delta^{18}O$) and −73.8‰ (δ^2H) for Bourke in Jan 2004 to +16.22‰ ($\delta^{18}O$) downstream at Burtundy in Feb of 2004 and +84.4‰ (δ^2H) at Lake Copi Hollow in April 2004. Surface waters were generally found to be highly enriched with an average of +3.40‰ ($\delta^{18}O$) and +26.8‰ (δ^2H).

Darling River samples are generally enriched in heavy isotopes and follow a distinctive enrichment trend in the δ-space with increasing distance from the headwaters of the river (Fig. 6.3). The data plots on a much lower slope than the Global Meteoric Water Line (GMWL: $\delta^2H = 8\delta^{18}O + 10$) with a slope of 5, forming the Local Evaporation Line (LEL) for the Darling River catchment ($\delta^2H = 5.00\delta^{18}O - 4.99$). This is what would be expected for a low flow system that is exposed to large amounts of solar energy (Darling *et al.*, 2005) and hence evaporation where waters will plot typically along lines with a slope of 4 to 8 in the $\delta^{18}O$-δ^2H space (Gat, 1996: Gibson *et al.*, 2000).

6.4.3 Temporal variability in isotopic signatures

Basin-wide trends are observed in the complete dataset but to identify catchment processes temporal variations in SWI signatures compared to flow conditions will be evaluated. This was completed for Brewarrina, Bourke and Menindee (Fig. 6.4). It can be observed that SWI signatures are generally governed by stream discharge conditions with the larger flow events being the primary influence. A strong isotopic enrichment

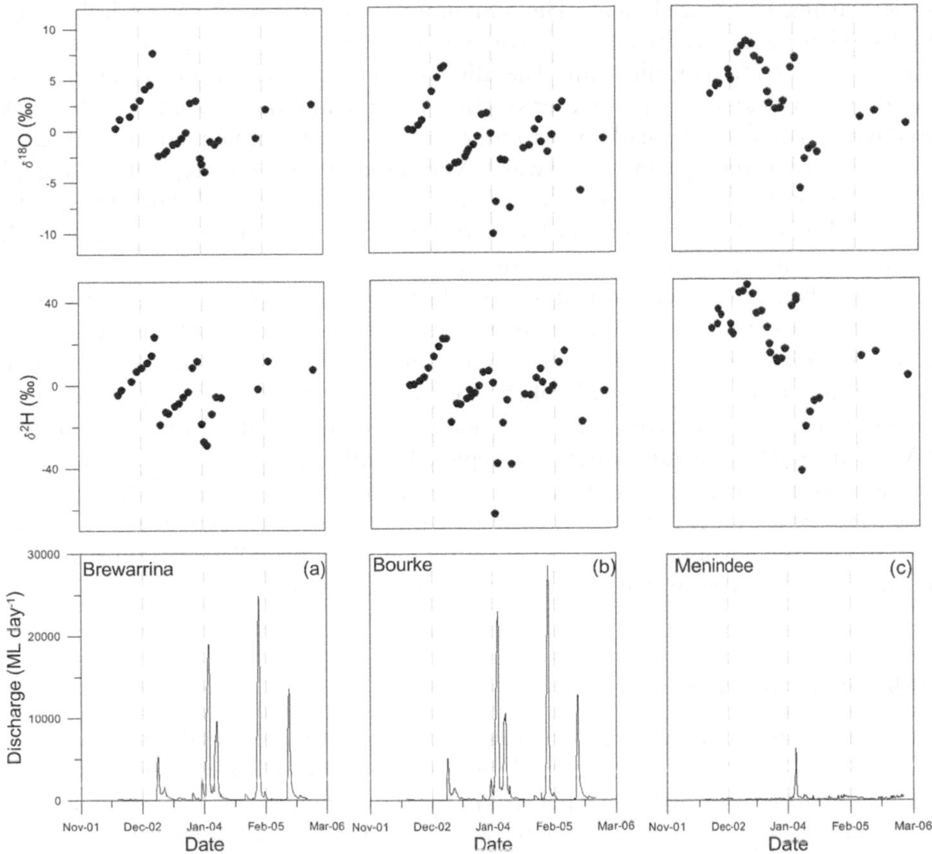

Figure 6.4 Stream discharge data compared with SWI values for Brewarrina, Bourke and Menindee along the Darling River (July 2003 to January 2006).

trend is evident from upstream (Brewarrina) to the more arid western gauging station (Menindee). The cyclic decrease and increase in SWI signatures with flow events and evaporation, respectively, dominate the hydrology of this system. An example is where evaporitic enrichment occurs until January 2004, when a highly depleted rainfall event was introduced into the system between Brewarrina and Bourke. This was also recorded at Menindee. This event was much more depleted at Bourke indicating these waters originated after Brewarrina and are likely to represent tributary floodwaters. Flow variability, seasonal evaporation trends and ensuing drought conditions in the region are variable controls on the SWI signatures of surface waters.

6.4.4 Groundwater

Groundwaters in the Glen Villa area were sampled from various intervals from the saturated zone ranging from 20 m (the top of the water table) to 140 m (base of the alluvium) below ground surface. Three unconsolidated alluvial aquifers are present in the Darling River catchment. The alluvium is separated from the underlying confined sandstone aquifer by an extensive aquitard. The chemistry of this system is not related to the overlying alluvium. The alluvial units are part of an extensive, closed and internally draining groundwater system. Alluvial fans are associated with rivers in the DRDB. The fans are made up of sequences of coarse sediments up to 150–200 m thick, becoming finer grained westward. Very limited groundwater data is available for this region because it is the most arid and remote area in the MDB (Ife and Skelt, 2004) and groundwater resources have not been considered important because they are not suitable for domestic or irrigation needs.

Groundwaters ranged from fresh ($0.7\,dS\,m^{-1}$) to extremely saline ($>50\,dS\,m^{-1}$) with the majority of groundwaters being very saline. It appears that the source of salinity is from evapo-concentration of atmospheric derived salts in the shallow alluvium but no comprehensive study of the source of salinity has been completed as yet. Groundwater chemical types were commonly Na-Cl-rich, with lesser concentrations of Mg^{2+} and SO_4^{2-}. Groundwaters are depleted in all heavy stable isotopes relative to seawater with average values of −3.7‰ and −29.4‰ for $\delta^{18}O$ and δ^2H, respectively. In contrast, river waters are generally fresh with enriched SWI signatures. The difference in groundwater and river water salinities together with SWI signatures allows us to trace groundwater-surface water interaction in this system.

6.4.5 Run-of-river

The ensuing drought conditions experienced in the Darling River Basin have resulted in the reduction of river connectivity along the river channel. Higher interval sampling was undertaken to identify whether groundwater exchange was occurring along the Mungindi to Burtundy stretch of the Barwon-Darling River in January and May 2007. River water EC values are generally low ($<<1\,dS\,m^{-1}$) except for a peak in salinity observed around Glen Villa (Fig. 6.5). This increase is also observed in major ion chemistry data (not shown) (Meredith *et al.*, 2009). Similarly, a decrease is observed in SWI values (^{18}O not shown but the same trend is observed). Even though other areas of high EC values are not observed, trends in stable isotope values are observed. Low values at Myrtle Vale and Milpa (May 2007) suggest the influx of fresh depleted groundwater into the system which may originate from bank storage release that has a stable isotope signature to flood waters. This depleted bank storage component is likely to be released gradually and would be identified in the stable isotope values during low flow conditions. These results alone show a multi-parameter approach is needed for understanding the hydrology of this system.

6.5 GROUNDWATER/SURFACE WATER EXCHANGE

The presence of saline groundwater exchange is not easily identified in the time-series stable isotope data because evaporation and catchment rainfall signals generally dominate. Run-of-river sampling performed during drought conditions clearly show an area of saline groundwater exchange at Glen Villa with elevated EC and a decrease

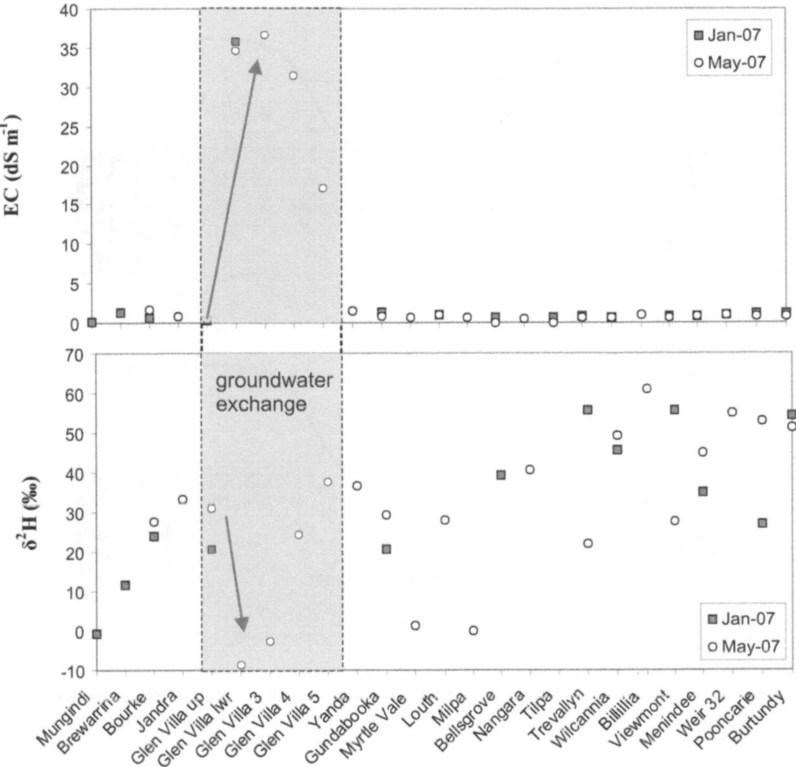

Figure 6.5 Run-of-river results (δ^2H and EC) for January and May 2007 for the Barwon-Darling River.

in stable isotope values. Hypothetical flow scenarios for the Glen Villa study area during "normal" flow conditions (Fig. 6.6a) and low flow conditions are presented (Fig. 6.6b). Results suggested that large flood events recharge the shallow aquifer with fresh-enriched river waters in the form of bank storage and during low flow conditions, bank storage is released and saline-depleted groundwaters rebound towards the river.

During low or zero flow conditions, the underlying saline groundwaters (with depleted isotopic signatures and high salinities) mix within the river channel, forming saline river water. Therefore, the influx of groundwater is dependent on flow conditions of the river. Fractional contributions of groundwater and river water end-members to those river waters were calculated using ^{18}O and Cl$^-$ concentrations from the May 2007 run-of-river sample event. It was found that 60–99% of the river water was composed of saline groundwater within the Glen Villa site during zero flow conditions (Meredith *et al.*, 2009). Low flow conditions are more common during drought periods such as those experienced in the Murray-Darling Basin during this study period. The flux of saline groundwaters into the river during low flow periods have the potential to create ideal environmental conditions for cyanobacteria blooms to develop such as those discussed by Bowling & Baker (1996) and Davis & Koop (2006). Indeed, during

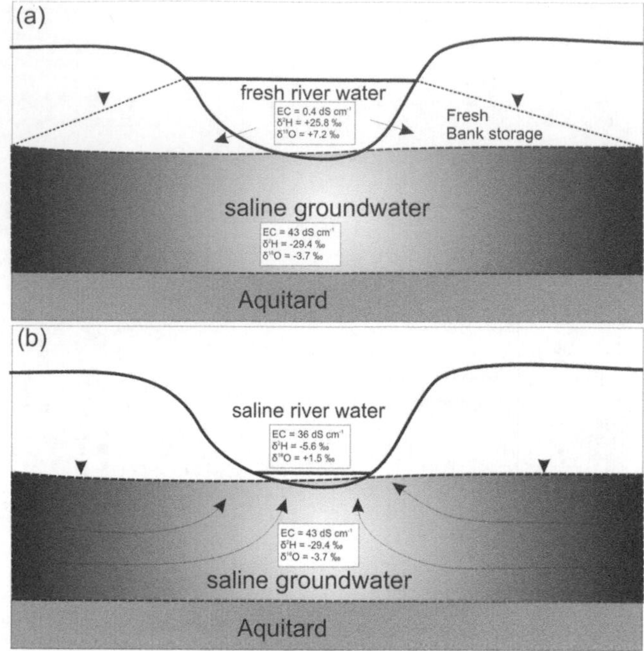

Figure 6.6 Hypothetical scenarios for groundwater/surface water exchange during (a) "normal" flow conditions and (b) low flow conditions in the Darling River near Glen Villa.

the May 2007 sampling event, fish kills and cyanobacteria blooms were evident at the site.

6.6 CONCLUSION

Stable isotope signatures of river water from both monthly and run-of-river samples were used to monitor the hydrological response of the Barwon-Darling River, to stream discharge, evaporation and delineate areas of groundwater interaction that are generally not readily identified using traditional stream hydrograph separation methods. The difference in EC and SWI signatures of river waters and groundwaters allowed for the identification of groundwater/surface water interaction along the Barwon-Darling River. It was also found that during 'normal' flow conditions, surface waters in the lower reaches of the river are primarily influenced by the isotopic signature of upstream waters that have been exposed to progressive evaporation during passage down the river. Groundwater exchange only becomes evident during zero flow conditions when up to 99% of the surface water is composed of groundwater. The hydrological response of the river to drought has had detrimental affects on the surface water system providing a pathway for saline groundwaters to discharge into the river system. The presence of underlying saline groundwaters not only influences the isotopic signature of the river waters but also increases the salinity of the surface waters.

ACKNOWLEDGEMENTS

The authors wish to thank Department of Water and Energy (DWE) for providing groundwater data and collecting monthly samples for SWI analysis. A special mention to Jim McCartney and Hari Haridharan for providing results and to Mick Allen from State Water for his local knowledge of the Darling River area. We would also like to thank Stuart Hankin (ANSTO) for his technical assistance in the field and Nathan Butterworth for extracting the stream flow data from PINEENA.

REFERENCES

Barnett, S., Caronsone, F., Dyson, P., Evans, R., Hillier, J., Ife, D., Jensen, G., Macumber, P., Marvenek, S., Morris, M., Skelt, K. & Wooley, D. (2004) *Murray–Darling Basin Groundwater Status 1990–2000: Technical Report*. Canberra, Murray-Darling Basin Commission.

Bowling, L.C. & Baker, P.D. (1996) Major Cyanobacterial bloom in the Barwon–Darling River, Australia, in 1991, and underlying limnological conditions. *Marine Freshwater Research*, 47, 643–657.

Burrell, M. & Ribbons, C. (2006) *NSW Drought Analysis 2006: Comparison of the current drought sequence against historic droughts*. Department of Natural Resources.

Darling, W.G., Bath, A.H., Gibson, J.J. & Rozanski, K. (2005) *Isotopes in Water*. In: Leng, M.J. (Ed.) *Isotopes in Palaeoenvironmental Research*. Netherlands, Springer. pp. 1–66.

Davis, J. & Koop, K. (2006) Eutrophication in Australian rivers, reservoirs and estuaries – a southern hemisphere perspective on the science and its implications. *Hydrobiologia*, 559, 23–76.

DeRose, R.C., Prosser, I.P., Weisse, M. & Hughes, A.O. (2003) *Patterns of Erosion and Sediment and Nutrient Transport in the Murray–Darling Basin*. CSIRO Land and Water, Technical Report 32/03.

Department of Natural Resources (DNR) NSW (2005) Pinneena 9: New South Wales Surface Water Archive.

Gat, J.R. (1996) Oxygen and Hydrogen isotopes in the hydrologic cycle. *Annual Review Earth and Planetary Science*, 24, 225–262.

Gat, J.R. & Airey, P.L. (2006) Stable water isotopes in the atmosphere/biosphere/lithosphere interface: Scaling-up from the local to continental scale, under humid and dry conditions. *Global and Planetary Change*, 51, 25–33.

Gibson, J.J., Price, J.S., Aravena, R., Fitzgerald, D.F. & Maloney, D. (2000) Runoff generation in a hypermaritime bog-forest upland. *Hydrological Processes*, 14, 2711–2730.

Gibson, J.J., Prepas, E.E. & McEachern, P. (2002a) Quantitative comparison of lake throughflow, residency, and catchment runoff using stable isotopes: modelling and results from a regional survey of Boreal lakes. *Journal of Hydrology*, 262, 128–144.

Gibson, J.J., Aggrawal, P., Hogan, P., Kendall, C., Martinelli, L., Stichler, A., Rank, D., Goni, I., Choudury, M., Gat, J., Bhattacharya, S., Sugimoto, A., Fekete, B., Pietroniro, A., Maurer, T., Panarello, H., Stone, D., Seyler, P., Maurice-Bourgoin, L. & Herzceg, A. (2002b) Isotope studies in large rivers basins: a new global research focus. *EOS*, 83 (52), 613–617.

Gibson, J.J., Edwards, T.W.D., Birks, S.J., St Amour, N.A., Buhay, W.M., McEachern, P., Wolfe, B.B. & Peters, D.L. (2005) Progress in isotope tracer hydrology in Canada. *Hydrological Processes*, 19, 303–327.

Gibson, J.J., Sadek, M.A., Stone, D.J.M., Hughes, C., Hankin, S., Cendon, D.I. & Hollins, S.E. (2008) Evaporative isotope enrichment as a constraint on reach water balance along a dryland river. *Isotopes in Environment and Health Studies*, 44 (1), 83–98.

Ife, D. & Skelt, K. (2004) *Murray–Darling Basin Groundwater Status 1990–2000: Summary Report*. Murray-Darling Basin Commission, Canberra.

Kendall, C. & Coplen, T.B. (2001) Distribution of oxygen-18 and deuterium in river waters across the United States. *Hydrological Processes*, 15, 1363–1393.

Meredith, K., Hollins, S., Hughes, C., Cendón, D., Hankin, S. & Stone, D. (2009) Temporal variation in stable isotopes (^{18}O and ^{2}H) and major ion concentrations within the Darling River due to saline groundwater exchange. *Journal of Hydrology*, 378, 313–324.

Sheldon, F. & Thoms, M.C. (2006) In-channel geomorphic complexity: The key to the dynamics of organic matter in large dryland rivers? *Geomorphology*, 77, 270–285.

Simpson, H.J. & Herczeg, A.L. (1991a) Stable isotopes as an indicator of evaporation in the River Murray. *Water Resources Research*, 27 (8), 1925–1935.

Simpson, H. & Herczeg, A. (1991b) Salinity and evaporation in the River Murray Basin, Australia. *Journal of Hydrology*, 124, 1–27.

Stone, D., Henderson-Sellers, A., Airey, P. & McGuffie, K. (2003) Murray Darling Basin Isotope Observations: an essential component of the Australian CEOP. *EOS Transactions, AGU*. AGU Fall Meeting Supplement. H22I-08, USA.

Thoms, M.C. & Sheldon, F. (2000) Water resource development and hydrological change in a large dryland river: the Barwon–Darling River, Australia. *Journal of Hydrology*, 228, 10–21.

Thoms, M.C., Southwell, M. & McGinness, H.M. (2005) Floodplain–river ecosystems: Fragmentation and water resources development. *Geomorphology*, 71, 126–138.

White, M.E. (2000) *Running Down Water in a Changing Land*. Sydney, NSW, Kangaroo Press.

Young, W., Brandis, K. & Kingsford, R. (2006) Modelling monthly streamflows in two Australian dryland rivers: Matching model complexity to spatial scale and data availability. *Journal of Hydrology*, 331, 242–256.

Chapter 7

A geochemical approach to determining the hydrological regime of wetlands in a volcanic plain, south–eastern Australia

Annette B. Barton[1], *Andrew L. Herczeg*[2], *Peter G. Dahlhaus*[3] & *James W. Cox*[4]

[1] *Bureau of Meteorology, Kent Town, Australia*
[2] *CSIRO Land and Water and CRC LEME, PMB 2 Glen Osmond, Australia*
[3] *University of Ballarat, Mt Helen, Australia*
[4] *The University of Adelaide, PMB1 Glen Osmond, Australia*

ABSTRACT

The Corangamite region in south-eastern Australia contains a large number of lakes and wetlands within an extensive, basaltic plain. To assess the impact of land-use change and groundwater pumping on wetland ecosystems, there is a need to develop a better understanding of their hydrology. This paper describes an approach using groundwater and surface water chemistry and stable isotopes to determine the extent that they are surface or groundwater dominant, and whether they are through-flow or terminal in nature. The ionic ratio HCO_3^-/Cl^- is higher in surface waters than groundwater, and lakes plot on a continuum between these two water types. Deuterium "excess" (δ_{xs}, where $\delta_{xs} = \delta^2H - 8 * \delta^{18}O$) reflects the deviation of a given sample from the global meteoric water line, with lower values indicating the increasing influence of evaporation, which in turn reflects longer water residence time (terminal lakes) compared with high δ_{xs} lakes that represent through-flow lakes.

7.1 INTRODUCTION

The Western Plains in the Corangamite region, south-west Victoria, Australia, are characterised by an abundance of freshwater and saline wetlands that are to varying degrees dependent on groundwater. The low, undulating country of the Plains, lies between the Central Highlands of the Great Dividing Range to the north and the Otway Ranges to the south (Fig. 7.1). The plain slopes generally toward the east, where there is an outlet to the sea. Its geographic features include cinder cones, maars, lava shields and the 'stony rise' landscapes.

The regional groundwater gradient of the Corangamite region moves from the Central Highlands generally south towards the Western Plains. The shallow groundwater moves through gravels, sands and silts of buried palaeochannels, and through the overlying fractured basalt (Cox *et al.*, 2007). On the Western Plains, the shallow groundwater moves under a very low regional gradient through an extensive, thin (<10 m) confined sand aquifer, underlying the basalt, and through sheets of fractured

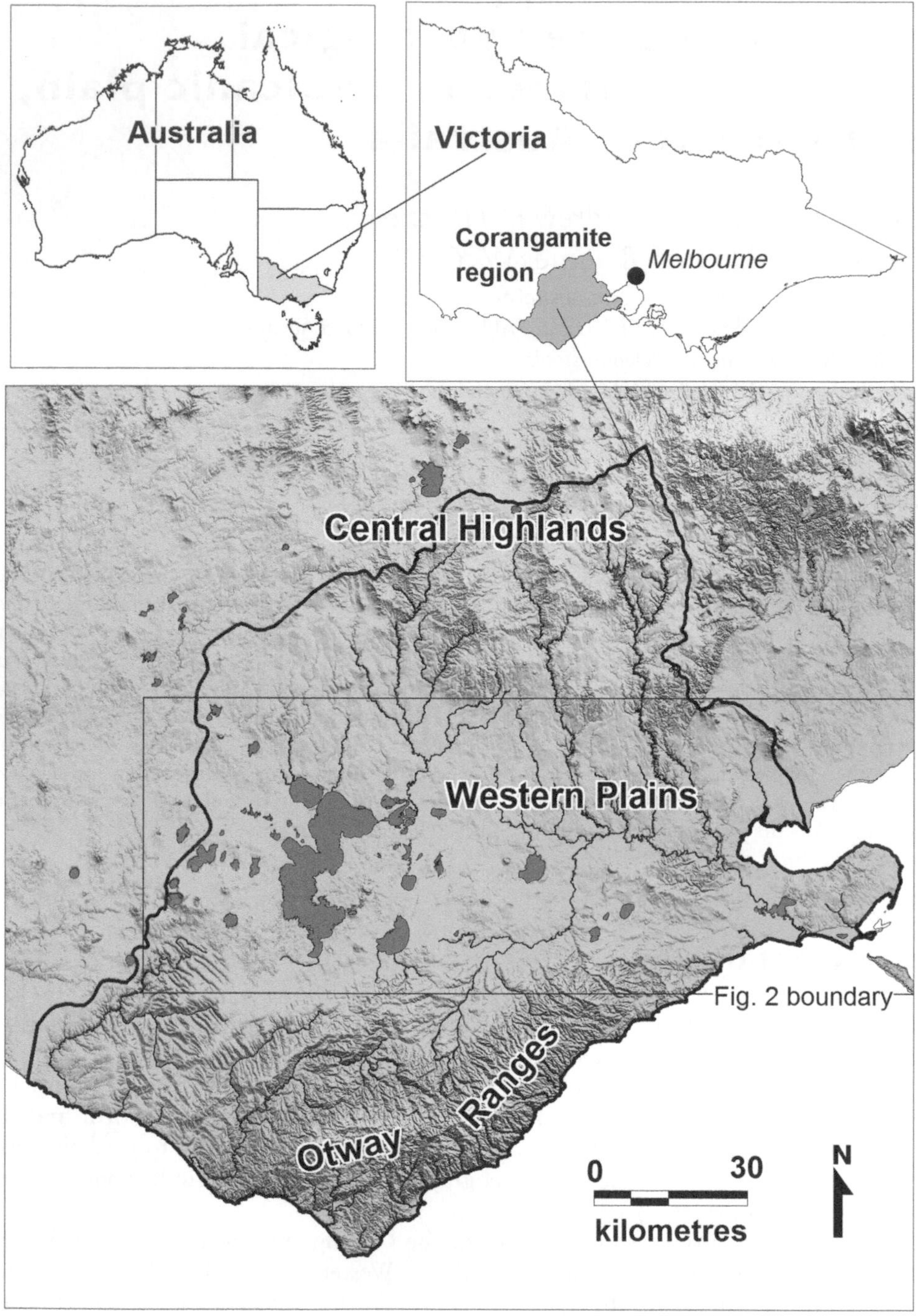

Figure 7.1 Location maps showing the Corangamite catchment management area, Victoria, Australia and the Western Plains (lakes) region between the Central Highlands and the Otway Ranges.

basalt towards the east and south east. Above the plains, more localised flows occur in the pyroclastic rocks of the eruption points, the fractured rocks of the youngest lava flows and the surficial Quaternary alluvial and aeolian sediments.

Low topographic gradients have resulted in sparsely distributed linear drainage channels usually terminating in shallow lakes. The lakes have a variety of landforms, such as crater lakes, maars, sag ponds, lakes formed in the depressions of the volcanic plains, lakes formed behind basalt barriers, and areas with impounded drainage. Some of the larger lakes are terminal lakes for both surface water and groundwater flow, whereas others are terminal for surface water flow but throughflow for groundwater (Coram *et al.*, 1998). Smaller lakes, especially those formed in volcanic craters and maars are groundwater sinks with little surface water input.

Annual rainfall varies from over 1000 mm in the elevated areas of the Central Highlands to less than 500 mm on the eastern edge of the Western Plains. The vast majority of the volcanic plains in the Corangamite region receive between 600 and 700 mm annually, with winter and spring as the dominant wet seasons. In times of abnormally high rainfalls, the lakes have been known to fill and merge, flooding large tracts of farmland. To manage the flooding, two major drainage channels were built in the 1950s and 60s to drain the larger lakes into the river systems of the plain.

Many of the more than 1400 declared wetlands of the Corangamite region are recognised as being of international ecological value, in particular the 27 Ramsar listed wetlands which include Lake Corangamite, Australia's largest, inland, permanent, saline lake. Sustained rises in lake salinity levels over the last 50 years has focussed attention on the impacts of human activities on wetland health (Williams, 1995, 1999; Adler & Lawrence, 2004; Nicholson *et al.*, 2006). Since the commencement of European settlement 160 years ago, widespread land-use and water-use changes have occurred. The extensive cultivation of pastures and crops has involved the clearing of natural vegetation and irrigation using groundwater resources. Flood mitigation/drainage works have also brought hydrological changes through the diversion of surface water flows from the lakes.

In order to assess the impact of groundwater pumping and changes to surface flow systems on the wetlands, research has been instigated to quantify the groundwater and surface water inputs to the lakes and hence improve understanding of their sensitivity to changes in groundwater recharge and discharge. In this paper the groundwater and surface water chemistry, including stable isotopes, is presented and a method described to distinguish between surface water, groundwater and evaporation dominated lakes.

7.2 APPROACH AND METHOD

Sampling of lake surface waters was undertaken with the aim of investigating a method for the rapid assessment of the groundwater dependence of the lakes of the region. It was hypothesised that a combination of isotopic and chemical parameters could be used to assess the groundwater dependence of the wetlands. The approach taken was to firstly obtain an understanding of the types of wetlands present before undertaking further, more in-depth research, into a few select "typical" lakes.

Grab sampling of the lake surface waters was undertaken in July 2006. Forty-six lakes were visited on the initial reconnaissance field trip, when it was ascertained that

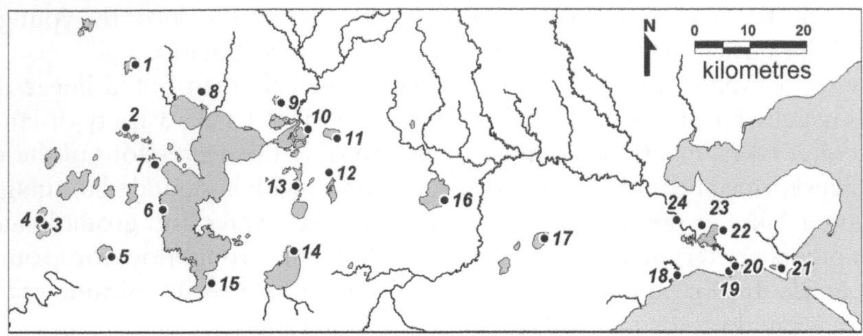

Figure 7.2 Location of lake sampling sites within the Western Plains (refer to Table 7.1).

Table 7.1 List of sampled lakes and location of sampling sites.

Site	*Lake*	*Longitude*	*Latitude*	*Site geology*
1	Lake Tooliorook	143.2826	−37.9772	Quaternary volcanics (basalt)
2	Kooraweera Lakes	143.2669	−38.0785	Neogene sand
3	Lake Bullenmerri	143.1080	−38.2376	Quaternary volcanics (crater lake)
4	Lake Gnotuk	143.0983	−38.2285	Quaternary volcanics (crater lake)
5	Lake Purrumbete	143.2432	−38.2861	Quaternary volcanics (maar lake)
6	Lake Corangamite	143.3449	−38.2098	Quaternary volcanics (basalt)
7	Lake Terangpom	143.3267	−38.1377	Neogene sand
8	Lake Struan	143.4169	−38.0192	Neogene sand
9	Lake Rosine	143.5784	−38.0342	Quaternary volcanics (basalt)
10	Cundare Pool	143.6350	−38.0759	Quaternary alluvium
11	Lake Weering	143.6952	−38.0887	Quaternary volcanics (basalt)
12	Upper Lough Calvert	143.6813	−38.1439	Quaternary alluvium
13	Lake Cundare	143.6126	−38.1669	Quaternary volcanics (basalt)
14	Lake Colac	143.6132	−38.2704	Quaternary volcanics (basalt)
15	West Basin	143.4475	−38.3249	Quaternary volcanics (crater lake)
16	Lake Murdeduke	143.9168	−38.1823	Quaternary volcanics (basalt)
17	Lake Modewarre	144.1194	−38.2381	Quaternary volcanics (basalt)
18	Breamlea Wetlands	144.3904	−38.2895	Quaternary alluvium
19	Barwon River mouth	144.4964	−38.2793	Quaternary alluvium
20	Barwon River estuary	144.5094	−38.2712	Quaternary alluvium
21	Lake Victoria	144.6044	−38.2714	Quaternary alluvium
22	Lake Connewarre	144.4817	−38.2156	Neogene marl
23	Reedy Lake	144.4373	−38.2088	Quaternary alluvium
24	Barwon River	144.3855	−38.2013	Quaternary alluvium

only a small fraction (around 2%) of the more that 1400 declared lakes and wetlands in the region (Wetland_1994) still contained some water. Like much of Australia, this area had been experiencing a prolonged period of below average rainfall, and in 2006 the 10 year rainfall average was at a record low (BOM, 2007).

On this occasion samples were obtained from 24 wetlands (Fig. 7.2, Table 7.1). Electrical conductivity (EC), pH, dissolved oxygen (DO) and temperature were measured on site using a calibrated field kit. The surface water grab samples were subsequently analysed in the CSIRO laboratories in Adelaide, South Australia, for chemistry and stable isotopes of water.

The lakes were revisited three months later, in October 2006, and a second set of samples were collected. On this occasion it was found that some of the lakes previously sampled had dried out and could not be sampled.

7.3 RESULTS

7.3.1 Chemistry

The salinity of the lakes sampled range over nearly three orders of magnitude (total dissolved solids (TDS) of 520 to ~310 000 mg/l) and are alkaline, with pH between 7.8 and 9.9. Dissolved solutes of the Corangamite lakes are in general dominated by Na^+ and Cl^- ions, although HCO_3^- makes up a significant fraction of anions at the lower salinity (Fig. 7.3). The dominance of Na^+ and Cl^- over the other ions increases linearly as a function of TDS. Calcium and HCO_3^- remain low throughout the entire salinity range indicating control of these dissolved ions through precipitation of carbonate minerals. The low salinity waters (TDS < 2500 mg/l) have higher proportion of HCO_3^- and alkaline earth ions (Mg^{2+} and Ca^{2+}) relative to other ions than the more saline waters (>2500 mg/l) indicating that these lakes probably have a significant fraction of their solutes derived from mineral weathering.

7.3.2 Stable isotopes

The stable isotopes of water ($^2H/^1H$ and $^{18}O/^{16}O$) are one of the most useful tracers in establishing a lake water balance, particularly with respect to the subsurface components (Rozanski *et al.*, 2000). Evaporation leads to increases in the ratio of $^2H/^1H$ and $^{18}O/^{16}O$ of residual waters with the isotopic concentrations evolving linearly in $\delta^2H - \delta^{18}O$ space. The degree of evaporative enrichment is dependent on atmospheric relative humidity over the lake and the surface water temperature.

Results for the stable isotopes of water for the two sets of lake samples collected are shown in $\delta^2H - \delta^{18}O$ space in Figure 7.4. Those lakes that lie further to the right along the trend line have undergone a greater degree of evaporation relative to the rate of inflow. The lake water balance can be simply represented by a balance between the relatively light isotopic composition of inflow, and the tendency of evaporation to remove the lighter isotope preferentially to the heavier isotope thereby enriching the remaining water in the heavier isotope. In a semi-quantitative way the lakes increase in residence time (residence time = total volume/total input rate) the further they lie to the right of the isotopic trend line.

The data have been presented together with isotope values for groundwater sampled from shallow (<25 m) bores in the vicinity of the lakes as well as the global meteoric water line (GMWL) and the local meteoric water line (LMWL) for Melbourne rain. Clearly the groundwater samples lie within the domain of values for Melbourne rainfall, hence the isotopic composition of groundwater is virtually indistinguishable from that for rainfall. This means the evaporation trend observed for the lakes would be almost identical for the two types of inflow, *viz* groundwater and surface water.

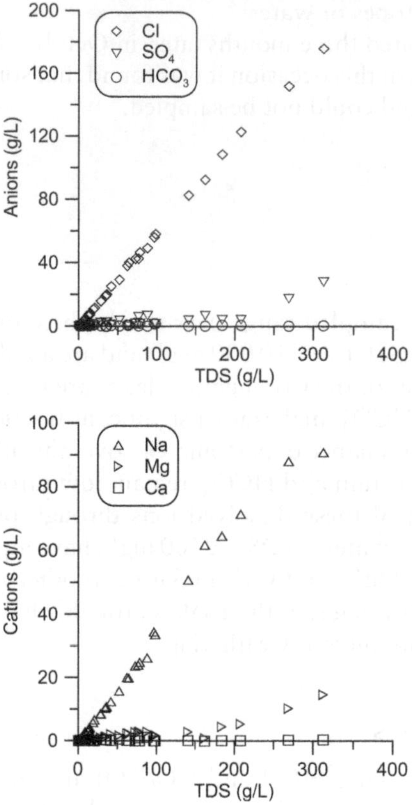

Figure 7.3 Plots showing lake water anions and cations as a function of total dissolved solutes.

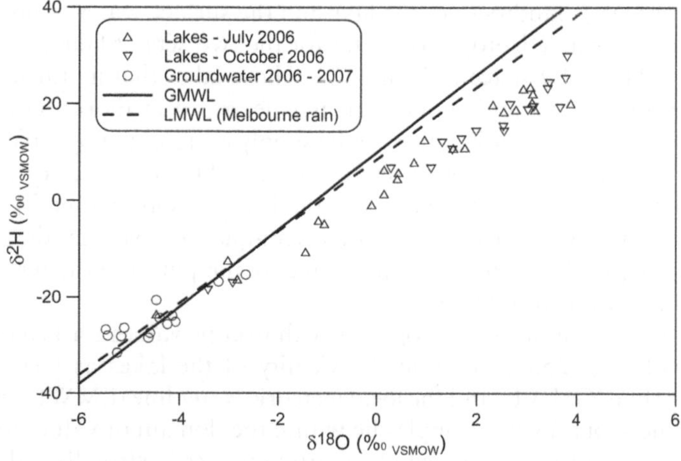

Figure 7.4 Plot of lake and groundwater stable isotopes of water showing similarities in the isotopic signatures of groundwater and rainwater.

7.4 DISCUSSION

7.4.1 A preliminary evaluation of lake types

The composition of the more saline lake waters (TDS > 2500 mg/l) is similar to that of seawater and the more saline groundwaters. Therefore, it is believed that the source of most of the dissolved ions for the both the lakes and groundwaters are derived from marine aerosols deposited by rainfall. The molar ratios of Cl^-/Br^- in all lakes are slightly higher than that of seawater (~695) which demonstrates the dominance of the marine origin of Cl^- and, by inference, Na^+ (Fig. 7.5). The linear relationship for the majority of the ions points to evaporation processes as the dominant control of salinity levels in the lakes (Fig. 7.3).

In plotting the ratio of HCO_3^-/Cl^- as a function of TDS (Fig. 7.6) it can be seen that the ratio ranges over some three orders of magnitude with an overall decrease with increasing TDS. The freshest lakes (TDS < 2500 mg/l) tend to have higher HCO_3^-/Cl^-

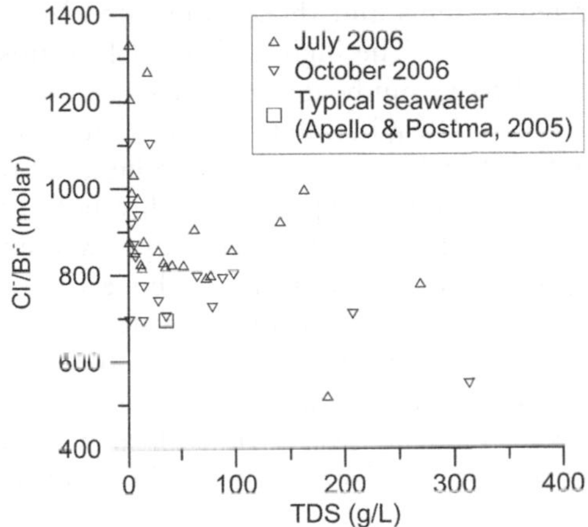

Figure 7.5 Cl^-/Br^- ratio plotted against TDS for the lake water, and compared to typical seawater.

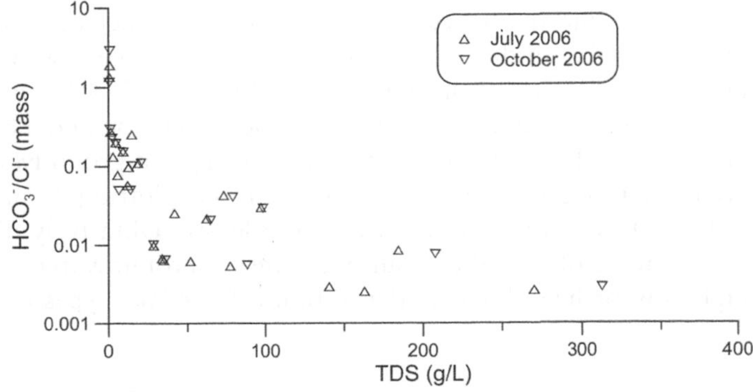

Figure 7.6 HCO_3^-/Cl^- as a function of TDS.

which is indicative of surface and inter-flow runoff components which tend to have a higher component of HCO_3^- due to mineral-solution reactions that produce HCO_3^- as a by-product. Lakes with salinities >2500 mg/l show $HCO_3^-/Cl^- < 0.08$ reflecting a higher saline groundwater component to the water balance. Therefore one may be able to separate the lakes into two groups: low salinity (<2500 mg/l) and high HCO_3^-/Cl^- (>0.08) that are surface water and inter-flow dominated; and higher salinity (>2500 mg/l) and low HCO_3^-/Cl^- (<0.01) that are groundwater dominated. The intermediate group may represent mixing between the two end-members.

While HCO_3^- is not conservative over the entire salinity range, the HCO_3^-/Cl^- values of groundwater samples are generally low due to evapo-transpiration and precipitation of carbonate within the soil zone. On the other hand, the short residence time of surface water runoff and interflow in the zone where CO_2 is produced, and consequent weathering of volcanic minerals results in much higher HCO_3^- relative to Cl^-. This leads to the contrast in HCO_3^-/Cl^- ratios which is used to infer the initial lake source water (e.g. high values = surface water; low values = groundwater). This approach is more sensitive than using salinity alone which cannot distinguish between groundwater fed systems and those dominated by surface water within a moderate degree of evaporation. The higher HCO_3^-/Cl^- ratios characteristic of surface in-flows (as opposed to groundwater in-flows) are only maintained until the onset of high salinities (e.g. 100 g/l), after which calcite saturation and precipitation occurs, reducing HCO_3^-/Cl^- ratios to low values, as observed in all of the lake waters with salinities above this approximate threshold value (Fig. 7.6). In general, such waters will also likely be characterised by low d-excess values, reflecting long residence time and intensive evaporation. Above the salinity threshold, it is not possible to distinguish the initial lake source water on the basis of the method. Further sampling over a longer period, covering periods immediately after substantial inflow of groundwater or surface water to the lakes may help to distinguish the dominant source waters (groundwater or surface water); as lake water composition will have not yet been substantially modified by evaporite precipitation reactions.

Accepting the HCO_3^-/Cl^- ratio as an indicator of surface water and groundwater dominance, the lakes can be further categorise by plotting the HCO_3^-/Cl^- ratio with respect to the deuterium excess (Fig. 7.7).

The deuterium excess, $\delta_{xs} = \delta^2H - 8 \times \delta^{18}O$, reflects the deviation of a given sample from the meteoric water line and hence is an indicator of residence time and a measure of evaporation. Lower values indicate the increasing influence of evaporation. Most of the groundwater samples have a deuterium excess of between 7–12, which is slightly less than the local meteoric water values of 13, while the lake waters have a δ_{xs} generally between 5 and −5. If there was a large flow-through of groundwater, then the δ_{xs} would approach that of groundwater δ_{xs} values. Hence, plotting the δ_{xs} values data as a function of HCO_3^-/Cl^- provides a means of at least qualitatively distinguishing between the water source of the lakes – surface water or groundwater – and whether they are through-flow or have long residence times. Four lake types can be defined (Fig. 7.8):

1 groundwater dominated, through-flow (high δ_{xs}, low HCO_3^-/Cl^-);
2 groundwater dominated, long-residence (low δ_{xs}, low HCO_3^-/Cl^-);

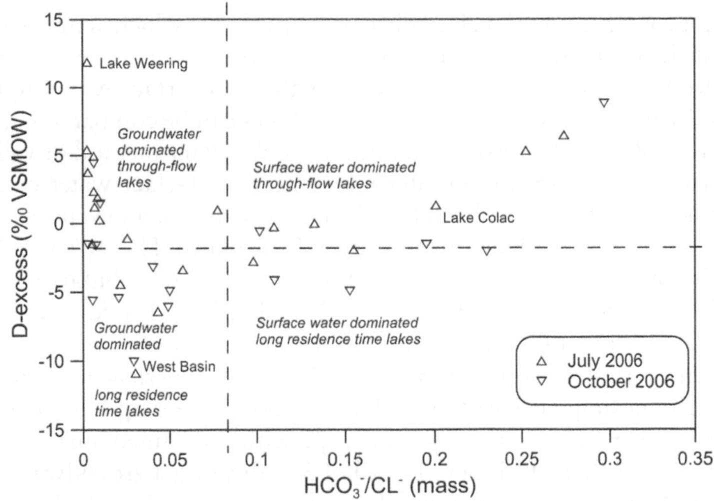

Figure 7.7 Deuterium "excess" versus HCO_3^-/Cl^-.

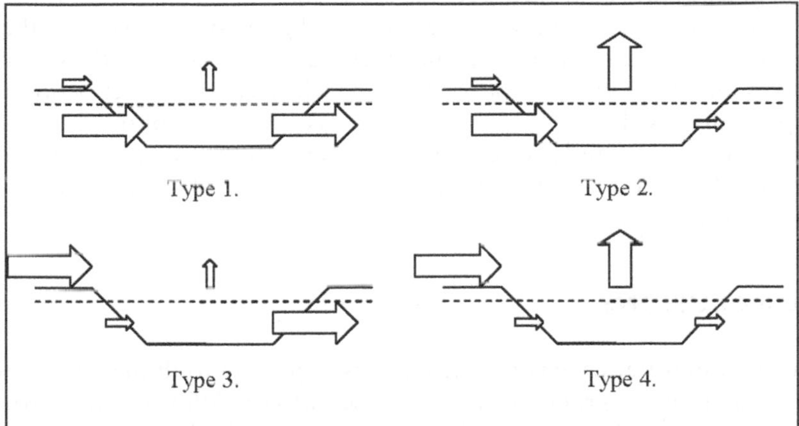

Figure 7.8 Conceptual representation of the four hydrological lake types.

3 surface water dominated, through-flow (high δ_{xs}, high HCO_3^-/Cl^-);
4 surface water dominated, long-residence (low δ_{xs}, high HCO_3^-/Cl^-).

As lakes evolve into the terminal type (low δ_{xs}) the differentiation between groundwater and surface water dominance is not possible due to the mineral precipitation controls on HCO_3^- concentration.

Most lakes do not fall clearly within one type and lie along a continuum between two or more groups. Furthermore variation in hydrologic condition (e.g. rising or falling groundwater tables, or changes in surface runoff) may result in a change to the lake's status.

7.4.2 Method assessment

This method provides an initial classification of the lakes which is not possible otherwise without detailed gauging of surface runoff and installing groundwater bores to obtain quantitative description of the groundwater and surface water inputs. Known physical and hydrogeological aspects of specific lakes can be compared with where the lakes lie on the HCO_3^-/Cl^- versus δ_{xs} plot, as a preliminary method of validation.

West Basin, for example, is a crater lake with little surface water catchment and no outlet. It is logical that that this lake falls in the region of an evaporating or terminal lake ($\delta_{xs} = -10.9$) with high groundwater dependence ($HCO_3^-/Cl^- = 0.03$). Lake Colac, on the other hand, is a larger, shallower, lake with tributaries on the south and an outlet to its north. It plots ($HCO_3^-/Cl^- = 0.2$, $\delta_{xs} = 1.3$ & -1.6) as a surface water dominated lake and between the terminal and through-flow types. This is consistent with the period of sampling when the lake was experiencing a prolonged dry period with diminishing surface flows. A third lake is Weering Lake which is essentially a shallow depression in the flat landscape with no defined inlet and outlet. The HCO_3^-/Cl^- ratio for this lake is very low, 0.003, indicating a groundwater dominated lake. Between the first and second sampling events δ_{xs} varied markedly between 11.9 and -1.6. This possibly indicates a decoupling of the lake from the groundwater table as it continued to dry out.

Fortnightly sampling of the three aforementioned lakes (West Basin, Lake Colac and Weering Lake) has since been undertaken and groundwater levels in the vicinity of these lakes monitored. These data will provide a basis to estimate the water balance of the lakes and make comparison with the geochemical data. The period in which the sampling discussed in this paper was undertaken was one of the driest on record and came at the end of an historic 10 year average low (BOM, 2007). The on-going drying of the lakes points to the fact that inflows, either of groundwater or surface water, have been limited.

7.5 CONCLUSIONS

The method uses water chemistry and stable isotope data to distinguish between the dominant water source (groundwater or surface water) and flow regime (through-flow or evaporating/terminal) of a series of wetlands in a basaltic plain of SE Australia. High HCO_3^-/Cl^- are indicative of surface water dominance, while low HCO_3^-/Cl^- reflect groundwater dominance, Deuterium "excess" (δ_{xs}, where $\delta_{xs} = \delta^2H - 8 \times \delta^{18}O$) is an indicator of lake water residence time, with low values indicating terminal or evaporation dominated lakes. The wetlands lie on a continuum between four end member lake types, *viz*: groundwater dominated, through-flow; groundwater dominated, long-residence time; surface water dominated, through-flow; and surface water dominated, long-residence time.

ACKNOWLEDGEMENTS

This project is funded through a National Action Plan for Salinity and Water Quality grant administered by the Corangamite Catchment Management Authority. We thank

Dr Matthew Currell (RMIT University) for his assistance and comments on aspects of the paper.

REFERENCES

Adler, R. & Lawrence, C.R. (2004) The drying of the Red Rocks Lakes Complex, Australia and its implications for groundwater management. *Paper presented at the XXXIII Congress of the International Association of Hydrogeologist*, Zacatecas, Mexico.

Appelo, C.A.J. & Postma, D. (2005) *Geochemistry, Groundwater and Pollution*. New York, Leiden, 649p.

BOM (2007) Rainfall data for weather station number 090147 (Colac Shire Office). Australian Bureau of Meteorology. http://www.bom.gov.au/silo/ "patched point data" set cited November 2006.

Coram, J.E., Weaver, T.R. & Lawrence, C.R. (1998) Groundwater–surface water interactions around shallow lakes of the Western District Plains, Victoria, Australia. *Paper presented at the International Groundwater Conference of the International Association of Hydrogeologists*, University of Melbourne, Melbourne, Australia, 8–13 February 1998.

Cox, J., Dahlhaus, P., Herczeg, A., Crosbie, R., Davies, P. & Dighton, J. (2007) Defining groundwater flow systems on the volcanic plains to accurately assess the risks of salinity and impacts of changed landuse. Final report to the Corangamite Catchment Managment Authority, Draft edn, CSIRO Land and Water, Urrbrae, SA.

Nicholson, C., Dahlhaus, P.G., Anderson, G., Kelliher, C.K. & Stephens, M. (2006) *Corangamite Salinity Action Plan: 2005–2008*. Colac, Victoria, Corangamite Catchment Management Authority.

Rozanski, K., Froehlich, K. & Mook, W.G. (2000) *Surface Water*. Technical Documents in Hydrology. Paris, UNESCO.

Wetland_1994 (n.d.) *Victorian Wetland Environments and Extent–up to 1994*. Victoria, Australia, Corporate Geospatial Data Library, Department of Sustainability and Environment.

Williams, W.D. (1995) Lake Corangamite, Australia, a permanent saline lake: conservation and management issues. *Lakes & Reservoirs: Research and Management*, 1, 55–64.

Williams, W.D. (1999) Conservation of wetlands in drylands: A key global issue. Aquatic conservation. *Marine and Freshwater Ecosystems*, 9 (6), 517–522.

Chapter 8

Mapping surface water-groundwater interactions and associated geological faults using temperature profiling

Matthijs Bonte[1], Josie Geris[2], Vincent E.A. Post[3], Victor Bense[4], Harold J.A.A. van Dijk[5] & Henk Kooi[6]

[1] *KWR Watercycle Research Institute, Nieuwegein, The Netherlands*
[2] *Newcastle University, School of Civil Engineering and Geosciences, Newcastle upon Tyne, United Kingdom*
[3] *Flinders University, Faculty of Science and Engineering, School of the Environment, Adelaide, Australia*
[4] *University of East Anglia, School of Environmental Sciences, Norwich, England*
[5] *Waterboard Aa en Maas, 's-Hertogenbosch, The Netherlands*
[6] *VU University, Faculty of Earth and Life Sciences, Amsterdam, The Netherlands*

ABSTRACT

In this study temperature profiling was used as a mapping tool to investigate surface water-groundwater interactions associated with low-permeability faults in the south-eastern part of the Netherlands. Data obtained from temperature profiling were simulated using a simple water and energy balance model. The model was used to quantify groundwater seepage rates near the faults. The model simulated the surface water temperature transects very well, but the model-based seepage rates are much higher than those reported in the literature for comparable hydrological settings. Of all the assumptions made in the model, the adopted values for the parameters that control the heat exchange between the water and the atmosphere are believed to be the most uncertain. Despite these uncertainties, temperature profiling proved to be a very useful tool to investigate surface water-groundwater interactions and to map groundwater seepage zones. The strength of using temperature as a tracer is that it is measured directly in the field with relative ease and using low cost equipment.

8.1 INTRODUCTION

Groundwater flow and geological structures and processes often show a strong two way linkage: groundwater flow can drive geological processes such as diagenesis, earthquakes and metamorphism (Ingebritsen & Sanford, 1998) while geological structures such as faults and folds can strongly control groundwater flow. There is a large body of literature that discusses the impact of geological faults on groundwater flow patterns. In hard rock terrains, faults act as conduits through which groundwater flow occurs (Conrad *et al.*, 2004; Ge *et al.*, 2008; Gudmundsson, 2000; Shalev & Yechieli, 2007; Wolaver & Diehl, 2010). This allows the use of springs or travertine deposits to map the spatial extent of these features (Crossey *et al.*, 2006; Springer & Stevens, 2009; Wolaver & Diehl, 2010). In high yielding aquifers of unconsolidated sediments, faults can act as groundwater flow barriers. Examples include the San Andreas fault system where head differences up to 60 m across the fault have been reported (Catchings *et al.*, 2009; Mayer *et al.*, 2007) and the Lembang fault in Indonesia, which minimises recharge of the downstream aquifer (Delinom, 2009).

A characteristic of both flow-obstructing and flow-conducting faults is the important role they play in controlling upward seepage of groundwater. This upward groundwater seepage can be discharged to surface water bodies sustaining aquatic ecosystems or it can be transpired directly by terrestrial vegetation. Such groundwater dependent ecosystems (GDEs) are highly dependent on groundwater seepage to provide: i) base-flow to streams, especially during dry periods (Sophocleous, 2002), ii) water with distinct physical and chemical characteristics (Claret & Boulton, 2009; Hancock *et al.*, 2005), iii) habitat for fish in streams (Malcolm *et al.*, 2009; Olsen & Young, 2009; Seilheimer & Fisher, 2010).

This paper focuses on mapping of fault-induced seepage areas and associated GDEs located in the southeast of the Netherlands. These seepage areas are colloquially referred to as *wijst* areas, and are historically known for poor farming properties due to water logging (Visser, 1948). Because of this aspect, these seepage areas have been intensively drained and the outcrop and the first few metres of the subcrop of the low-permeability fault zones have been removed to allow horizontal groundwater through flow. Such deterioration of GDEs is no exception in the Netherlands: Runhaar *et al.* (1996) estimated that at least 50% of the Dutch GDEs have been negatively-impacted by agricultural drainage, groundwater extraction and water quality issues. In recent decades, there has been a re-appraisal, mostly from an ecological perspective, of the value of groundwater seepage. The discharging groundwater in the south of the Netherlands is characterised by a high Fe^{2+} concentration (Mendizabal *et al.*, 2011). Dissolved ferrous iron immobilises phosphate and reduces eutrophication of surface water (Griffioen, 2006). This causes the discharging groundwater to sustain valuable vegetation types such as the seepage indicator Water Violet (*Hottonia palustris*), phreatic grassland and swamp vegetation (*Calthion palustris*, *Filipendulion*, *Caricion elatae*), and groundwater dependent tree species (*Alnion glutionosae*) (Meuwissen & van den Brand, 2003; Verwijst, 1982).

Another distinct aspect of the groundwater seepage zones is their visual appearance: water logging occurs on relatively high areas upstream from the fault, while downstream, topographical lower areas are relatively dry. Where groundwater discharges, the dissolved iron reacts with oxygen and precipitates as iron-oxides with its characteristic reddish-brown colour. The localised precipitation of iron has resulted in the development of iron concretions along the fault lines (ranging in size from several cm^3 to m^3). The permanent wet conditions have locally resulted in the formation of peat bogs.

Because of the ecological and environmental value of these areas, the regional water management authority, the Waterboard Aa and Maas (here after called 'the Waterboard'), commissioned a project to map the seepage areas and exact location of the faults on a regional scale. A second objective of the project was to assess the technical opportunities to restore or increase the ecological health of the seepage areas.

There are numerous techniques to map fault-related groundwater seepage (Brodie, 2007; Hillier *et al.*, 2002) such as: i) seepage meters, ii) geological mapping of faults and facies changes, iii) geophysical surveys, iv) base-flow separation methods,

v) groundwater flow-net analyses, vi) hydrochemical tracers, v) water budgets and numerical modelling, and vii) using heat as a tracer. This latter method is based on the temperature contrast between groundwater and surface water.

For this project, temperature profiling of surface water and shallow groundwater (0.4 m depth below the stream surface) was undertaken. The rationale for this approach is that temperature can be determined in situ with relative ease, making it relatively cheap compared to many of the other above mentioned methods. This aspect, combined with the recent development of new measurement techniques, have resulted in a gaining popularity of the use of heat as a natural and omnipresent tracer, which is reflected by an increase in the number of publications in the literature (Saar, 2011; Anderson, 2005; Fairley & Nicholson; 2006, Bense & Kooi, 2004; Forster & Smith, 1989).

The objective of the study was to evaluate the feasibility of temperature measurements in streams and shallow groundwater for mapping the exact location of low-permeability faults. The results are presented in the form of profiles across the fault zone. A quantitative interpretation of the measurements using two analytical models is attempted in order to determine seepage flux magnitudes.

8.2 STUDY AREA

The study area encompasses the water management area of the Waterboard, which has a surface area of about 1600 km^2. The area is located in the southeast of the Netherlands (Fig. 8.1), in the Lower Rhine Graben (LRG) system, which is the northwestern branch of the Rhine Graben Rift System (Ziegler, 1994). The LRG is the most prominent neotectonically active area in the Netherlands (van Balen *et al.*, 2005), which is reflected by earthquakes (for example, in Roermond (1994) in Geluk *et al.* (1994)), fault-scarps and in topographic elevation differences.

The LRG system comprises a number of tectonic elements, three of which traverse the Waterboard area in a NW-SE direction, including (from west to east, see Fig. 8.1): the Roer Valley Graben (RVG, also known as Central Graben), the Peel Block (PB) and the Venlo Block (VB) (Geluk *et al.*, 1994; van Balen *et al.*, 2005). The RVG is separated from the PB by the Peel Boundary Fault Zone and the PB is separated from the VB by the Tegelen Fault Zone. These faults have a low permeability due to clay smearing and iron precipitation (Bense & Van Balen, 2004) and have a pronounced influence on groundwater flow.

The thickness of the unconsolidated sediments and aquifers varies strongly between these three tectonic units. The thickness of Tertiary and Quaternary sediments in the RVG ranges between 900 and 1400 m, while at the PB the thickness ranges between 200 and 300 m (Geluk *et al.*, 1994).

The surface water levels in the area are highly regulated by use of an intense drainage network consisting of drainage pipes (generally 5 to 10 cm diameter), drainage channels (one to several metres width and ranging in depth from less than one to several meters), and larger channels or small streams. Water level regulation during wet periods is achieved by gravity drainage and weirs, while during dry periods water is imported from major rivers such as the Meuse. Groundwater levels closely follow the topographical relief of the area, with a water divide present at the centre of the

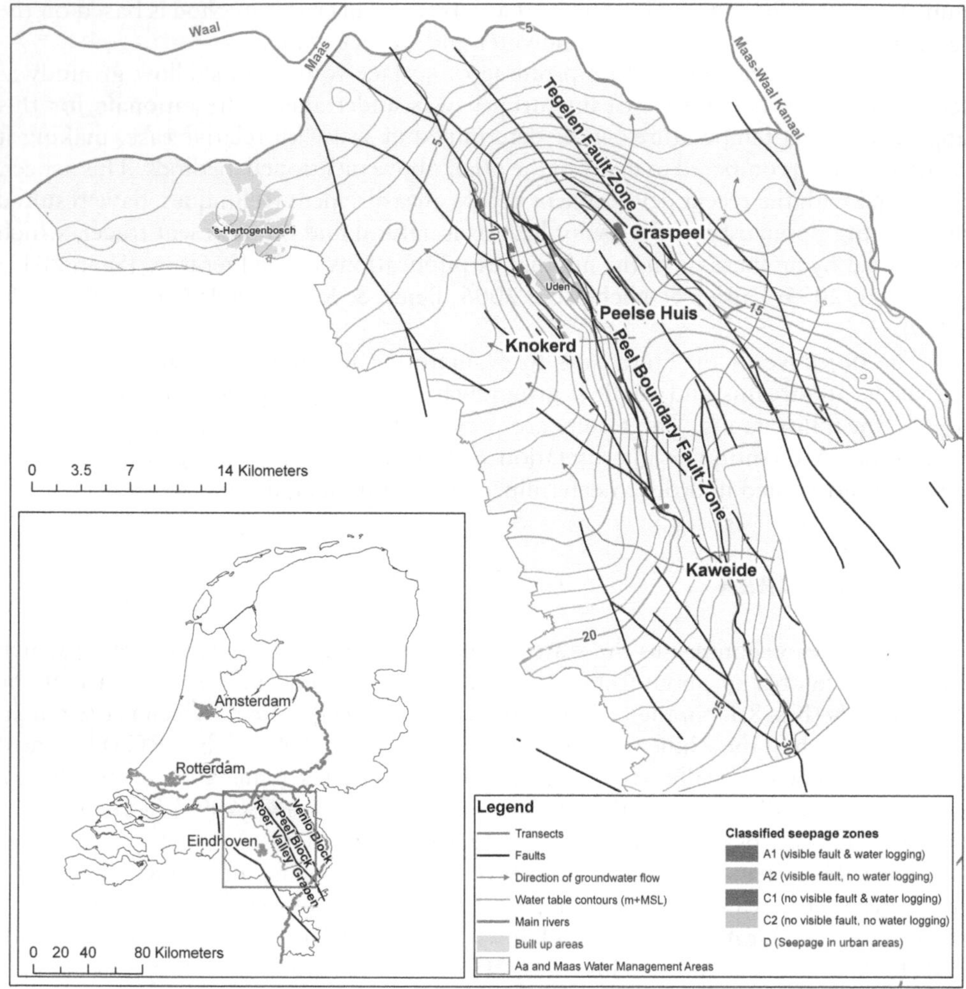

Figure 8.1 Study site showing the location of the thermal profiling transects, major faults and classified *wijst* seepage areas.

PB and direction of groundwater flow in a south-west and north-west direction away from the water divide.

8.3 FIELD METHODS

The selection of locations where fault-induced groundwater discharge was expected to occur was based on an desk study of data on groundwater levels, soil type, fault locations inferred from seismic and drilling data, historical maps and local expert knowledge. Thirty-nine locations were identified and visited between October and

December 2006. At each site a drainage channel perpendicular to the inferred fault was selected in which the temperature of the surface water and the groundwater below the drain bottom was measured along a transect. Measurements were taken at intervals varying between 5 to 25 meters, depending on the observed spatial variability in temperature. These measurements were made using a rod with a thermistor (YSI 400 series) built into its cone-shaped tip. The measurements in surface water were taken at half the water-filled channel depth. Groundwater temperatures were measured at 0.4 m below the channel bottom, at which daily temperature variations have dampened out (Conant, 2004). A practical consideration is that this depth is still accessible with relative ease using the thermistor-rod, which experiences strong friction with the sandy bottom sediments.

At each site the color of surface water, the electrical conductivity (EC), Fe^{2+} and NO_3^- were recorded. The latter two parameters were determined using colorimetric test strips (Merckoqant®) and the EC was measured using a handheld EC-meter (Greissinger GMH-3410). The results were directly stored in a palmtop fitted with a GPS receiver and Arcview® GIS software. The results were overlain on aerial photographs that gave an indication of the regional extent of the groundwater seepage zones.

8.4 RESULTS

Typical examples of temperature profiles observed at four different study sites are shown in Figure 8.2. The measured Fe^{2+} and NO_3^- concentrations are also indicated.

At the Graspeel site, groundwater temperatures measured in the streambed varied between 10 and 12°C and were consistently higher than surface water temperatures. Based on the temperature distribution in the streambed, two seepage zones were distinguished: between 140 and 240 m and 490 and 580 m. Ferrous iron concentration ranges between 5 and 15 mg/l and no nitrate was observed.

At the Knokerd site, the drainage channel is dry at the upstream side of the fault. The fault induces groundwater discharge into the channel, which starts to carry water just upstream of the fault. The location of the fault is indicated by a pronounced decrease in groundwater temperature. NO_3^- concentrations remain constant in the seepage zone and decrease downstream of the fault. Iron concentrations increase in the seepage zone and downstream of the fault, and then level off at a constant value approximately 50 m downstream of the fault.

At the Kaweide site, the temperature profile measured in both surface water and shallow groundwater showed a very clear anomaly which can be related to groundwater discharge to the surface water. At the start of the transect, at 0 m, groundwater and surface water temperature are in the range 9–11°C, which is slightly higher than the atmospheric temperature at the time of collection (9°C). At a distance between 100 and 245 m from the start of the profile a zone is observed where groundwater and surface water temperatures are 2 to 3°C higher than the atmospheric temperature. Considering that there are no other heat sources present in the channel, this temperature anomaly is attributed to groundwater seepage. Upstream of the seepage zone, both the Fe^{2+} and NO_3^- concentrations are low. The highest NO_3^- concentration is observed just inside the seepage zone. The highest Fe^{2+} concentration is measured just downstream

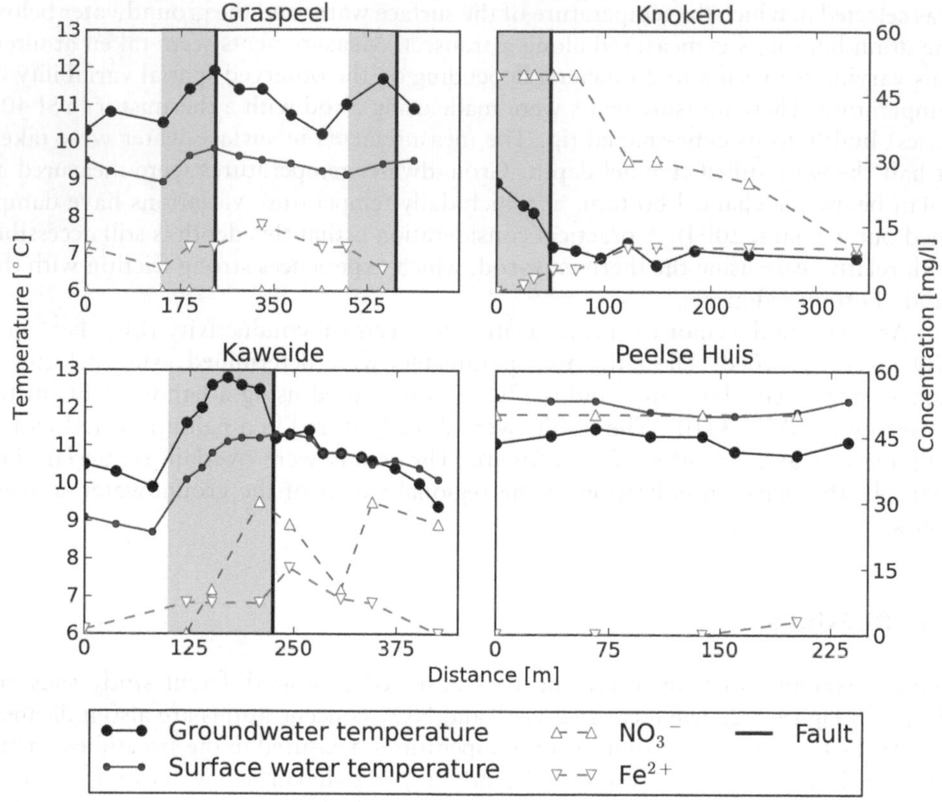

Figure 8.2 Temperature profiles at four transects in the study area in groundwater and surface water (at Knokerd only groundwater was measured) and measured ferrous iron and nitrate concentrations. Channel flow is always from left to right and transects start at the upstream end. Seepage areas are shown with grey shading; the faults are indicated by the black lines at the downstream end of the seepage zones.

the seepage zone, but given the use of graduated indicator test strips to measure Fe^{2+}, the significance of this observation is questionable. The elevated NO_3^- concentrations show that shallow groundwater is contaminated by agricultural activities.

At the Peelse Huis site, no indications of groundwater seepage were found. No clear temperature anomaly was observed, which is interpreted to signify that there is little or no groundwater seepage at this site. The low Fe^{2+} and NO_3^- concentrations provide further indication that groundwater seepage is negligible.

The autumn of 2006 was relatively mild, with temperatures of on average 2 to 5°C above the annual average. The absence of a temperature anomaly (e.g. at the Peelse Huis site) could, therefore, also be due to a lack of a clear temperature contrast between groundwater and surface water. It is expected that the resolution of the measurements will improve when they are taken well into or towards the end of prolonged cold or hot periods when the contrast of surface water and groundwater temperatures is greatest. In this sense the temperature and chemical concentration measurements are complementary, as the water quality parameters may still show an anomaly in the absence of a temperature contrast. If, however, both temperature and water quality parameters do not show any significant variation, as with the Peelse Huis site, this may be taken as an indication that there is no groundwater seepage.

8.5 MODELLING OF THE TEMPERATURE PROFILES

Various studies have shown that spatial temperature variations, both in the vertical (as depth profiles below the channel bottom) and horizontal (as transects along the channel), can be used for quantification of groundwater discharge rates (Saar, 2011; Bense *et al.*, 2008; Fairley & Nicholson, 2006; Anderson, 2005; Bense & Kooi, 2004; Forster & Smith, 1989). A detailed approach to simulate the temperature profiles in groundwater under the channel bed sections was presented by Bense & Kooi (2004) who used a transient groundwater flow and heat transport model to predict seepage at the Peel Boundary Fault near the village of Uden. Bense & Kooi (2004) were able to reproduce the temperature anomalies measured in shallow groundwater reasonably well and found seepage fluxes in the *wijst* zone between 15 and 80 mm/d. A disadvantage to using this method to quantify the seepage flux is that the construction and calibration of the transient groundwater model required for the analysis is a time consuming effort.

Anibas *et al.* (2009, 2011), Arriage & Leap (2006) and Schmidt *et al.* (2007) used a steady state solution of the advection-conduction equation for heat transport based on the work by Bredehoeft & Papaopulos (1965) which relates the seepage flux to the surface water temperature, the temperature in the streambed and the deeper (> several metres) temperature in groundwater. The simplicity of a steady state equation makes it easily applicable in many situations to assess the flux magnitude. In the present study, however, it proved not possible to apply this analytical model because we lacked local groundwater temperature data at depth.

Leach & Moore (2011) and Westenhof *et al.* (2007) used surface water temperature transects to assess water inflows into streams. However, they addressed lateral inflows and their models cannot be directly used to constrain groundwater seepage. Therefore, it was decided to develop a simple one-dimensional water and energy balance model to simulate the observed surface water temperatures. The model assumes: 1) a quasi-steady state in the channel temperatures during the time that the temperature transect is measured (i.e. the temperature change during the measurement time is negligible compared to the temperature differences measured along the transect); 2) complete vertical and lateral mixing in the channel; and 3) the surface water temperature is determined solely by heating of channel water due to upward groundwater inflow and cooling of channel water due to atmospheric exchange. Conductance of heat to the streambed is neglected, as this is expected to be a minor contribution of the total heat transport in strongly advection-dominated system. Based on these assumptions, the water and energy balance is defined by a set of two ordinary differential equations:

$$\frac{dq(x)}{dx} = \frac{q_z}{D} \tag{8.1}$$

$$\frac{d(q(x)T(x))}{dx} = \frac{q_z T_z - \frac{k_{air}}{c_w \rho}(T(x) - T_e)}{D} \tag{8.2}$$

where $q(x)$ is the horizontal flow velocity (m day^{-1}) in the channel at distance x (m), q_z is the seepage flux density in the channel (m d^{-1}) and D is the channel depth (m), $T(x)$ is the temperature in the surface water (°C) at distance x (m), T_z is the temperature of the streambed (°C), k_{air} is the atmospheric heat transfer coefficient (J m^{-2} d^{-1} °C^{-1}) , T_e is the equilibrium air temperature (°C), if the water has temperature there is no energy exchange between water and atmosphere, ρ_f and c_w are the density (g m^{-3}) and specific heat of water (J g^{-1} °C^{-1}), respectively. The use of this type of linear atmospheric cooling model to simulate energy exchange between water and atmosphere is a simplified way to describe the complete energy balance of surface water, avoiding explicit use of solar and long wave radiation energy and evaporation and conduction energy (Williams, 1963; Wright *et al.*, 2009). Sweers (1976) derived the following equation for k_{air} by solving the heat balance of surface water:

$$k_{air} = (4.48 + 0.049T) + f(u)(1.12 + 0.0180T + 0.00158T^2) \tag{8.3}$$

where $f(u)$ is the wind function (J s^{-1} m^{-2} mbar^{-1}) which is used to calculate the evaporative energy losses from water as a function of the actual and saturated water vapour in the air. Sweers (1976) uses the following wind function derived by McMillan (1973) for a 6 m deep lake of 5 km^2 in Wales:

$$f(u) = 3.6 + 2.5u_3 \tag{8.4}$$

where u_3 is the wind speed (m/s) measured at 3 m above the water's surface. When all heat balance terms are available T_e can be calculated iteratively (Boderie & Dardengo, 2003), but because these date were unavailable it was assumed here that T_e equals the average air temperature during the day of sampling. This assumption was also made by Williams (1963) in one of the first papers on water cooling by atmospheric exchange and yielded good results for the lakes in his study. Data on average daily measured wind speed and temperature were downloaded from http://www.knmi.nl for a weather station in Eindhoven (circa 20 km from the study area).

The set of equations (1) and (2) was solved using Python's ODEINT (based on the fortran collection of solvers "ODEPAK" by Hindmarsh, 1983) for the surface water temperature data for Kaweide and Graspeel. Data obtained at Peelse Huis and Knokerd were not simulated because of the lack of seepage (Peelse Huis) and surface water temperature data (Knokerd). An objective function based on the squared errors between measured and simulated surface water temperatures was minimized using python's LEASTSQ module (which is a wrapper around the MINPACK's lmdif and lmder algorithms by Burton *et al.*, 1980) based to find: (i) the inflow at the upstream end of the transect, and (ii) the upward water flux in the seepage zone. The seepage zone was delineated based on the measured streambed temperatures as described in the previous section. Table 8.1 presents an overview of the model parameters, Figure 8.3 shows the simulated temperature transects. The model yields seepage fluxes ranging between 510 and 1400 mm/d and is very well able to reproduce the observed temperature anomalies in the surface water.

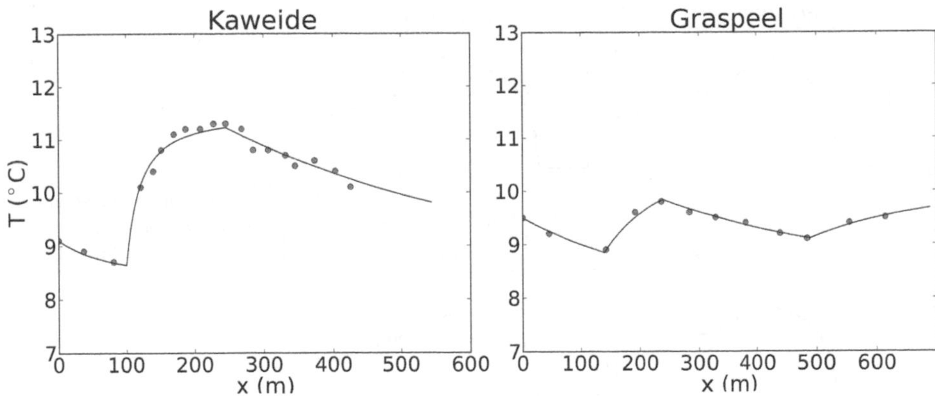

Figure 8.3 Simulated surface water temperatures at the Kaweide and Graspeel sites. Best fits were found at Kaweide with a seepage flux of 1400 mm/d between 100 and 245 m distance from the profile start and at Graspeel with seepage fluxes of 810 and 510 mm/d between respectively 140 and 240 m and beyond 490 m from the profile start.

Table 8.1 Parameters applied to analytical model defined by equations (8.1) to (8.4).

Parameter	*Unit*	*Site*		*Source*
		Graspeel	*Kaweide*	
c_w	J g^{-1} °C^{-1}	4.19		
ρ	kg m^{-3}	999.7		Lide (1990) at 10°C
U	m s^{-1}	6	6	Weather station Eindhoven, average daily data
k_{air}	J m^{-2} d^{-1} °C^{-1}	2.5	2.5	Method by Sweers (1976)
T_e	°C	8	9.1	Assumed to be equal to air temperature sourced from weather station Eindhoven, average daily data
T_z	°C	12.7	11.5	Measured in streambed
D	m	0.15	0.15	Measured in field

8.6 DISCUSSION

The thermal transects at the sites presented here as well as the transects collected at other investigated sites showed that seepage zones are readily identified in the field by mapping temperature anomalies. Using temperature transects for a quantitative assessment of seepage fluxes is a simple task at sites where sufficient groundwater temperature data are available to warrant the use of a steady state model as derived by Bredehoeft & Papaopulos (1965). Numerical modelling to reproduce a transient situation dramatically increases the amount of work required to make an assessment, making this method for estimation of seepage fluxes more cumbersome. In our study a simple water and energy balance model was developed to simulate surface water temperatures. The model considers the effects of channel flow, seepage flux and temperature and equilibration with air temperature. Although the observed temperature

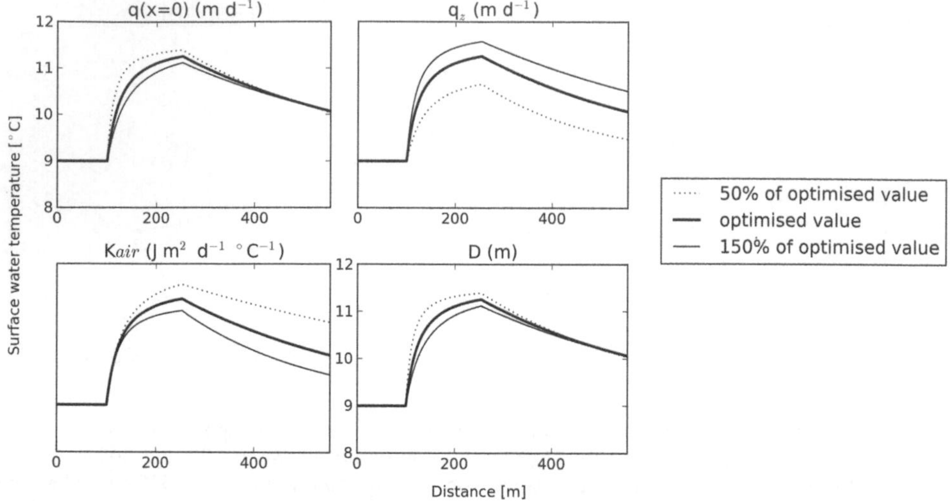

Figure 8.4 Sensitivity analyses of simulated surface water temperature transects to variations in the main model parameters.

transects can be simulated very well, the calculated seepage fluxes are high compared to values reported in literature. For example, Bense & Kooi (2004) found values ranging between 15 to 80 mm/d based on transient numerical modelling of a temperature transect measured in the streambed near the PBF in the nearby village of Uden. Other reported seepage fluxes in reasonably similar hydrological conditons range between: 22 to 77 mm/d measured in an unconsolidated aquifer comprising heterogeneous fine sand with clay in Belgium (Anibas *et al.*, 2011), −10 (infiltration rather than seepage) to 455 mm/d in an unconsolidated aquifer comprising glacio-fluvial sandy gravels in Germany (Schmidt *et al.*, 2006) and 0.03 to 446 mm/d with one outlier of 7040 mm/d in an aquifer comprising mainly very fine to fine sands with some areas with gravel and cobbles in Ontario, Canada (Conant, 2004).

The key uncertain factor in the model is the atmospheric heat exchange coefficient. The value for average daily wind and temperature conditions was derived using a relationship for a much larger and deeper water body than studied here. It is likely that elongated, relatively narrow drain water bodies are characterised by a microclimate with a lower atmospheric heat exchange coefficient than that based on the model by Sweers (1976). In order to investigate the sensitivity of the model outcomes to the different parameters in the model, a sensitivity analysis was performed in which the parameter values were increased and decreased by 50%. The results show that k_{air} is the most sensitive parameter and that lowering k_{air} has a similar impact on the simulated temperature transect as increasing the seepage flux (Fig. 8.4). There is a clear linear relation ship between k_{air} and the optimised seepage flux and initial flow velocity (Fig. 8.5).

The model cannot be used without adequate determination of the value of k_{air}. If the range in k_{air} values can be narrowed, or can be related to easily measured channel

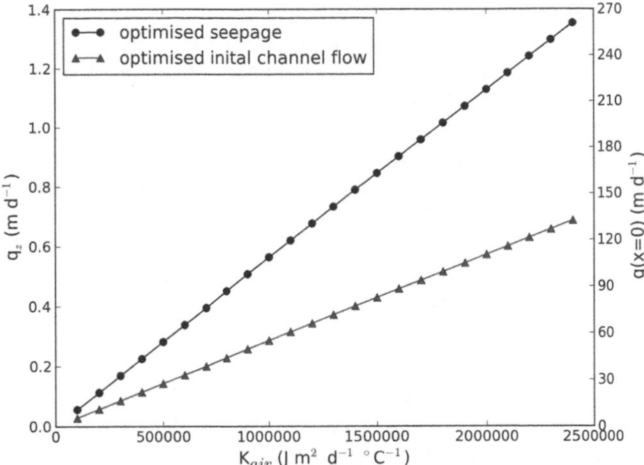

Figure 8.5 Sensitivity analyses of optimised seepage flux and initial channel flow velocity to variations in k_{air} value.

characteristics, the model developed here may prove to be a useful additional tool for hydrological investigations.

8.7 CONCLUSION

Mapping temperature and solute concentrations in stream transects perpendicular to faults was shown to be an accurate, fast and cost-efficient method for the delineation of groundwater seepage zones. The collected field data identify the locations of major groundwater seepage zones that can be used to constrain regional groundwater flow conditions. At the local scale, the amplitude of the surface water temperature anomalies constrains the relative magnitude of the groundwater seepage rates. A simple surface water and energy balance model was developed to quantify groundwater seepage rates. The model simulated the surface water temperature transects very well, but the model-based seepage rates are much higher than those reported in the literature for comparable hydrological settings. Of all the assumptions contained in the model, the adopted values for the parameters that control the heat exchange between the water and the atmosphere are believed to be the most uncertain. Better estimates of the site-specific values for these parameters and further testing of the adopted approach are required to test the applicability of the model under these conditions.

ACKNOWLEDGEMENTS

This project was primarily funded by the waterboard Aa en Maas. Additional funding to complete this manuscript was provided by the joint research project of the Dutch water companies (BTO).

REFERENCES

Anderson, M.P. (2005) Heat as a ground water tracer. *Ground Water*, 43, 951–968.

Anibas, C., Buis, K., Verhoeven, R., Meire, P. & Batelaan, O. (2011) A simple thermal mapping method for seasonal spatial patterns of groundwater–surface water interaction. *Journal of Hydrology*, 397, 93–104.

Bense, V.F. & Kooi, H. (2004) Temporal and spatial variations of shallow subsurface temperature as a record of lateral variations in groundwater flow. Journal of Geophysical Research, 109, B04103. doi:10.1029/2003jb002782.

Bense, V.F. & Van Balen, R. (2004) The effect of fault relay and clay smearing on groundwater flow patterns in the Lower Rhine Embayment. *Basin Research*, 16, 397–411. doi:10.1111/j.1365-2117.2004.00238.x.

Bense, V.F., Person, M.A., Chaudhary, K., You, Y., Cremer, N. & Simon, S. (2008) Thermal anomalies indicate preferential flow along faults in unconsolidated sedimentary aquifers. *Geophysical Research Letters*, 35, L24406. doi:10.1029/2008gl036017.

Boderie, P. & Dardengo, L. (2003) *Heat Discharge in Surface Water and Atmospheric Exchange* [in Dutch] WL Delft Hydraulics, Delft, pp. 110.

Bredehoeft, J.D. & Papaopulos, I.S. (1965) Rates of vertical groundwater movement estimated from the Earth's thermal profile. *Water Resources Research*, 1, 325–328. doi:10.1029/WR001i002p00325.

Brodie, R., Sundaram, B., Tottenham, R., Hostetler, S. & Ransley, T. (2007) An adaptive management framework for connected groundwater–surface water resources in Australia Bureau of Rural Sciences, Canberra.

Garbow, B.S., Hillstrom, K.E. & More, J.J. (1980) Documentation for MINPACK subroutine LMDIF Double precision version Argonne National Laboratory. Report available online: http://www.math.utah.edu/software/minpack/minpack/lmdif.html.

Catchings, R.D., Rymer, M.J., Goldman, M.R. & Gandhok, G. (2009) San Andreas fault geometry at Desert Hot Springs, California, and its effects on earthquake hazards and groundwater. *Bulletin of the Seismological Society of America*, 99, 2190–2207.

Claret, C. & Boulton, A. (2009) Integrating hydraulic conductivity with biogeochemical gradients and microbial activity along river–groundwater exchange zones in a subtropical stream. *Hydrogeology Journal*, 17, 151–160. doi:10.1007/s10040-008-0373-3.

Conant, B. (2004) Delineating and quantifying ground water discharge zones using streambed temperatures. *Ground Water*, 42, 243–257.

Conrad, J., Nel, J. & Wentzel, J. (2004) The challenges and implications of assessing groundwater recharge: A study – Northern Sandveld, Western Cape, South Africa. *Water SA*, 30, 623–629.

Crossey, L.J., Fischer, T.P., Patchett, P.J., Karlstrom, K.E., Hilton, D.R., Newell, D.L., Huntoon, P., Reynolds, A.C. & de Leeuw, G.A.M. (2006) Dissected hydrologic system at the Grand Canyon: Interaction between deeply derived fluids and plateau aquifer waters in modern springs and travertine. *Geology*, 34, 25–28.

Delinom, R.M. (2009) Structural geology controls on groundwater flow: Lembang fault case study, West Java, Indonesia. *Hydrogeology Journal*, 17, 1011–1023.

Fairley, J.P. & Nicholson, K.N. (2006) Imaging lateral groundwater flow in the shallow subsurface using stochastic temperature fields. *Journal of Hydrology*, 321, 276–285.

Forster, C. & Smith, L. (1989) The influence of groundwater flow on thermal regimes in mountainous terrain: A model study. *Journal of Geophysical Research*, 94, 9439–9451.

Ge, S., Wu, Q.B., Lu, N., Jiang, G.L. & Ball L. (2008) Groundwater in the Tibet Plateau, western China. *Geophysical Research Letters*, 35.

Geluk, M.C., Duin, E.J.T., Dusar, M., Rijkers, R.H.B., Van Den Berg, M.W. & Van Rooijen, P. (1994) Stratigraphy and tectonics of the Roer Valley Graben. *Geologie en Mijnbouw*, 73, 129–141.

Griffioen, J. (2006) Extent of immobilisation of phosphate during aeration of nutrient-rich, anoxic groundwater. *Journal of Hydrology*, 320, 359–369. doi:10.1016/j.jhydrol.2005.07.047.

Gudmundsson, A. (2000) Active fault zones and groundwater flow. *Geophysical Research Letters* 27: 2993–2996.

Hancock, P.J., Boulton, A.J. & Humphreys, W.F. (2005) Aquifers and hyporheic zones: Towards an ecological understanding of groundwater. *Hydrogeology Journal*, 13, 98–111. doi:10.1007/s10040-004-0421-6.

Hillier, J., Middlemis, H., Bonte, M. & Ross, J. (2002) *A Review of Stream-Aquifer Interaction Assessment Methods*. Unpublished report to the Murray Darling Basin Commission. Resource and Environmental Management, Adelaide. 73p.

Hindmarsh, A.C. (1983) ODEPACK, A systematized collection of ODE solvers. In: Stepleman, R.S. *et al.* (eds.) *Scientific Computing*. Amsterdam, North-Holland (vol. 1 of *IMACS Transactions on Scientific Computation*). pp. 55–64.

Ingebritsen, S.E. & Sanford, W.E. (1998) *Groundwater in Geologic Processes*. Cambridge University Press (Sciences).

Leach, J.A. & Moore, R.D. (2011) Stream temperature dynamics in two hydrogeomorphically distinct reaches. *Hydrological Processes*, 25, 679–690.

Lide, D.R. (1990) *CRC Handbook of Chemistry and Physics*. 70th edition. Boca Raton, FL, CRC Press.

Malcolm, I., Soulsby, C., Youngson, A. & Tetzlaff, D. (2009) Fine scale variability of hyporheic hydrochemistry in salmon spawning gravels with contrasting groundwater-surface water interactions. *Hydrogeology Journal*, 17, 161–174. doi:10.1007/s10040-008-0339-5.

Mayer, A., May, W., Lukkarila, C. & Diehl, J. (2007) Estimation of fault-zone conductance by calibration of a regional groundwater flow model: Desert Hot Springs, California. *Hydrogeology Journal*, 15, 1093–1106.

McMillan, W. (1973) Cooling from open water surfaces: Final report, Part 1: Lake Trawsfynydd cooling investigation. Scientific Services Department, CEGB, Manchester.

Mendizabal, I., Stuyfzand, P. & Wiersma, A. (2011) Hydrochemical system analysis of public supply well fields, to reveal water-quality patterns and define groundwater bodies: The Netherlands. *Hydrogeology Journal*, 19, 83–100. doi:10.1007/s10040-010-0614-0.

Meuwissen, I.J.M. & van den Brand, L. (2003) *Wijst Grounds in the Picture* [in Dutch] Waterboard Aa, Forestry Service, Boxtel. 39 p.

Olsen, D. & Young, R. (2009) Significance of river–aquifer interactions for reach-scale thermal patterns and trout growth potential in the Motueka River, New Zealand. *Hydrogeology Journal*, 17, 175–183. doi:10.1007/s10040-008-0364-4.

Runhaar, J., vanGool, C.R. & Groen, C.L.G. (1996) Impact of hydrological changes on nature conservation areas in the Netherlands. *Biological Conservation*, 76, 269–276.

Saar, M. (2011) Review: Geothermal heat as a tracer of large-scale groundwater flow and as a means to determine permeability fields. *Hydrogeology Journal*, 19, 31–52. doi:10.1007/s10040-010-0657-2.

Schmidt, C., Bayer-Raich, M. & Schirmer, M. (2006) Characterization of spatial heterogeneity of groundwater-stream water interactions using multiple depth streambed temperature measurements at the reach scale. *Hydrology and Earth System Sciences Discussion*, 3, 1419–1446. doi:10.5194/hessd-3-1419-2006.

Seilheimer, T.S. & Fisher, W.L. (2010) Habitat use by fishes in groundwater-dependent streams of Southern Oklahoma. *American Midland Naturalist*, 164, 201–216.

Shalev, E. & Yechieli, Y. (2007) The effect of Dead Sea level fluctuations on the discharge of thermal springs. *Israel Journal of Earth Sciences*, 56, 19–27.

Sophocleous, M. (2002) Interactions between groundwater and surface water: The state of the science. *Hydrogeology Journal*, 10, 52–67. doi:10.1007/s10040-001-0170-8.

Springer, A. & Stevens, L. (2009) Spheres of discharge of springs. *Hydrogeology Journal*, 17, 83–93. doi:10.1007/s10040-008-0341-y.

Sweers, H.E. (1976) A nomogram to estimate the heat-exchange coefficient at the air-water interface as a function of wind speed and temperature; A critical survey of some literature. *Journal of Hydrology*, 30, 375–401.

van Balen, R.T., Houtgast, R.F. & Cloetingh, S.A.P.L. (2005) Neotectonics of The Netherlands: A review. *Quaternary Science Reviews*, 24, 439–454.

Verwijst, T. (1982) *Ecology of Soils at Wijst (North Brabant): Study on Relationships Between Geology, Water Management, Soil and Vegetation Along the Breaches in the Northern Peelhorst* [in Dutch] National Forestry Service (Staatsbosbeheer), Utrecht. 162 p.

Visser, W.C. (1948) The problem of the wijst areas [in dutch: Het probleem van de wijstgronden]. *Tijdschrift Koninklijk Nederlands Aardkundig Genoodschap*, 65, 798–823.

Westhoff, M.C., Savenije, H.H.G., Luxemburg, W.M.J., Stelling, G.S., van de Giesen, N.C., Selker, J.S., Pfister, L. & Uhlenbrook, S. (2007) A distributed stream temperature model using high resolution temperature observations. *Hydrology and Earth System Sciences*, 11, 1469–1480. doi:10.5194/hess-11-1469-2007.

Williams, G.P. (1963) Heat transfer coefficients for natural water surfaces. *General Assembly of Berkeley International Association of Scientific Hydrology Publication* No. 62, 203–212.

Wolaver, B.D. & Diehl, T.M. (2010) Control of regional structural styles and faulting on Northeast Mexico spring distribution. *Environmental Earth Sciences*, 1–15.

Wright, S.A., Anderson, C.R. & Voichick, N. (2009) A simplified water temperature model for the Colorado River below Glen Canyon Dam. *River Research and Applications*, 25, 675–686. doi:10.1002/rra.1179.

Ziegler, P.A. (1994) Cenozoic rift system of Western and Central Europe: An overview. *Geologie en Mijnbouw*, 73, 99–127.

Chapter 9

Typology of groundwater-surface water interaction (GSI typology) – with new developments and case study supporting implementation of the EU Water Framework and Groundwater Directives

Mette Dahl[1] *& Klaus Hinsby*[2]

[1]*Sejeroe, Denmark*

[2]*Geological Survey of Denmark and Greenland, Copenhagen, Denmark*

ABSTRACT

The EU Water Framework and Groundwater Directives outline an approach to water administration in which interactions between groundwater bodies, groundwater dependent terrestrial ecosystems and surface water bodies (associated aquatic ecosystems) take on a central role. To facilitate monitoring and status assessment the directives allow grouping according to typologies. A typology used for evaluating interdependency between groundwater bodies, riparian areas and streams must be based on processes controlling flow, contaminant transport and attenuation. The typology of Groundwater – Surface water Interaction (GSI typology) has been developed as a process oriented scaled framework classifying interactions between the three hydrological components. The controlling processes are characterised using an eco-hydrological approach based on geomorphology, hydrogeological setting and flow paths on gradually smaller scales. As part of a programme of measures to obtain good ecological status of associated coastal waters (including a Natura 2000 site) an assessment was conducted in a 50 km^2 catchment area in a moraine landscape by combined use of new developments of the GSI typology and a field campaign of nitrate concentration measurements in groundwater, tributaries and streams. One purpose was to delineate sub-catchments in which nitrate concentrations in a shallow local groundwater body exceeded threshold values derived from good status objectives for the coastal ecosystems. Another purpose was to assess in which selected sub-catchments riparian area hydrology restorations or reductions in nitrate input from fields were appropriate measures to reduce nitrate loads to streams and ultimately to the associated coastal waters.

9.1 INTRODUCTION

The EU Water Framework Directive (WFD) and Groundwater Directive (GWD) stipulate that groundwater threshold values derived from good ecological status objectives for dependent terrestrial ecosystems and associated surface water bodies have to be established for the protection or restoration of these ecosystems. Recognition of the interdependency between components of the hydrological continuum is a major strength of the directives as a management framework.

A groundwater body is defined as a distinct volume of groundwater within an aquifer or aquifers. A surface water body is defined as a discrete element of surface water. These are subdivided into rivers, lakes, transitional waters, and coastal waters. Despite its significance, a terrestrial ecosystem is not well defined in the WFD, but the term covers areas where the water table is at or near ground surface. Riparian areas and wetlands are encompassed within this term.

A main purpose of the WFD and GWD is to promote sustainable water use, which protect and ensure water requirements of terrestrial ecosystems and good ecological status of associated aquatic ecosystems.

To obtain good quantitative and chemical status of a groundwater body, the directive requires that groundwater level and chemical composition of the groundwater body must cause neither the associated surface water bodies to fail in achieving their environmental objectives, nor cause any significant damage to groundwater dependent terrestrial ecosystems. A typology primarily addressing the quantitative status must reflect processes that control spatial, temporal and quantitative linkages with dependent ecosystems.

The WFD directive calls upon two features of riparian areas to be evaluated: (1) their water requirements, and (2) their capability of maintaining high water quality of adjacent surface water bodies. To evaluate the water requirements it is of equal importance to conceptualise how a site works hydrologically and by what mechanisms water requirements of the ecosystems are met. Consequently, a typology must characterise processes controlling riparian water sources. The water requirements form part of the status assessment of the contributing groundwater body.

Finally, assessment of ecological status of streams and rivers requires evaluation of biological, hydromorphological and physico-chemical elements of the stream and it's riparian area, if the structure and condition of the riparian area is relevant for achieving the environmental objectives of the stream. In this context the riparian area water quality function form part of the status assessment of the adjacent stream or river.

To facilitate monitoring and status assessment, the WFD directive allows grouping of both groundwater and surface water bodies according to typologies. Previous typologies or classification systems, however, commonly characterise only one component of the hydrological continuum as reviewed by Dahl *et al.* (2007). Therefore, a need exists to develop an integrated typology characterising functional linkages and controlling flow processes between groundwater bodies, riparian areas and streams.

The GSI typology has been developed and tested for use within temperate climates and hydrogeological settings dominated by loose sediments and matrix flow.

The typology provides a conceptual tool to assess which associated stream or river reaches and groundwater dependent terrestrial ecosystems exchange groundwater with a given groundwater body. It thereby provides a framework for understanding which dependent ecosystems are affected by groundwater abstraction and nitrate load from the groundwater body (Dahl *et al.*, 2007; Dahl, 2008; Dahl *et al.*, 2010; Hinsby & Dahl, 2010).

The GSI typology also provides a conceptual framework to assess the water quality function of riparian areas which to a large extent is controlled by flow paths through the riparian area (Dahl *et al.*, 2007; Dahl *et al.*, 2010). The typology thus provides very general path-specific nitrate reduction capacities for defined flow paths through the riparian area. It is expected that these flow paths will also control riparian area reduction of other pollutants. However, path-specific reduction capacities have not yet been developed for other substances.

9.2 TYPOLOGY OF GROUNDWATER – SURFACE WATER INTERACTION

In the GSI typology, controlling processes are characterised using an eco-hydrological approach based on geomorphology, hydrogeological setting and flow paths on a gradually smaller scale (Dahl *et al.*, 2007). The hierarchy includes three scales illustrated in Figure 9.1.

As the deposits of riparian areas are extremely heterogeneous and often organic-rich caused by a changing flow regime, water logging, sediment delivery, and channel displacement over time, these characteristics give rise to a perception of the deposits as a unique type of aquifer termed Riparian Area Aquifer in the typology.

9.2.1 Landscape type

On a catchment scale of more than 5 km, the Landscape Type classifies the groundwater flow systems and the groundwater system as a regional frame controlling the complexity of flow processes and discharge patterns. A groundwater flow system is a discrete, closed three-dimensional system containing flow paths from recharge point to a topographically lower discharge point. Development of flow systems are primarily controlled by topography. A groundwater system constitutes a hydrogeological setting with a characteristic distribution of aquifers and confining layers.

The classification criteria are thus regional geomorphology and regional hydrogeological setting, respectively. The typology includes an unlimited number of regional geomorphologies forming landscapes characterized by a typical pattern of regional and local slopes. The typology distinguishes between four regional hydrogeological settings:

- A single dominant unconfined aquifer
- Two interconnected aquifers of equal importance
- A three-unit system consisting of an unconfined aquifer, a confining layer, and a confined aquifer
- A complexly interbedded sequence of aquifers and confining layers with no dominant aquifer.

The position, magnitude and confinement of these aquifers give way to perceive them as specific groundwater body types defined below.

Through landscape forming processes, the geomorphology is often linked to geology and, subsequently, to the hydrogeological setting of the groundwater system. Figure 9.2 illustrates an example of a regionally sloping and locally undulating moraine landscape with a hydrogeological setting comprising a complexly interbedded sequence of sand aquifers and confining layers of clayey till.

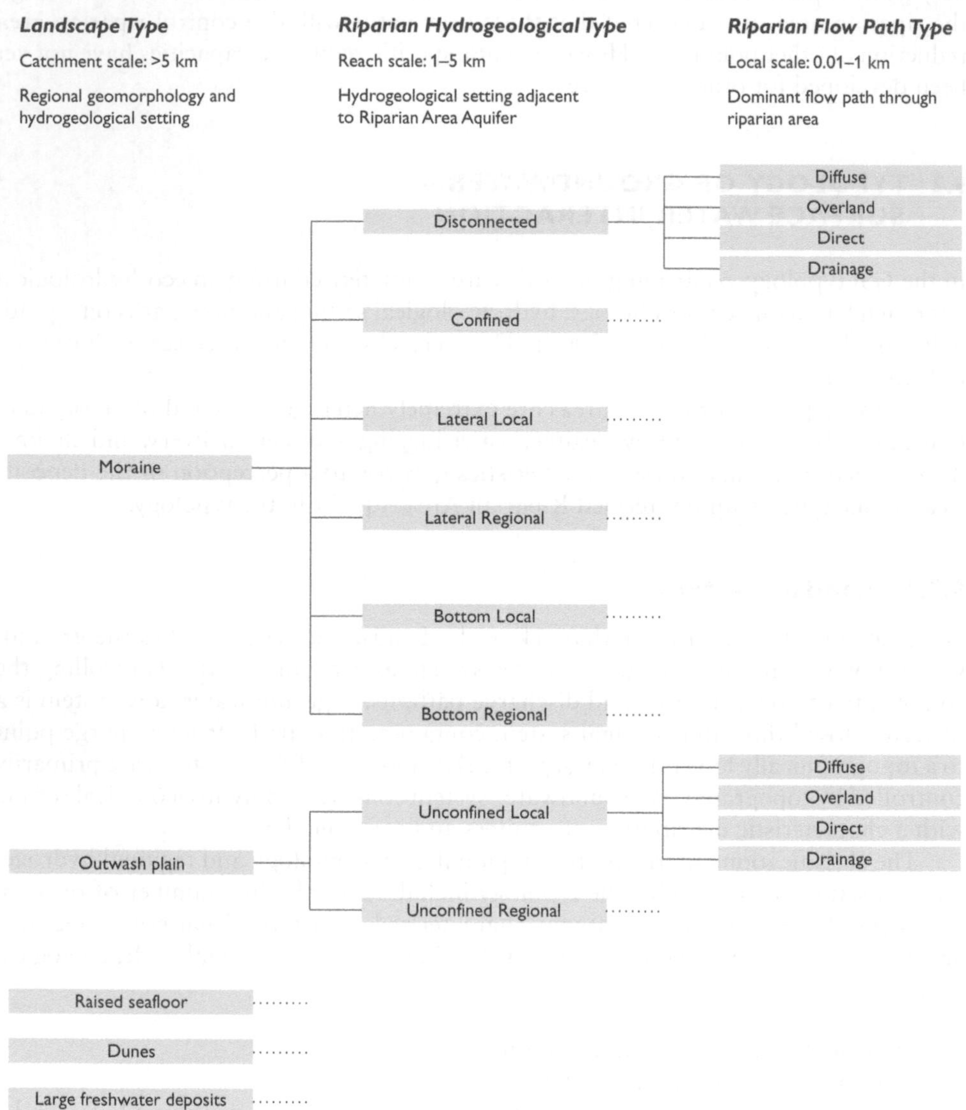

Figure 9.1 Terminology, scale hierarchy and classification criteria of the GSI typology. Applying the typology in Denmark results in five most important Landscape Types, eight Riparian Hydrogeological Types, and four Riparian Flow Path Types. Dotted lines following Landscape Types indicate that a number of Riparian Hydrogeological Types may be associated to specific Landscape Type. Dotted lines connected to Riparian Hydrogeological Types indicate that all four Riparian Flow Path Types are common to all Riparian Hydrogeological Types (Dahl *et al.*, 2007).

9.2.2 Riparian hydrogeological type

On a reach scale of 1–5 km the Riparian Hydrogeological Type classifies the hydrogeological setting adjacent to a Riparian Area Aquifer. This scale characterises the

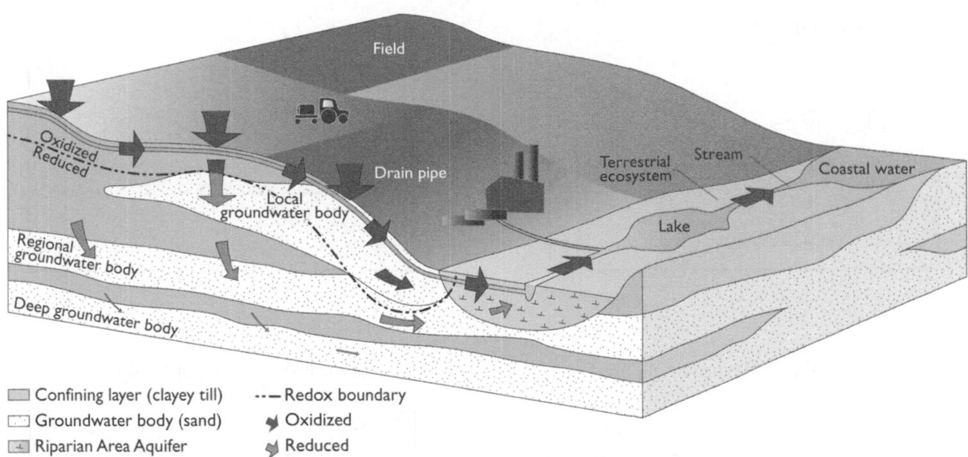

Figure 9.2 Moraine landscape with a groundwater system comprising a complexly interbedded sequence of aquifers and confining layers. The regionally sloping and locally undulating geomorphology creates local, intermediate and regional flow systems directing groundwater through Local, Regional or Deep Groundwater Bodies and Riparian Area Aquifers to various surface water bodies (Dahl *et al.*, 2007).

exchange of groundwater between a groundwater body and a Riparian Area Aquifer. The hydrogeological settings are defined by a combination of two features, namely the Contact Type and the Groundwater Body Type.

The typology distinguishes between five Contact Types (Fig. 9.3) characterising the physical contact between a groundwater body and a Riparian Area Aquifer. The Contact Types are arranged in order from no contact to full contact. They determine groundwater ability to enter the Riparian Area Aquifer and, to a large extent, the entry point:

- Disconnected. An impermeable confining layer between a groundwater body and a Riparian Area Aquifer permits no groundwater to enter the Riparian Area Aquifer.
- Confined. A low-permeable confining layer between a groundwater body and a Riparian Area Aquifer permits some groundwater to enter the Riparian Area Aquifer.
- Lateral. A Riparian Area Aquifer is in contact with a groundwater body at a hill slope adjacent to the Riparian Area Aquifer. Groundwater enters the Riparian Area Aquifer at the hill slope.
- Bottom. A Riparian Area Aquifer is in contact with a groundwater body at the base of the Riparian Area Aquifer. Groundwater is conveyed to the Riparian Area Aquifer along its base.
- Unconfined. A Riparian Area Aquifer is in contact with an unconfined groundwater body on all three sides. Groundwater is easily conveyed into the Riparian Area Aquifer along all sides.

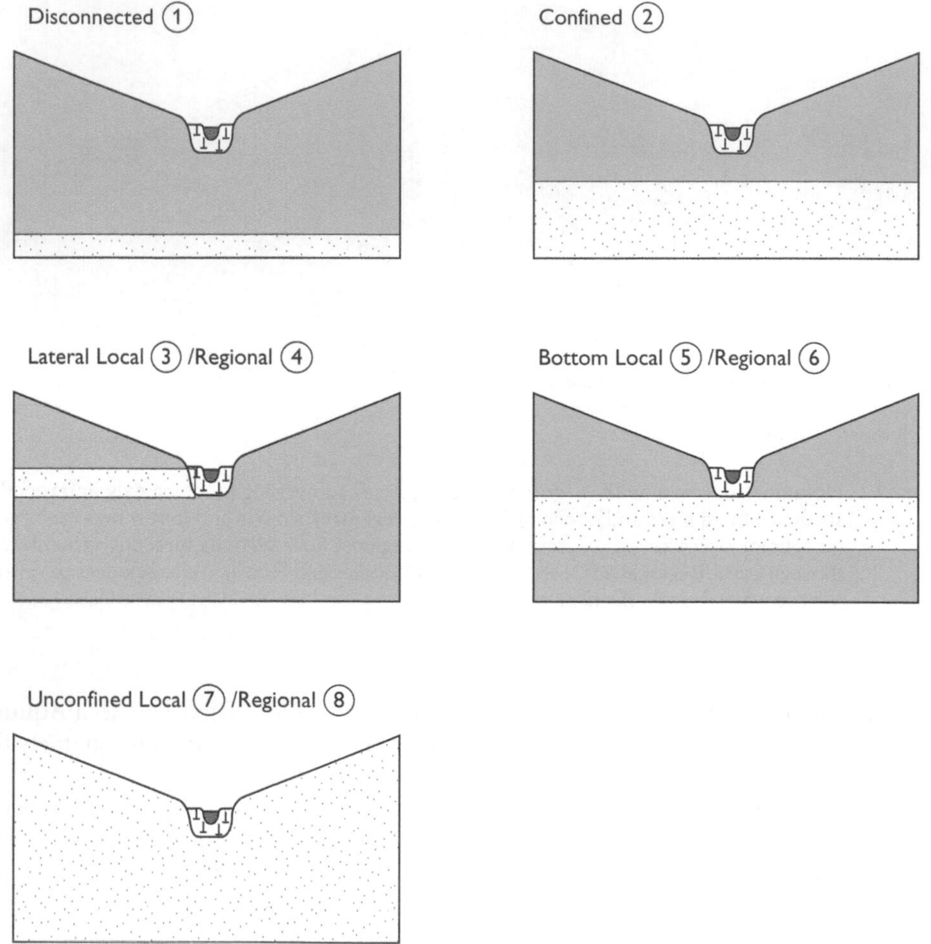

Figure 9.3 Riparian Hydrogeological Types. The names at numbers in circles refer to types in Figure 9.1. The hydrogeological legend corresponds to Figure 9.2 (Dahl *et al.*, 2007).

The typology encompasses two Groundwater Body Types (Figs. 9.2 and 9.3). They control the temporal contact, stability and flux of groundwater to the Riparian Area Aquifer:

- Local Groundwater Body. A shallow and generally small groundwater body in which groundwater tends to connect with an adjacent riparian area and stream through a local flow system. This groundwater body generally has the largest recharge rate, shallowest flow penetration depth, shortest flow path and residence time, and it discharges seasonally unsteady amounts of water. Consequently, baseflow is quite small and may even cease in the dry season.

- Regional Groundwater Body. A deeper and generally larger groundwater body in which groundwater tends to connect with riparian areas along medium-sized streams and large rivers through a larger scale flow system. Generally, this groundwater body has a smaller recharge rate, a deeper flow penetration depth as well as a longer flow path and residence time. Discharge from this type of groundwater body is continuous and seasonally steadier. The discharge flux, however, depends on size of recharge area and leakage properties of possible confining layers overlying the groundwater body.

When an extensive groundwater body lies at depth, and groundwater flows along a regional flow system only connected to transitional or coastal waters, the groundwater body is defined as a Deep Groundwater Body (Fig. 9.2). Generally, this groundwater body has the smallest recharge rate, deepest penetration depth, and longest flow path and residence time. Because this kind of groundwater body does not exchange water with a Riparian Area Aquifer along a stream, it is not included in the GSI typology.

Eight Riparian Hydrogeological Types illustrated in Figures 9.1 and 9.3 are defined through combining Contact Types and Groundwater Body Types: Disconnected, Confined, Lateral Local, Lateral Regional, Bottom Local, Bottom Regional, Unconfined Local, and Unconfined Regional. In the case of a Disconnected or Confined type there is no need to characterise the corresponding Groundwater Body Type because the discharge is nil or very small, respectively. However, this must be done in case of a Lateral, Bottom or Unconfined type.

9.2.3 Riparian flow path type

On a local scale of 10–1000 m the Riparian Flow Path Type classifies the dominant flow path through the riparian area to the stream based on the flow path distribution in the riparian area.

Four flow paths have been identified as the most important transferring water through the riparian area to the stream (Fig. 9.4). They are defined based on type of flow (matrix, preferential, overland, channel or pipe), and contact time between water and riparian deposits with an organic content. To a large extent, these flow paths control transport and attenuation processes and, consequently, the capability of the riparian area to maintain high water quality of an adjacent stream:

- Diffuse Flow (Q_1). This flow path passes slowly through the Riparian Area Aquifer as matrix flow with a long contact time between water and deposits. Groundwater may recharge the aquifer upwards or laterally. Water may infiltrate into the aquifer from seepage faces or drains. Groundwater may also discharge to the riparian surface as return flow, or it may stay in the aquifer until it discharges into the stream. Residence time is expected to range from weeks to years.
- Overland Flow (Q_2). This flow path passes across the riparian area as overland flow. Water has a short contact time to riparian deposits. Water may enter the riparian area from drains ending at the hill slope or it may flood the riparian area from the stream. Water may also reach the riparian surface as springs in areas with preferential flow. Residence time is expected to range from hours to days.

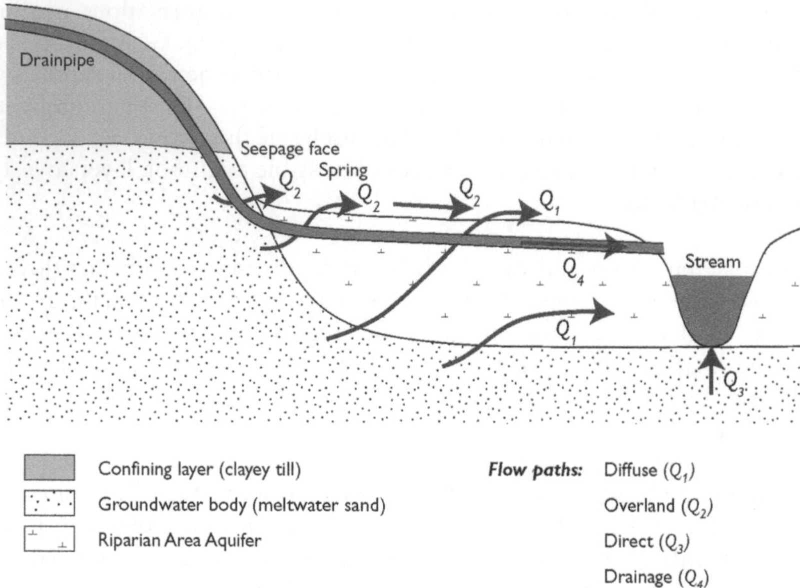

Figure 9.4 Flow paths through a riparian area to a stream (Dahl *et al.*, 2007).

- Direct Flow (Q_3). This flow path transfers groundwater beneath the Riparian Area Aquifer, through an underlying groundwater body, directly through the stream bed. Within the order of a few days water may or may not have contact with riparian deposits beneath the stream bed.
- Drainage Flow (Q_4). This flow path transfers water through drainage pipes or ditches bypassing the Riparian Area Aquifer, conducting it directly into the stream. The transfer is expected to take place in the order of hours.

9.3 DISCUSSION

9.3.1 Landscape type

Regional geomorphology to a large extent controls which groundwater flow systems operate in the catchment (Toth, 1963). Dependent ecosystems are affected by these systems because they control a number of important factors like location of recharge and discharge areas, flux and stability of discharge, flow penetration depth, and residence time within biogeochemical environments affecting quality of discharging water.

Subsurface heterogeneity, however, strongly influences water flux and travel time in the flow systems. The heterogeneity also influences groundwater ability to enter riparian areas and streams resulting in a more patchy discharge pattern. The classification of groundwater systems by Heath (1982) is applicable to characterise the heterogeneity through types of regional hydrogeological setting. The setting divides groundwater flow through specific Groundwater Body Types defined in the GSI typology. Subsequently, several functional aspects of a groundwater body can be evaluated such as response to climate input and abstraction, susceptibility to pollution, water quality, and interaction pattern with dependent ecosystems.

Therefore, both features are important classification criteria on catchment scale.

9.3.2 Riparian hydrogeological type

Hydraulic properties of aquifers and confining layers adjacent to riparian areas and streams significantly impact the spatial distribution of groundwater discharge on a reach scale. Riparian Hydrogeological Types provide a framework classifying this heterogeneity in greater detail through combining Contact Types and Groundwater Body Types.

From the Riparian Hydrogeological Types several functional aspects concerning the groundwater discharge to Riparian Area Aquifers and streams may be evaluated, i.e. physical contact between the contributing groundwater body and the Riparian Area Aquifer controlling the ability of groundwater to enter the Riparian Area Aquifer, and to a large extent the entry point. Temporal contact, stability and flux of groundwater discharge controlling the seasonal stability of the flow path distribution in the riparian area may also be evaluated. These factors are all critical for maintaining diverse ecosystems (Wheeler *et al.*, 2004) as they together with riparian permeability, depth and width determine the riparian hydrological regime.

Likely redox conditions and water quality of groundwater recharging a Riparian Area Aquifer may to some extent also be deduced. As illustrated in Figure 9.2 there is a tendency for groundwater originating from Regional (and Deep) Groundwater Bodies to be reducing and nitrate free. The opposite is expected for groundwater originating from shallow Local Groundwater Bodies, where water presumably is oxidised and may be nitrate rich depending on land use.

9.3.3 Riparian flow path type

Flow path distribution

Various researchers have studied which geomorphologic and hydrogeological parameters control the flow path distribution through riparian areas in glacial landscapes.

Based on results from Langhoff *et al.* (2006) it seems that in a Danish outwash plain landscape where the Riparian Area Aquifer is in full contact with a thick unconfined groundwater body and the permeability of the Riparian Area Aquifer is comparable throughout the catchment, the extent of the contributing groundwater body and the width of the riparian area are the major controlling parameters for the flow path distribution. In narrow riparian areas along upper stream reaches recharged by a Local Groundwater Body the Riparian Area Aquifer is able to transfer most groundwater through the aquifer diffusively (Q_1). Along middle stream reaches, where the contributing groundwater body becomes Regional, the direct flow (Q_3) through the stream bed also becomes important. The extent of the Regional Groundwater Body and width of the riparian area increase downstream. Along these reaches direct flow through the stream bed is dominant. However, a seepage face (Q_2) arises at the hillslope base because the groundwater flux exceeds the ability of the Riparian Area Aquifer and stream bed to transfer groundwater diffusively.

In Canada Vidon & Hill (2004) demonstrated that depth of a contributing local aquifer overlying an aquitard controls the duration of contact between upland and Riparian Area Aquifers. It also controls magnitude and seasonality of groundwater flux recharging the Riparian Area Aquifer and water table amplitude within the Riparian Area Aquifer. Local topographical slope controls the stability of the flow pattern within the Riparian Area Aquifer. Their field work demonstrated that the permeability of the Riparian Area Aquifer is also a very important controlling parameter. Low-permeable aquifers thus tend to create seepage faces and springs close to the hillslope base from where small rivulets evolve (Q_2). Opposed to this, the flow through high-permeable aquifers tends to pass diffusively through the Riparian Area Aquifer all the way to the stream (Q_1). Finally, the hydrogeological setting adjacent to the Riparian Area Aquifer is also important. Absence of a confining layer in shallow depth below the Riparian Area Aquifer thus favours groundwater bypass beneath the Riparian Area Aquifer directly through the stream bed (Q_3).

In Danish Weichselian moraine landscapes the heterogeneity of the Riparian Area Aquifer is large and the downstream aquifer quite deep (up to about 10 m). The deposits generally consist of sand, silty/limnic deposits and fen peat. Based on conceptual numerical simulations Banke (2005) demonstrated that the most important parameters controlling the riparian flow path distribution in this setting are:

- Predominantly bulk permeability of the Riparian Area Aquifer,
- Less pronounced the total extent of the contributing groundwater body and the riparian area,
- The presence or absence of a confining layer between the Riparian Area Aquifer and the contributing groundwater body, and
- To a small degree, the thickness of this confining layer.

Nitrate reduction capacities

Nitrate and other elements and compounds are gradually transformed through the changing redox zones observed during passage of a Riparian Area Aquifer. For this reason the entry point and flow paths to a large extent control where and to what degree transformation processes take place. As groundwater recharges the Riparian Area Aquifer at the hillslope base or at the base of the aquifer, this is where the important zone of denitrification is found. The hyporheic zone immediately beneath the stream bed is of comparatively minor importance in Denmark.

By reviewing nitrate reduction capacities of many riparian areas in Denmark Dahl *et al.* (2004) proposed very general path-specific reduction capacities. The capacity of Q_1 and Q_3 is close to 100% if the organic content of the riparian deposits exceeds 3%, and the capacity is close to 0% if the organic content is less than 3%. For Q_3 this implies that the water contacts riparian deposits beneath the stream bed. If it does not the reduction capacity is close to 0%. The reduction capacity of Q_2 is approximately 50%, and the capacity of Q_4 close to 0%. These capacities are the result of hydrogeochemical environments controlling different redox conditions, contact times with riparian deposits, and carbon availabilities along the flow paths. The denitrification capacity of a riparian area can be estimated through multiplying the flow path distribution by the respective path-specific reduction capacities.

The Riparian Flow Path Type thus provides a classification of the capability of riparian areas to maintain high water quality of adjacent streams based on flow paths controlling transport and attenuation through the riparian area.

9.4 NEW DEVELOPMENTS AND AAKAER STREAM CASE STUDY

As part of a programme of measures to obtain good ecological status of associated coastal waters an assessment was conducted in Aakaer Stream catchment area by combined use of new developments of the GSI typology and a field campaign of nitrate concentration measurements in groundwater tributaries and streams. The purpose was to:

- Delineate sub-catchments in which nitrate concentrations in a Local Groundwater Body exceeded threshold values derived from good status objectives for the associated coastal ecosystems.
- Assess in which selected sub-catchments riparian area hydrology restorations or reductions in nitrate input from fields were appropriate measures to reduce nitrate loads to streams and ultimately to the associated coastal ecosystems.

The catchment area is situated in a Weichselian moraine landscape with a hydrogeological setting comprising a complexly interbedded sequence of sandy aquifers and confining layers of clayey till in the eastern part of Jutland, Denmark. All streams within the 50 km^2 catchment area are recharged by a shallow sandy Local Groundwater Body, which is nitrate polluted due to intensive farming. Aakaer Stream is one of several surface water tributaries to the coastal water, Kolding Inner Fjord (estuary), and ultimately to Kolding Outer Fjord and the Belt Sea (including a Natura 2000 site). The coastal waters do not comply with the requirements of good ecological status according to the EU WFD and GWD due to an excessive nutrient load.

9.4.1 New developments of the GSI typology

The Riparian Hydrogeological Types classify the hydrogeological setting adjacent to a Riparian Area Aquifer on reach scale and characterise the physical exchange of groundwater between a groundwater body and a Riparian Area Aquifer.

However, as the hydrogeological setting all the way up to the groundwater divide to a large extent determines the chemical composition of groundwater interacting with a dependent ecosystem, the Riparian Hydrogeological Types were further developed into Groundwater-Surface water Interaction Response Units (GSI Response Units or GSI RUs) (Dahl, 2008; Dahl *et al.*, 2010). The purpose of these GSI RUs is to characterize the distribution of flow paths between drainage (oxidied, potentially nitrate rich), shallow (oxidised, potentially nitrate rich) and deep (reduced, nitrate free) groundwater discharge into the Riparian Area Aquifer on reach scale.

The GSI RUs are defined by combining hydrogeological settings with types of groundwater bodies as described earlier in the paper. Eight types of hydrogeological

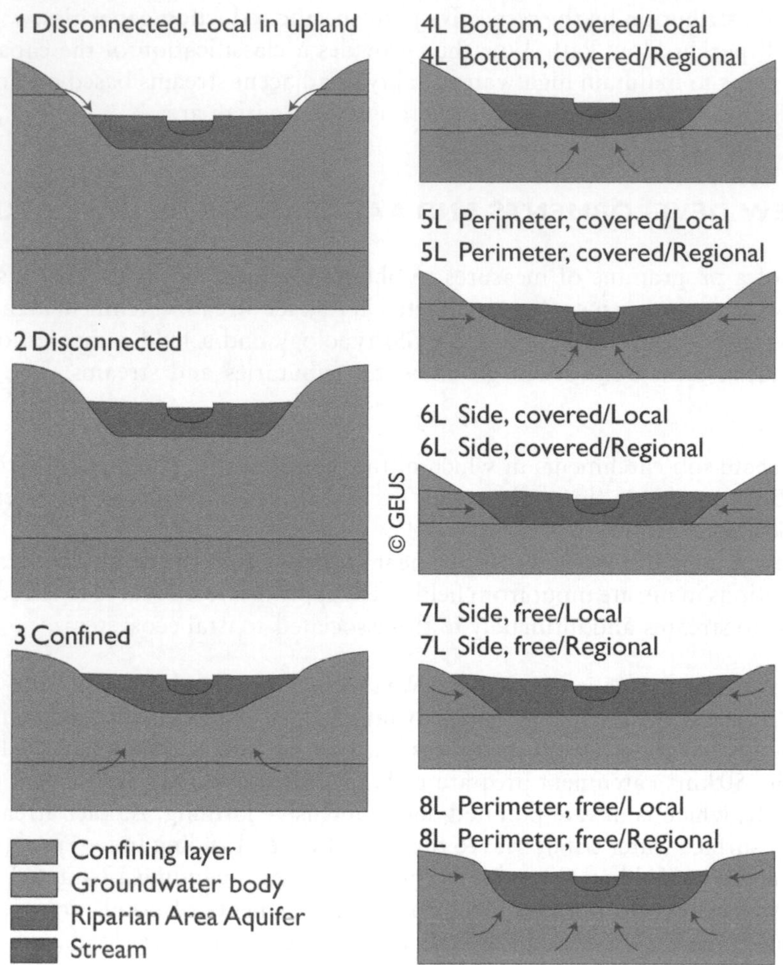

Figure 9.5 Groundwater-Surface water Interaction Response Units (GSI Response Units or GSI RUs). In type 1–3 the groundwater body has no physical contact with the Riparian Area Aquifer, whereas the opposite is the case in type 4–8. The "side", "bottom" and "perimeter" types characterise where the contact is located. In the "covered" settings the groundwater body is partly or completely covered by a low permeable layer in the upland area, whereas the groundwater body in the "free" settings reaches the ground surface in the upland area.

settings (1–8) are thus combined with either a shallow Local (L) or a Regional (R) groundwater body defining thirteen types of GSI RUs illustrated in Figure 9.5.

Through mapping of GSI RUs a catchment area is thus divided into sub-catchments comprising a specific hydrogeological setting with complete flow systems from recharge to discharge and an expected distribution between drainage, shallow and deep groundwater discharge into a dependent ecosystem.

9.4.2 Mapping of GSI response units

In the Aakaer Stream catchment area GSI RUs were mapped based on existing geological (surface lithology and borehole data), hydrological (hydraulic heads, stream stages and synchronous median minimum discharges) and topographical data (Dahl, 2008; Dahl *et al.*, 2010). In a field campaign nitrate concentrations in streams and in groundwater tributaries like ditches, drains and springs, mainly characterising shallow groundwater discharge, were measured (Hinsby & Dahl, 2010). The campaign was conducted in the winter when discharge and nitrate concentrations were annually highest.

Figure 9.6 illustrates the distribution of GSI RUs in Aakaer Stream catchment area. Eight types of GSI RUs were mapped, all except type 2 and 3 fed by the shallow Local Groundwater Body. The GSI RUs have been grouped based on similar hydrogeological settings, groundwater tributary nitrate concentrations and land use, roughly separating between agriculture and nature/forest.

The lowest nitrate concentrations (mainly <20 mg/l) were measured in groundwater tributaries from GSI RUs 2 and 3, mainly situated west of Aakaer Stream. These are characterised by a thick confining layer completely covering the groundwater body causing no physical contact between the groundwater body and the Riparian Area Aquifer or stream. An undulating topography makes these areas difficult to drain. Groundwater is, therefore, forced to pass through the low-permeable clayey till layer where nitrate is reduced below the redox boundary (Fig. 9.2). These landscapes are often forested. In this case land use was half forest, half agriculture. Presumably, these features cause the relatively low nitrate concentrations measured in groundwater tributaries in the sub-catchments. Depending on topography, however, these GSI RUs may be drained and farmed causing high groundwater tributary nitrate concentrations discharging into dependent ecosystems. In the Aakaer Stream case they do, however, offer a good protection and are the least critical to nitrate discharge into dependent ecosystems.

Medium nitrate concentrations of 11–50 mg/l were measured in groundwater tributaries from GSI RUs 4L, 5L and 6L. The contributing shallow sandy Local Groundwater Body is partly or fully covered by a thin low-permeable clayey till layer, but the groundwater body is in physical contact with the Riparian Area Aquifer. In this setting shallow groundwater discharge and drainage discharge are likely. The land use in the sub-catchments was mainly agriculture. Data thus indicate that GSI RUs 4L, 5L and 6L offer medium protection. They are, therefore, medium critical to nitrate discharge into dependent ecosystems.

Very high nitrate concentrations of predominantly 51–102 mg/l were measured in groundwater tributaries from GSI RUs 1, 7L and 8L. The contributing sandy Local Groundwater Body has no covering layer and it is in good contact with the Riparian Area Aquifer. These settings favour a large shallow groundwater discharge in springs and rivulets. The land use in these sub-catchments was mainly agriculture. The high nitrate concentrations indicate a relatively low-lying redox boundary (Fig. 9.2). Data thus indicate that GSI RUs 1, 7L and 8L offer poor protection. Consequently, they are the most critical to nitrate discharge into dependent ecosystems.

In Aakaer Stream the nitrate concentration was 30–35 mg/l which indicates that nitrate from the upper oxidised part during passage of the deeper reduced part of the

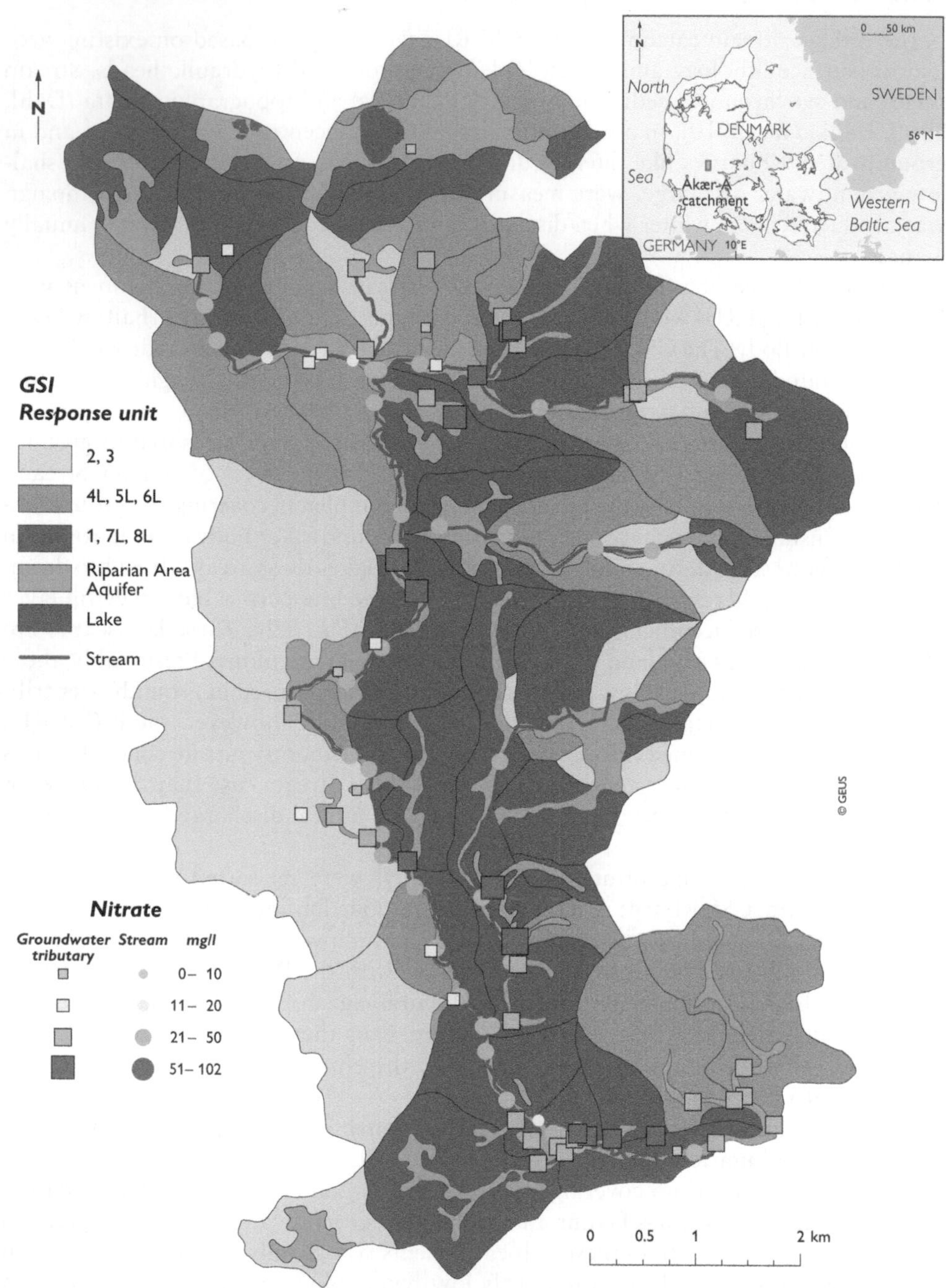

Figure 9.6 Distribution of GSI Response Units and nitrate concentrations in groundwater tributaries (squares) and streams (points) in Aakaer Stream catchment area.

groundwater body and passage of Riparian Area Aquifers is partly reduced, and that the streams also receive nitrate poor groundwater from forested areas.

9.4.3 Chemical status

According to the EU WFD and GWD a groundwater body is considered to be of good chemical status when both the groundwater quality standard and the relevant threshold value, the establishment of which is based on the extent of interaction between groundwater and associated aquatic ecosystems (surface water bodies) and dependent terrestrial ecosystems (wetlands), are not exceeded in that particular groundwater body.

For nitrate the groundwater quality standard is 50 mg-NO_3/l. Case studies show that threshold values derived from good status objectives of associated ecosystems are often significantly lower than the groundwater quality standard, i.e. in the range of 9–30 mg/l in typical Danish settings (Hinsby *et al.*, 2008; Hinsby *et al.*, 2012) and possibly even lower in others (Camargo & Alonso, 2006). As both criteria must be met the lowest value should be used to assess the chemical status of a groundwater body.

In three Danish case studies in Weichselian moraine landscapes nitrate threshold values derived from associated coastal waters were estimated to 20 mg/l for shallow oxidized groundwater and 10 mg/l for associated streams (Hinsby & Dahl, 2010). A new study involving marine and freshwater ecologists further demonstrate that derived groundwater threshold values depend on technically and politically feasible measures and management options, as the required reductions in nitrogen loads may be distributed with different amounts between different sources (Hinsby *et al.*, 2012). They recommend to establish threshold values for streams and to include stream quality monitoring to support groundwater chemical status assessments. The GSI typology would be a helpful tool to plan reliable integrated groundwater and surface water monitoring programmes.

The Danish parliament has decided that the groundwater quality standard of 50 mg/l nitrate constitutes the threshold value for good chemical status of both groundwater and associated streams. In Figure 9.6 the largest squares and points thus denote poor chemical status. The threshold value is only exceeded in GSI RUs 1, 7L and 8L in which the groundwater body is free (uncovered in the upland area) and has a good contact with the Riparian Area Aquifer, and the land use is agriculture. These sub-catchments of the groundwater body thus have poor chemical status. In the rest of the GSI RUs the chemical status of the contributing groundwater body is good. As a whole, the shallow Local Groundwater Body of Aakaer Stream catchment area thus has a poor chemical status.

If the threshold values derived from three Danish associated coastal waters in Weichselian moraine landscapes are applied in the case of Aakaer Stream as well, the nitrate threshold value for groundwater tributaries is estimated to 20 mg/l and for streams 10 mg/l assuming all nitrogen sources contribute to the required reduction with the same relative amount. If the only management option would be to decrease the load from diffuse sources alone, the derived groundwater threshold values would be even lower. However, when the groundwater threshold value is 20 mg/l almost all

GSI RUs, and thus the whole groundwater body of Aakaer Stream catchment area, have poor chemical status.

A very similar result would probably display if a threshold value derived from dependent terrestrial ecosystems were used instead, as Ejrnæs *et al.* (2009) report environmental quality standards in the range of 4–13 mg/l nitrate for Danish terrestrial ecosystems.

9.4.4 Programme of measures

The mapped GSI RUs have, in combination with nitrate concentration data from groundwater tributaries and streams, as well as the classification of Riparian Flow Path Types, been applied in Aakaer Stream catchment area to propose a programme of measures to decrease nitrate loads to the stream, and the coastal water, Kolding Inner Fjord (Dahl, 2008; Hinsby & Dahl, 2010). For selected GSI RUs the most appropriate measure was proposed. Two out of six originally planned restorations of riparian hydrology would probably cause no significant nitrate reduction due to a very small groundwater nitrate discharge. Contrary to this, a long reach was delineated where the groundwater body discharged high nitrate loads. As the hydrology of the corresponding dependent riparian areas were undisturbed and dominated by overland flow (Q_2), no restoration was possible or needed here. Along these reaches, where the GSI RU was classified as type 1, an appropriate measure would be to reduce agricultural nitrate input instead.

9.5 CONCLUSION

In relation to the EU WFD and GWD the GSI typology may be applied to group interacting groundwater bodies, riparian areas and streams in a functional way, to support reliable monitoring, groundwater risk and status assessments as well as appropriate programmes of measures within temperate climates and hydrogeological settings dominated by loose sediments and matrix flow. Applying the underlying concepts, the GSI typology may be further developed to cover other climates, hydrogeological settings and scales.

The GSI typology, related to eco-hydrological interaction, may provide an important nested framework overcoming difficulties in dealing with different time and space frames within the two fields. The GSI Response Units may on reach scale be applied to conceptually assess a number of aspects such as physical and temporal contact, stability, flux, and nitrate load in the discharge from a groundwater body to a Riparian Area Aquifer or a stream. These aspects to a large extent control abiotic site factors determining riparian plant community distributions. They also affect stream ecology through sustaining baseflow, moderating surface water level and temperature, and supplying nutrients and other substances. Consequently, GSI Response Units may serve as a framework for assessing the vulnerability to water requirements of dependent ecosystems.

The GSI Response Units support delineation of riparian areas receiving nitrate free or nitrate polluted groundwater. The Riparian Flow Path Type and path-specific nitrate reduction capacities, on the other hand, support assessment of denitrification capacities of the riparian areas. This knowledge may be applied in programmes of measures to protect and ensure the ecological status of dependent terrestrial and associated aquatic ecosystems, which is a main purpose of the directives.

However, there is a need to develop methods estimating the flow path distribution through riparian areas situated in settings with complex groundwater systems, as the flow path distribution to a large extent controls the quality of groundwater passing through them. Subsequently, path-specific reduction capacities could be established for a variety of substances.

The Riparian Hydrogeological Types and the GSI Response Units also provides a conceptual tool to assess the susceptibility of a given stream or river reach or a dependent terrestrial ecosystem to groundwater abstraction in the contributing groundwater body. The better the contact is between the groundwater body and the dependent ecosystem, and the more shallow the groundwater body is, the more affected the dependent ecosystem is expected to be.

In regional hydrological models Riparian Hydrogeological Types or GSI Response Units may be used to distribute leakage properties between groundwater bodies and surface water systems. However, if the modelling purpose is to simulate contaminant transport and attenuation the controlling processes on various scales point to the need of applying flexible models to simultaneously quantify regional-scale groundwater body and local-scale Riparian Area Aquifer processes.

Finally, the typology may act as a framework for comparison between sites, a method of regionalising features, and a tool to point out data and research needs.

REFERENCES

Banke, M. (2005) *Method for Estimating Flow Path Distribution* in *Stream Valleys* (in Danish). M.Sc. thesis. University of Copenhagen, Denmark.

Camargo, J.A. & Alonso, A. (2006). Ecological and toxicological effects of inorganic nitrogen pollution in aquatic ecosystems: a global assessment. *Environment International*, 32, 831–849.

Dahl, M. (2008) *GSI Mapping and Protection Strategy in Wetland Restoration Areas in the Aakaer Stream Catchment Area* (in Danish). Report from the Geological Survey of Denmark and Greenland (GEUS), Copenhagen, Denmark. No. 20.

Dahl, M., Hinsby, K. & Refsgaard, J.C. (2010) *Application Possibilities of the GSI Typology* (in Danish). Report from Ministry of the Environment, Copenhagen, Denmark. ISBN 978-87-92617-82-8.

Dahl, M., Langhoff, J.H., Kronvang, B., Nilsson, B., Christensen, S., Andersen, H.E., Hoffmann, C.C., Rasmussen, K.R., Platen-Hallermund, F.V. & Refsgaard, J.C. (2004) *Progress in Development of the GSI Typology* (in Danish). Report from the Environmental Protection Agency No. 16, Copenhagen, Denmark.

Dahl, M., Nilsson, B., Langhoff, J.H. & Refsgaard, J.C. (2007) Review of classification systems and new multi-scale typology of groundwater–surface water Interaction. *Journal of Hydrology*, 344, 1–16.

Ejrnæs, R., Andersen, D.K., Baattrup-Pedersen, A., Christensen, B.S., Damgaard, C., Hygaard, B., Nilsson, B., Johanson O.M. & Dybkjær, J.B. (2009) *Hydrological and Hydrochemical Requirements of a Good Ecological Status of Groundwater Dependent Terrestrial Ecosystems* (in Danish). The National Environmental Research Institute, University of Aarhus, Denmark.

Heath, R.C. (1982) Classification of ground-water systems of the United States. *Ground Water*, 20 (4), 393–401.

Hinsby, K., Markager, S., Kronvang, B., Windolf, J., Sonnenborg, T. O., and Thorling, L. (2012) Threshold values and management options for nutrients in a catchment of a temperate estuary with poor ecological status, *Hydrology and Earth System Sciences*, 16, 2663–2683.

Hinsby, K. & Dahl, M. (2010) *Delineation of Critical Sub-Catchments for Nitrate Load to Groundwater Dependent Ecosystems* (in Danish). Report from the Ministry of the Environment, Copenhagen, Denmark. ISBN 978-87-92617-83-5.

Hinsby, K., Melo, T.C. & Dahl, M. (2008) European case studies supporting the derivation of natural background levels and groundwater threshold values for the protection of dependent ecosystems and human health. *Science of the Total Environment*, 401, 1–20.

Langhoff, J.H., Rasmussen, K.R. & Christensen, S. (2006) Quantification and regionalisation of groundwater-surface water interaction along an alluvial stream. *Journal of Hydrology*, 320, 342–358.

Toth, J. (1963) A theoretical analysis of groundwater flow in small drainage basins. *Journal of Geophysical Research*, 68, 4785–4812.

Wheeler, B.D., Gowing, D.J.G., Shaw, S.C., Mountford, J.O. & Money, R.P. (2004) Ecohydrological for Lowland Wetland Plant Communities. In: Brooks, A.W., Jose, P.V., Whiteman, M.I. (eds.) Peterborough, United Kingdom, Environment Agency (Anglian Region).

Vidon, P.G.F. & Hill, A.R. (2004) Landscape controls on hydrology of stream riparian zones. *Journal of Hydrology*, 292, 210–228.

Chapter 10

Conservation of trial dewatering discharge through re-injection in the Pilbara region, Western Australia

Lee R. Evans[1] *& Jed Youngs*[2]
[1]*Pilbara Iron, Tom Price, Australia*
[2]*MWH, Perth, Australia*

ABSTRACT

Groundwater is an important resource supporting many ecosystems in the Pilbara region of Western Australia. Large scale dewatering programmes are a widespread feature of iron ore mining in the Pilbara. Pilbara Iron, a member of the Rio Tinto Group, with the aid of MWH, recently completed a closed trial dewatering and re-injection programme which demonstrated how ecosystem protection can be furthered by innovative approaches to mine dewatering. Groundwater mounding and depression in the dewatered aquifer was limited in extent and a rapid recovery was achieved. With no net water usage, water levels within an aquifer associated with potential groundwater dependent ecosystems were relatively unaffected. Comprehensive monitoring of aquifer response during the programme led to improved parameterisation of the existing groundwater model to develop it as a tool for future groundwater management at the site.

10.1 INTRODUCTION

The climate of the Pilbara region of Western Australia is classified as hot desert and grassland ranging from arid to semi-arid using the Köppen scheme (BOM, 2007; Trewin, 2006). Local annual rainfall recorded over nine years ranges from 187 mm to 697 mm, and is highly seasonal. Rainfall typically occurs mostly during the hot wet season months of October to March and is dominated by cyclonic or low-pressure events, however, winter rain events are not uncommon (Table 10.1). Pilbara rivers are ephemeral, however, the region contains recognised extensive supplies of good quality groundwater (Johnson & Wright, 2001).

Mining operations and nearby communities are primarily dependent on groundwater for water supply in the Pilbara. Growing requirements for mine dewatering in the Pilbara as mining increasingly progresses below the water table (BWT) mean that many operations will have short term water excesses in the near future. Finding appropriate management techniques for these excesses at each site is an important issue faced by the industry (RTTS, 2006).

Water is recognised as a serious environmental issue in Western Australia, and in much of Australia (Grierson *et al.*, 2007). There is a growing understanding of the relationship between some ecosystems and groundwater in Australia (NGC, 2004), and that mining has the potential to threaten groundwater dependent ecosystems (GDEs) (SKM, 2001). Pilbara spring, karstic, and arid zone groundwater calcrete ecosystems

Table 10.1 Rainfall events (>10 mm/day) at Marandoo, July 2005–June 2006.

Tropical Cyclone	*From*	*To*	*Rainfall (mm)*
	10/07/2005	10/07/2005	58.0
	30/12/2005	30/12/2005	37.0
	06/01/2006	06/01/2006	18.2
Clare	10/01/2006	10/01/2006	60.8
	14/01/2006	14/01/2006	20.2
Daryl	26/01/2006	27/01/2006	40.8
	06/02/2006	06/02/2006	31.2
	12/02/2006	12/02/2006	17.4
Glenda	31/03/2006	31/03/2006	15.4
Hubert	08/04/2006	08/04/2006	11.2

have all been identified as entirely dependent on groundwater and of high conservation value (*ibid*). Mine dewatering has been categorised as a potential threatening process to these GDEs by altering groundwater levels, flux, or quality (*ibid*).

Innovative techniques such as aquifer re-injection of mine dewatering discharge are being widely explored in the Pilbara region as a method for preserving groundwater and GDEs. Aquifer re-injection has great potential for the mining industry, but will not be an option for all sites as appropriate receptor aquifer conditions and business case must exist. Chief among these conditions is the existence of a suitable aquifer, accessible both physically and in terms of jurisdiction, at a distance which is economic to infrastructure installation but sufficient to prevent significant water recirculation. In addition, aquifer re-injection in unconfined aquifers has the potential to impact vegetation by over supply of water and alteration of the natural groundwater regime.

10.2 BACKGROUND

In 2005, Pilbara Iron planned to extract a bulk sample of BWT ore from the Marandoo iron ore mine. This required a temporary dewatering system to lower the water table across a 100 m by 60 m area by some eight metres to allow the sample drop cut to be blasted, excavated and subsequently backfilled with inert waste rock to the pre-blast level. Marandoo is operated as an above water table (AWT) mine and backfilling was needed to meet specific licensing conditions for the sampling programme.

Previous work at Marandoo (Youngs and Brown, 2005) had demonstrated the feasibility of re-injection in the area, through a project which used re-injection as a low impact groundwater investigation tool. In 2004, Pilbara Iron and Liquid Earth operated a pumping and re-injection trial in which 3.7 Ml/day was delivered for a continuous 44 day period (Liquid Earth, 2005; Youngs & Brown, 2005).

The hydrogeological challenges of the 2005 dewatering programme were significant but manageable, and similar to those faced in many mining operations. Less standard was the handling of the dewatering discharge. The Marandoo mine lease is bordered on all sides by Karijini National Park which calls for high standards of water management to maintain natural ecosystems. Prolonged surface discharge into the park had previously been discouraged by park managers, the Department of Environment and Conservation (DEC). The decision was made to retain all water from the programme on the mine lease by re-injecting it in order to overcome possible concerns from neighbouring stakeholders and further trial the application of aquifer re-injection in the Pilbara. Beyond the practical requirements of the programme, the opportunities it presented to Pilbara Iron included:

(i) re-injection provided the means for ensuring a net zero water usage from the aquifer, thus only locally altering the groundwater regime temporarily;
(ii) the dewatering and re-injection programme was essentially a large scale pumping test from static conditions to full recovery and with relatively constant flow rates and, therefore, ideal for aquifer analysis;
(iii) the aquifer of the Marra Mamba Iron Formation was known to be highly variable (Liquid Earth, 2005), thus the need for better resolution of spatial variation;
(iv) existing low extraction rates within the aquifer for water supply produced relatively no 'stress' on the groundwater regime providing little insight into the hydraulic properties of the aquifer;
(v) the scale of the dewatering and re-injection programme, in terms of both flow rates and time, was greater than any other yet undertaken within the aquifer, providing the largest anthropogenic 'stress' yet observed;
(vi) the programme could be used to validate and recalibrate the existing Marandoo numerical groundwater model (Liquid Earth, 2005) and thereby improve the effectiveness of the model as a management tool;
(vii) full scale mine dewatering in the Pilbara had previously only occurred within the high permeability aquifers of the Brockman Iron Formation orebodies and the channel iron deposit ore bodies. This programme presented an opportunity to gather experience on dewatering of Marra Mamba Iron Formation;
(viii) to fund and achieve governmental approval of a programme at this scale purely on its hydrogeological knowledge advancement merits would be improbable; and
(ix) the dewatering and re-injection programme provided a small scale dewatering and closure analogy as a potential water management option for future mining operations, given adequate conditions, economics and environmental constraints.

10.3 HYDROGEOLOGICAL SETTING

Investigation of the hydrogeologic setting of the Marandoo area has been accelerated since 1992 with development of the Marandoo mine (e.g., AGC Woodward Clyde 1992; Waterhouse & Howe 1994; Youngs 2004; Liquid Earth 2005; Youngs & Brown 2005; MWH, 2006a; MWH, 2006b). The Marandoo operation is situated on the southern fringe of a broad internally draining catchment referred to as the

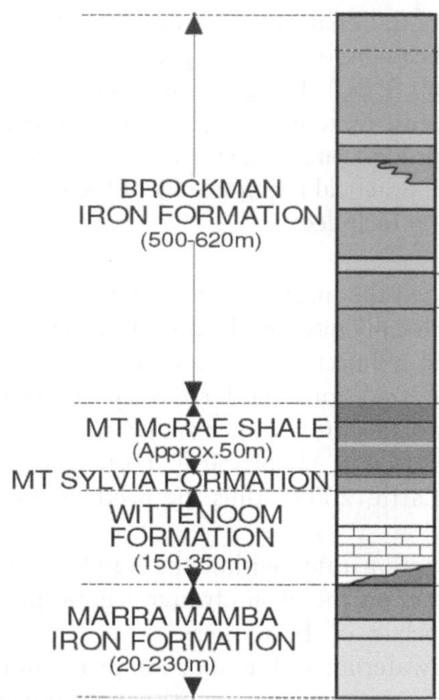

Figure 10.1 Stratigraphic sequence of the lower Hamersley Group (after HI, 2003).

Mt. Bruce Flats. The watersheds of the Southern Fortescue River and Turee Creek form at the margins of the Mt. Bruce Flats (Liquid Earth, 2005).

Quaternary and Tertiary sediments overlie much of the Proterozoic Hamersley Group on the Mt. Bruce Flats. In the area the Proterozoic Hamersley Groups constitutes the Brockman Iron Formation, Mt. McRae Shale, Mt. Sylvia Formation, Wittenoom Formation and Marra Mamba Iron Formation (Fig. 10.1).

The principal aquifer units, collectively referred to as the 'deep aquifer', are the karstic Wittenoom Dolomite and the mineralised Marra Mamba Iron Formation (Liquid Earth, 2005). The deep aquifer is overlain by a tertiary clay and gravel layer incorporating a calcrete horizon termed the 'intermediate aquifer'. A sequence of tight reactive clays hydraulically isolate these aquifers from an extensive calcrete horizon, referred to as the 'shallow aquifer' (Fig. 10.2). Locally the Brockman Iron Formation, Mt McRae Shale, and Mt. Sylvia Formation are generally above water table and present on the flanks of Mt. Bruce.

The groundwater regime is relatively consistent between the deep and intermediate aquifers, suggesting a strong connectivity, with a general hydraulic gradient to the west (Liquid Earth, 2005). In contrast groundwater in the shallow aquifer mounds near the centre of the internally draining basin of the Mt. Bruce Flats and flow radiates outwards (Liquid Earth, 2005), an antonym of the surface water flow.

A potentially groundwater dependent ecosystem occurs in the area of the shallow aquifer groundwater mound. This ecosystem, referred to locally as the Coolabah open

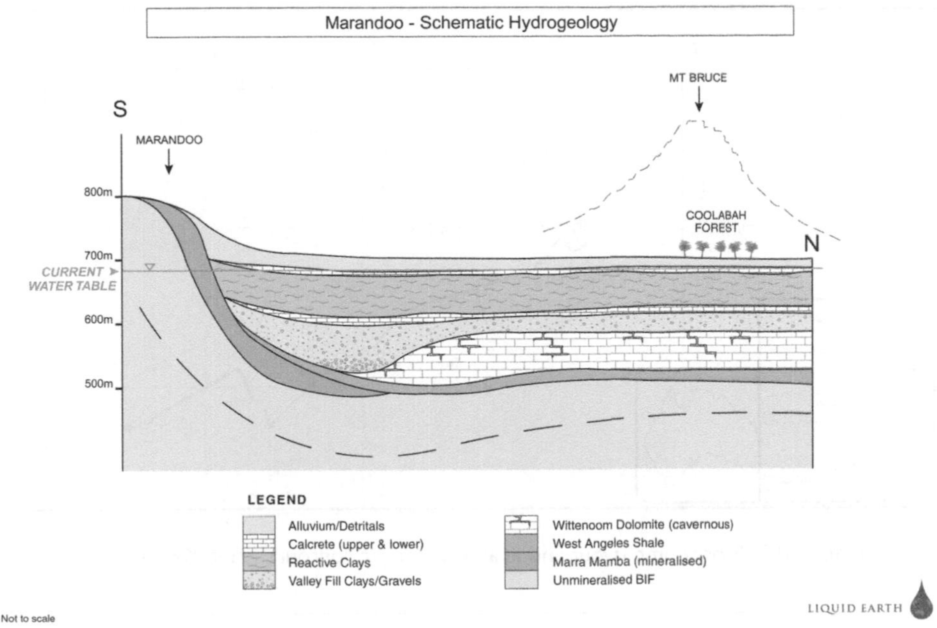

Figure 10.2 Conceptual model (after Liquid Earth, 2005).

woodland is considered a wetland of national importance (English, 1999) and is incorporated within an ongoing Pilbara wide collaborative study (Grierson *et al.*, 2007) supported in part by Pilbara Iron.

10.4 THE PROGRAMME

10.4.1 Operation

The primary aim of the dewatering programme was to extract a sample of approximately 100 000 tonnes 'dry' BWT ore. It was recognised that the programme could also provide a valuable insight into the aquifer dynamics within the Marandoo area. The approved trial programme was run within the terms of a licence to take water granted by the Water and Rivers Commission of the Government of Western Australia.

The programme comprised abstraction of water from five dewatering bores positioned around the drop cut, and re-injection of this water into five injection bores, located to the east of the drop cut, along the strike of the ore body. Dewatering bores were located as close to the drop cut as possible and re-injection bores were selected based upon yield and location within the aquifer. Figure 10.3 shows the location of the production bores, injection bores used for disposal water, and the associated monitoring bore network. The dewatering and re-injection programme is reported in MWH (2006a) and Evans (2006).

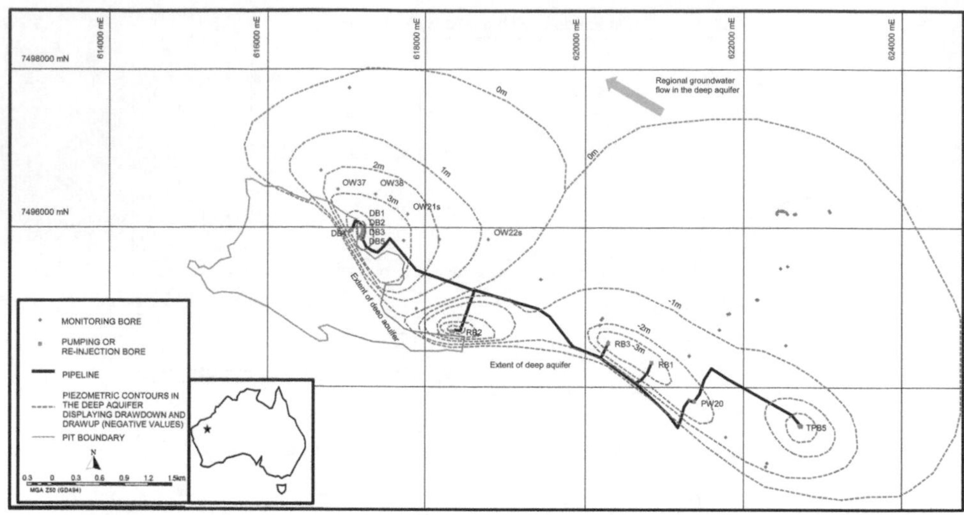

Figure 10.3 Programme layout and final piezometric contours (after MWH, 2006a).

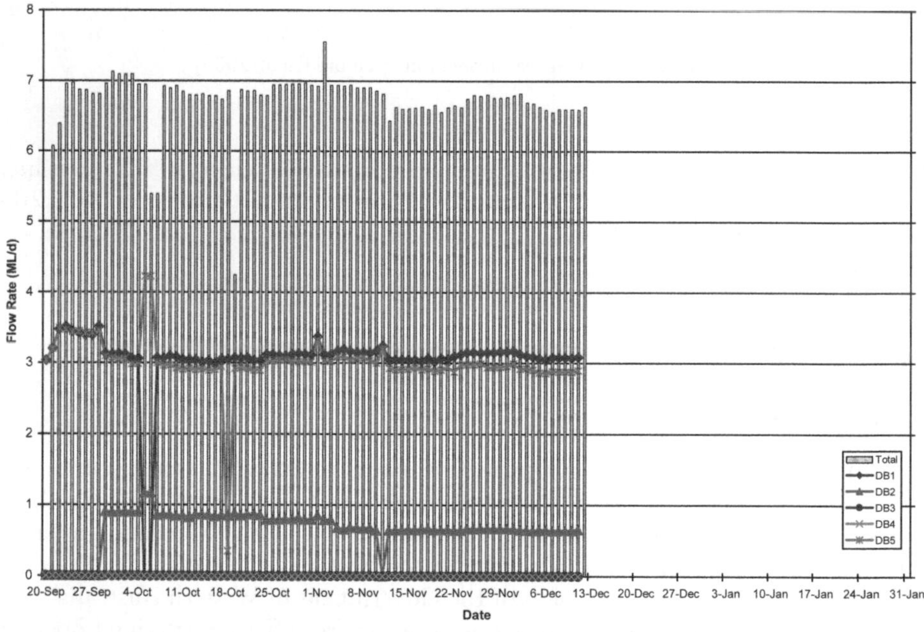

Figure 10.4 Abstraction (after MWH, 2006a).

A temporary drop cut was dewatered to allow blasting, bulk sampling and backfilling. The drop cut dewatering operated from 21 September 2005 to 11 December 2005, during which time a total of 503 Ml was pumped at an average rate of 6.4 Ml/d (Fig. 10.4). All water pumped was re-injected (Fig. 10.5) and a contingency in the

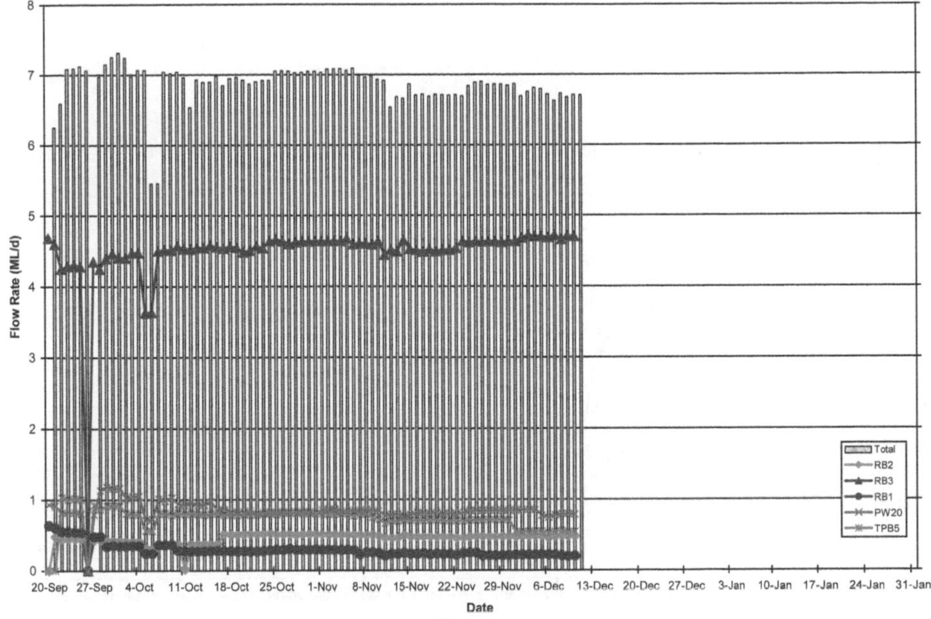

Figure 10.5 Re-injection (after MWH, 2006a).

operating strategy allowing three days surface discharge was not enacted. Minor discrepancies between totals in Figures 10.4 and 10.5 can be explained by operational requirements requiring minor discharges to surface for sampling, maintenance and in the filling of the volume of pipework. Once the drop cut open void was completely backfilled, the dewatering and re-injection programme was decommissioned, allowing the aquifer to rebound to static conditions naturally. Significant rain events also contributed to the rebound of the groundwater regime through recharge in the very late stages of aquifer recovery (Table 10.1).

10.4.2 Monitoring

Cumulative and instantaneous flows from each of the production, dewatering and re-injection bores were measured at least daily using a volumetric flow meter and compiled on a weekly basis. Automatic rainfall data was taken at 15 minute intervals and significant events are summarised in Table 10.1.

Prior to, during, and after the dewatering and re-injection operation, more than 80 monitoring bores were measured to record water levels. Most of the bores were manually dipped on a regular basis, with some being installed with automated water level recording probes that provided high resolution water level data. Water level responses in the dewatering and re-injection bores are given in Figures 10.6 and 10.7 respectively. The response in the shallow aquifer is shown in Figure 10.8. Piezometric contours representing drawdown and recovery at the end of the programme (before shutdown) are shown on the programme layout (Fig. 10.3).

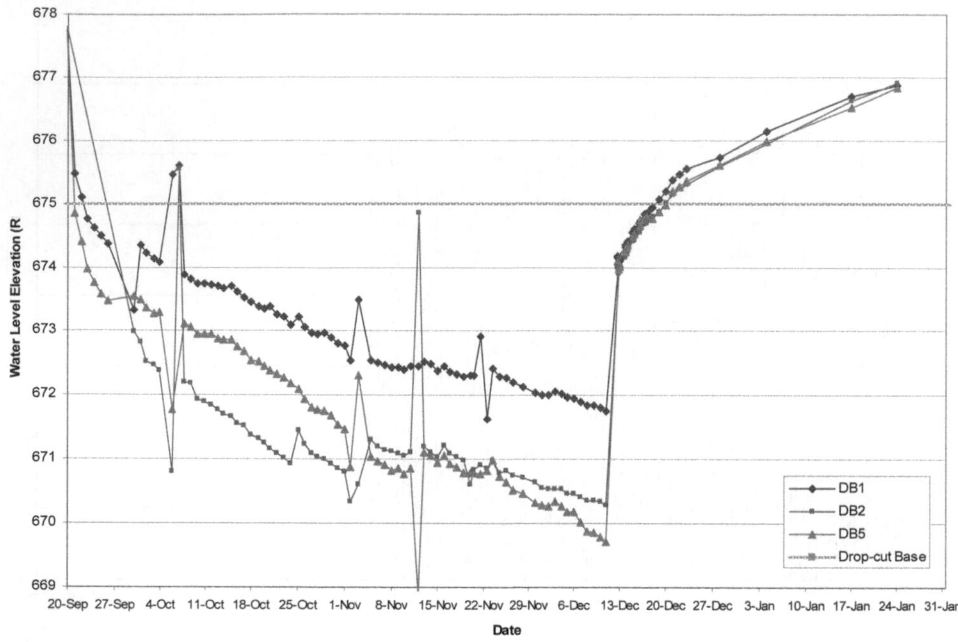

Figure 10.6 Selected dewatering bore hydrographs (after MWH, 2006a).

Figure 10.7 Selected re-injection bore hydrographs (after MWH, 2006a).

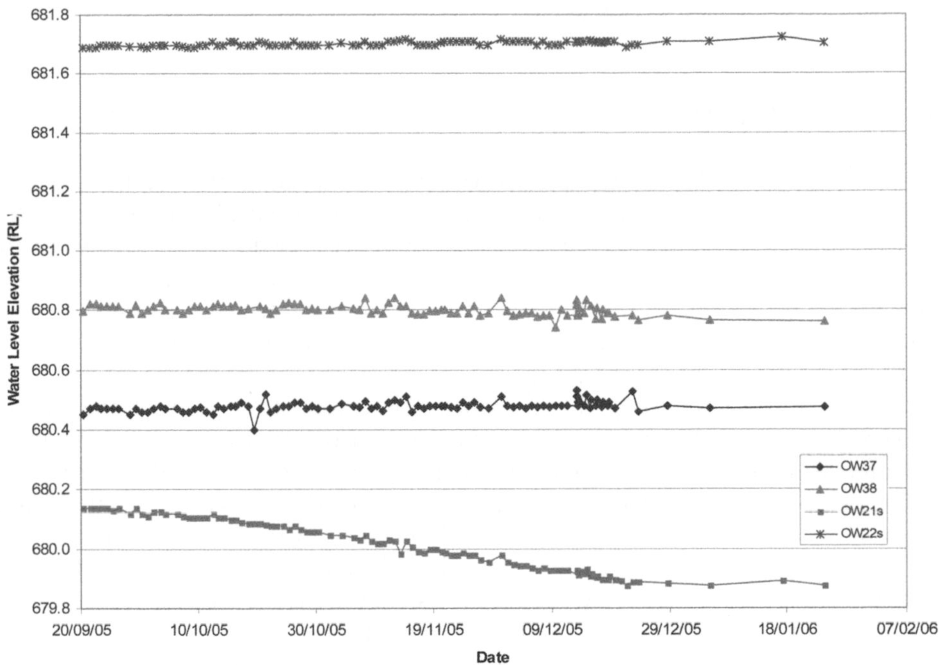

Figure 10.8 Shallow aquifer monitoring (after MWH, 2006a).

10.4.3 Modelling

The dewatering and re-injection programme provided the largest anthropogenic 'stress' on the local hydrogeological environment at Marandoo induced to date. The programme induced 'stresses' at the dewatering site and the five re-injection sites, all within the deep aquifer. The focus of the post programme modelling was to validate the model, refine parameter zonations and quantify aquifer properties. The monitoring delivered by Pilbara Iron 'provided the ideal dataset for the refinement of the model' (MWH, 2006b).

Dewatering and re-injection flow data was converted to model units for input to the Marandoo groundwater model. All of the available water level data were processed and used to construct data sets for validation and recalibration of the groundwater model. Neil Milligan (CyMod Systems) completed the post programme numerical groundwater flow modelling (MWH, 2006b), having been principal modeller for previous model construction (Liquid Earth, 2002 and Liquid Earth, 2005).

The principal outcomes from the modelling (MWH, 2006b) were:

(i) the representation of the Marra Mamba orebody aquifer by the Marandoo groundwater model was improved;
(ii) the intermediate aquifer may play a larger role in groundwater flows than previously thought;
(iii) the orebody along strike is hydraulically well connected; and

(iv) the variable nature of the hydraulic conductivity of the Marra Mamba was further reiterated.

10.5 DISCUSSION

The inferred recovery in the deep aquifer was about 6 km along strike and 4 km across, while the drawn down cone was about 4.5 km along strike and 2 km across (Fig. 10.3). The transition between net recovery (draw up) and net drawdown occurred some two kilometres east of the trial pit. The influence of recirculation from the re-injection bores on the dewatering operation was minimal. As all abstracted groundwater was re-injected into the same aquifer, the net volume of groundwater within the deep aquifer remained constant.

Although unusually heavy rainfall occurred late in the trial recovery, the rebound observed before the onset of the rains indicated that the recovery was not reliant on recharge. Groundwater level response to the programme was observed in one shallow aquifer monitoring bore (OW21s), as shown in Figure 10.8. OW21s is located within 0.5 km of the pumping area (Fig. 10.3) and drew down by 0.27 m. Following the shut down of pumping a recovery trend was well established in the bore before the onset of rains. The response indicated a degree of connectivity between the deep aquifer and shallow aquifer in the vicinity of the dewatering site. The thickness of the clay aquitard in this area is less than is found near most of the Marandoo orebody and this degree of connection has not been observed in other hydrogeological investigations. No measurable affects on water levels were observed in the shallow aquifer:

(i) at a distance of 1.5 km from the dewatered area (OW22s);
(ii) within the vicinity of the re-injection bores, other than locally at RB3 which is related to the construction of RB3; and
(iii) within the vicinity of a potential GDE (the Coolabah open woodland), some 5 km away from the dewatering site.

It was found that for most bores the specific capacity of the bore during re-injection approximately equalled the specific capacity during pumping. One notable exception was bore PW20, in which the re-injection specific capacity was only 20% of the pumping figure. In this case the bore was unconfined during pumping and confined during re-injection. The performance results indicate that re-injection response can be predicted with reasonable confidence from pumping tests, but also highlight the importance of considering site specific hydrogeological conditions in planning.

The dewatering and re-injection programme design was successful in minimising the impact on potential GDEs by:

(i) maintaining the net volume of groundwater in the Marra Mamba Iron Formation aquifer,
(ii) returning water to the pumped aquifer within close proximity to pumping,
(iii) negating the need to discharge groundwater to the surface,
(iv) removing the need for surface recharge or external sources to facilitate groundwater rebound, and

(v) not altering the groundwater levels in the shallow aquifer in the vicinity of a potential GDE.

The programme has broader benefits for local and regional GDEs due to the increased understanding of the hydrogeological regime. Comprehensive monitoring of aquifer response during the programme led to improved parameterisation of the existing groundwater model. Greater confidence in the Marandoo groundwater model obtained through the present study will aid in future water management decisions.

10.6 CONCLUSIONS

The Marandoo drop cut dewatering and re-injection programme demonstrated that with the right circumstances small scale trial mine dewatering and large scale aquifer testing can operate with no net water usage. This approach can help to minimise impacts on aquifers and GDEs. Although the programme highlights the benefits of the use of re-injection as a low impact water management option on this scale, the extent of impacts remain *a priori*. Further research is required before the approach presented here can be up scaled to a sustainable and viable full scale long term mine dewatering exercise.

ACKNOWLEDGEMENTS

The authors gratefully acknowledge the work of: the Pilbara Iron Hydrogeology Group, particularly Jason Hart, Ian Warner and Eddie Newell; John Box for making the dewatering and re-injection programme happen; and Neil Milligan in development of the modelling. The authors would also like to thank Pilbara Iron, Rio Tinto Iron Ore and MWH for enabling this paper to be written.

REFERENCES

AGC Woodward Clyde (1992) Marandoo Iron Ore Project – Investigation of the Hydrogeology of the Mount Bruce Flats and Environs, unpublished consultant report.

BOM (2007) Australian climate zones – major classification groups (based on the Koeppen classification system) Australian Government, Bureau of Meteorology, http://www.bom.gov.au/cgi-bin/climate/cgi_bin_scripts/clim_classification.cgi Cited 12 Feb 2007.

Evans, L.R. (2006) *Pilbara Iron*. Preliminary MMBWT Trial Operational Report, January 2006, unpublished report, 10 pp.

English, V. (1999) *A Directory of Important Wetlands in Australia*. Australian Government, Department of the Environment and Water Resources. http://www.environment.gov.au/cgi-bin/wetlands/report.pl. Cited 12 February 2007.

Grierson, P.F., Adams, M.A., Madden, S., White, S., van Leeuwen, S., Smith, B. & Evans, L.R. (2007) *Dynamics of Water and Woody Vegetation in Central Pilbara – Understanding and Managing for Environmental Change*. Australian Research Council Linkage Projects, Application for funding, 88 pp.

HI (2003) Hamersley Iron 2003 Resource Evaluation. Undeveloped Resource Handbook, unpublished report.

Johnson, S.L. & Wright A.H. (2001) *Central Pilbara Groundwater Study,Waters and Rivers Commission,Hydrogeological Record Series*. Report HG8. 102 pp.

Liquid Earth (2002) Marandoo Below Watertable Numerical Modelling Investigation: Stage 1, unpublished consultant report.

Liquid Earth (2005) Marandoo Hydrogeological Investigation Report, Pilbara Iron Pty. Limited, unpublished consultant report.

NGC (2004) *Improved Management and Protection of Groundwater Dependent Ecosystems*. Issue paper 2, National Groundwater Committee, Department of the Environment and Heritage, 2004. http://www.environment.gov.au/water/groundwater/committee/issue-2/index.html. Cited 12 February 2007.

MWH (2006a) Marra Mamba BWT, Dewatering and Reinjection Program, Pilbara Iron and RTIO Expansion Projects, unpublished consultant report prepared by Youngs, J.

MWH (2006b) Marandoo Model Development 2006, Pilbara Iron, unpublished consultant report prepared by Youngs, J. and Milligan, N.

RTTS (2006) Rio Tinto Technical Services, Surplus Water Management Strategy for RTIO Pilbara Operations, unpublished report.

SKM (2001) Environmental Water Requirements of Groundwater Dependent Ecosystems, Sinclair Knight Merz Pty Ltd, Technical Report Number 2, Environment Australia, November 2001, Australian Government, Department of the Environment and Water Resources. ISBN 0 642547696. http://www.environment.gov.au/water/rivers/nrhp/groundwater/chapter2.html#mining. Cited 12 February 2007.

Trewin, B. (2006) Australian Deserts, Climatic Aspects Of Australia's Deserts, 1301.0 – Year Book Australia, 2006 Previous Issue Released at 11:30 AM (Canberra time) 20/01/2006 Dr Blair Trewin, National Climate Centre, Australian Bureau of Meteorology, Melbourne.

Waterhouse, J.D. & Howe, P.J. (1994) An Environmental Hydrogeological Study near Mount Bruce, Pilbara Region of Western Australia. *Proceedings of Water Down Under '94,Congress of the International Association of Hydrogeologists and International Hydrology and Water Resources Symposium*, Adelaide.

Youngs, J. (2004) *Groundwater Level Behaviour at a Pilbara Mine*. M.Eng., University of Technology Sydney, Australia.

Youngs, J. & Brown, D. (2005) Aquifer re-injection as a low-impact groundwater investigation tool – a case study from the Pilbara Region, Western Australia. *Conference Proceedings of the 5th International Symposium on Management of Aquifer Recharge, Berlin, Germany, 11–16 June 2005*.

Chapter 11

Nitrogen cycle in gravel bed rivers: The effect of the hyporheic zone

Alessandra Marzadri[1] *& Alberto Bellin*[2]

[1] *Center for Ecohydraulics Research, University of Idaho, Boise, ID, United States*

[2] *Department of Civil, Environmental and Mechanical Engineering, University of Trento, Trento, Italy*

ABSTRACT

Several studies have shown the importance of microbial processes in removing nitrogen from rivers, with first-order gravel bed streams playing a major role in reducing the total load. Consequently, modelling transport of nutrients and contaminant in rivers calls for including temporary storage and nitrification-denitrification processes within hyporheic and riparian zones. In gravel bed rivers alternate zones of high and low pressure generated by bedforms induce a complex flow pattern within the hyporheic zone that interacts with the stream through downwelling and upwelling zones. The two transport equations for NH_4^+ and NO_3^- were solved, coupled with the chained first-order kinetics modelling nitrification and denitrification processes within the hyporheic zone, by using a Lagrangian approach and assuming that dispersion is negligible. With this simple, yet powerful, model the interplay between streambed morphology and nitrogen removal from hyporheic zone of a gravel bed river including the interplay between ammonium and nitrate were studied.

11.1 INTRODUCTION

The hyporheic zone is a saturated area surrounding the stream that provides the linkage between the river and the aquifer. It is a rich ecotone influencing solute transport and nutrient cycling in rivers (Alexander *et al.*, 2000). Hyporheic flow is generated by a uneven distribution of the water pressure at the riverbed, which is tied dynamically with bed topography (Elliot & Brooks, 1997). In gravel-bed rivers the pressure at the bed surface is non-uniform, with zones of high and low pressure at the upstream and downstream ends of the bedforms, respectively. This creates a complex three-dimensional flow pathway within the hyporheic zone that is connected to the streamflow through an alternate sequence of downwelling and upwelling zones for whose the spatial pattern is dictated by bed morphology. Focusing on a single wavelength of the bed form one should note that the mean residence time of water within the hyporheic zone is by orders of magnitude larger than the renewal time of the stream water between successive downwelling and upwelling zones. The different timing of solute transported in the stream and through the hyporheic zone creates tailing in the observed breakthrough curve (Haggerty *et al.*, 2002). Furthermore, biological activity within the hyporheic zone is chiefly controlled by the distribution of the residence time.

Most of the streams flowing in agricultural and urban areas suffer from the impact of excessive reactive nitrogen inputs under the form of ammonium (NH_4^+) and nitrate

(NO_3^-) (Alexander *et al.*, 2000; Peterson *et al.*, 2001). Under oxidized conditions, aerobic bacteria (Nitrosomonas and Nitrobacter) use oxygen as a terminal acceptor during the transformation of ammonium to nitrate (nitrification) which in turn is reduced to nitrous oxide (N_2O) or molecular nitrogen (N_2) by heterotrophic denitrifying bacteria (denitrification). However, as the environment becomes increasingly reduced, nitrification is inhibited while denitrification continues to reduce nitrate to N_2O or N_2. Along the flowpaths from downwelling to upwelling zones the environment becomes reduced as oxygen is consumed by aerobic bacteria. Consequently, the hyporheic zone may act as a source or a sink of nitrate, depending on the oxygen dynamics (Triska *et al.*, 1993), which is indirectly related to the stream travel time and ultimately to the dimension of the bed form (Edwards, 1998; Malard *et al.*, 2002).

Several methods have been proposed in order to quantify the hydrological exchange between streams and shallow ground water in the hyporheic zone (see Packman & Bencala (2000) for an extensive review). Mass exchange between the stream and the hyporheic zone is traditionally modelled by a first-order mass transfer equation, as in the Transient Storage Model (TSM) proposed by Bencala & Walters (1983). In this model the hyporheic zone is considered a well mixed zone of constant volume.

Although this simplified model may work under certain circumstances (Zaramella *et al.*, 2003), there are evidences that the hyporheic zone is far from being well mixed and that more sophisticated transport models are needed, in particular for modelling processes, such as nutrients dynamics, for whose the spatial pattern depends critically on the residence time distribution (see e.g. Packman & Bencala, 2000; Cardenas *et al.*, 2004; Tonina & Buffington, 2007). Numerical models are commonly used in rivers with complex three-dimensional topography (Wagenschein & Rode, 2008, Boano *et al.*, 2010). In this work we seek for an answer to the following questions:

What are the important factors controlling the export of ammonium and nitrate by the hyporheic zone?

Under which conditions does the hyporheic zone act as a source rather than a sink of nitrate for the stream water?

To answer these questions two transport equations for NH_4^+ and NO_3^- need to be solved, coupled with the chained first-order kinetics modelling nitrification and denitrification processes within the hyporheic zone. Transport is solved by using a Lagrangian approach (Dagan, 1989; Cvetkovic & Dagan, 1994; Rubin, 2003) and assuming that local dispersion is negligible. The Lagrangian approach is particularly appealing in this case because the transport equation is written in term of residence time which is the controlling parameter of both retention and nitrification/denitrificaton processes. With this simple, yet powerful, model we studied the interplay between streambed morphology and nitrogen export from the hyporheic zone of a gravel bed river including the interplay between ammonium and nitrate.

11.2 MODELLING APPROACH

11.2.1 Flow

Let us consider the hyporheic zone of a gravel bed river with an alternate sequence of pools and riffles (alternate bars). Two successive diagonal fronts delimitate the bar

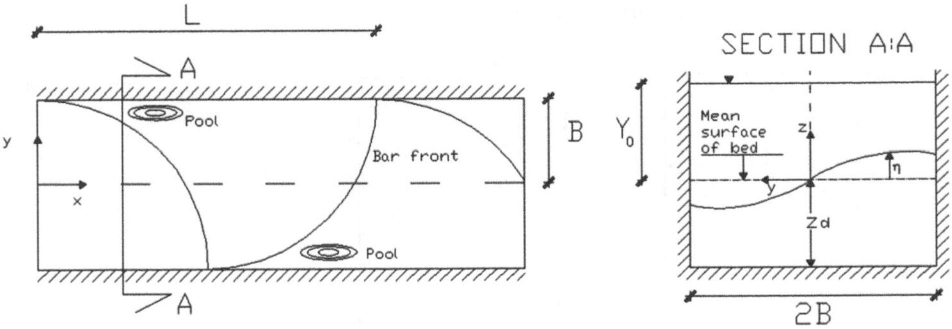

Figure 11.1 Sketch of the channel structure.

unit with a pool at the downstream end of each front in proximity of the channel banks (Fig. 11.1). For simplicity, we consider the case of a straight channel of constant width and constant streamflow. Marzadri *et al.* (2010) presented a three-dimensional analytical model of the hyporheic flow induced by pumping caused by the pressure head at the bottom of the stream. A similar analysis has been performed by Elliott & Brooks (1997) for a river developing dunes. Due to the morphological characteristics of the dunes the flow field is two-dimensional, while in our case it is fully three-dimensional. The stationarity hypothesis implies that the stream water discharge is constant, or slowly varying over times of the order of the particle residence time.

The dimensionless governing equation is (Marzadri *et al.*, 2010):

$$\frac{\partial^2 h^*}{\partial x^{*2}} + \frac{\partial^2 h^*}{\partial y^{*2}} + \frac{\partial^2 h^*}{\partial z^{*2}} = 0 \tag{11.1}$$

with the following boundary conditions (Fig. 11.1):

$$\left.\frac{\partial h^*}{\partial y^*}\right|_{y^*=\pm 1} = 0, \quad h^*(-L/2B, y^*, z^*) = h^*(L/2B, y^*, z^*), \quad \left.\frac{\partial h^*}{\partial z^*}\right|_{z^*=-z_d^*} = 0 \tag{11.2}$$

where $(x^*, y^*, z^*, z_d^*) = (x/B, y/B, z/B, z_d/B)$, $h^* = h/Y_0$ and h is the hydraulic head, $2B$ is the channel width, Y_0 is the mean flow depth, z_d is the thickness of the hyporheic zone, and L is the bar length. The boundary condition at the mean bed surface has been provided by Colombini *et al.* (1987) as superimposition of Fourier harmonics.

The terms of order up to the second in both directions are retained in the expansion, since according to Ikeda (1984) and Colombini *et al.* (1987) higher-order terms exert a negligible impact on the hydraulic head. Once the distribution of h^* within the hyporheic zone has been obtained the dimensionless velocity field is computed by the Darcy's equation:

$$\vec{u}^* = (u^*, v^*, w^*) = -\nabla h^* \tag{11.3}$$

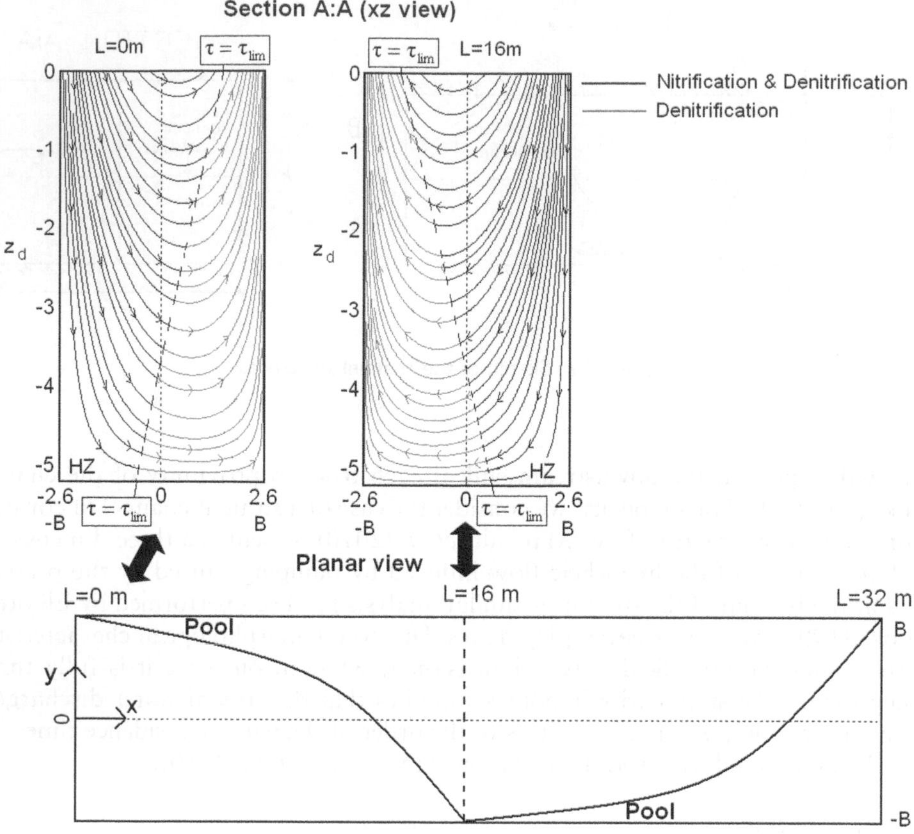

Figure 11.2 Sketch of flowpaths within the hyporheic zone and the associated redox conditions.

where $\vec{u}^* = (B\vartheta/KY_0)\vec{u}$ is the dimensionless velocity field, ϑ is the porosity, and K is the hydraulic conductivity. The resulting flow field is symmetric because of the symmetry of the forcing term and with a complex structure dictated by the pressure distribution at the bed surface. The magnitude of the velocity decays exponentially with depth as for the hydraulic head (Fig. 11.2).

11.2.2 Nitrogen cycle

Nutrient cycling within stream is controlled by several factors ultimately related to the proportion of geomorphological units creating the hyporheic flow and their size. It is widely recognised that within the hyporheic zone the biochemical reactions, and their spatial pattern, depend on the residence time distribution, which in turn depends on the river bed morphology other than the streamflow (see e.g. Malard *et al.*, 2002). For example, along a flowpath connecting the downwelling with the upwelling zone first oxygen is consumed by aerobic respiration that transforms ammonium to nitrate. As oxygen is depleted denitrification may remove nitrate produced in the upstream end of the flowpath. Consequently, hyporheic flow systems with dominance of short pathways may be a source of nitrate, whereas longer bars with longer pathways would be a sink of nitrate (Fig. 11.2).

Since NH_4^+ can be only depleted, while NO_3^- is subjected to more complex dynamics of enrichment, in aerobic zones, and depletion, in anaerobic zones, nutrient cycling is influenced by the ratio between ammonium and nitrate concentration in the downwelling area (Sjodin *et al.*, 1997):

$$\mathrm{RC} = \frac{[\mathrm{NH}_4^+]_0}{[\mathrm{NO}_3^-]_0} \tag{11.4}$$

At a scale larger than the single morphological unit more complex dynamics may emerge due to the nonlinear coupling between biological processes within the hyporheic zone and variations of RC along the stream. Other important factors controlling biological activity are water temperature and pH. Only the effect of oxygen consumption is considered, assuming that temperature and pH conditions remain the same at the level of the single morphological unit.

The concentration of dissolved oxygen (DO) into the water is chiefly controlled by temperature, pressure and salinity. In well aerated streams the concentration of DO is about 9–10 mg/l; when this concentration reduces to 2–4 mg/l conditions shift from aerobic to anaerobic (Böhlke & Denver, 1995).

The governing equations for transport of reactive solutes are the following:

$$\frac{\partial C_i}{\partial t} + \vec{u}\nabla C_i = f_i(C_i, \ldots, C_j, \ldots, C_n), \quad i = 1, \ldots, n \tag{11.5}$$

where C_i is the concentration of the i-th compound, and f_i is the reaction term. In equation (11.5) we neglected local (pore scale) dispersion which has been shown of secondary importance in advection dominated processes (e.g. Dagan, 1989; Rubin, 2003). Equation (11.5) can be written in a more convenient way for the following analytical treatment by applying mass conservation along a stream tube (Shapiro & Cvetkovic, 1988; Dagan *et al.*, 1992):

$$\frac{\partial C_i}{\partial t} + \frac{\partial C_i}{\partial \tau} = f_i(C_i, \ldots, C_j, \ldots, C_n), \quad i = 1, \ldots, n \tag{11.6}$$

In equation (11.6) the advective term has been written by means of the travel time τ of a targeted particle along the streamline which assumes the following expression:

$$\tau = \int_0^l \frac{d\zeta}{\vec{u}(\zeta)}. \tag{11.7}$$

where ζ represent the distance from the downwelling position and the integration is performed along the streamline (flowpath) of length l (Fig. 11.2). Solution of equations (11.6) and (11.7) provide the contribution of a stream tube to transport of the aqueous species i.

The interplay between ammonium ($i=1$) and nitrate ($i=2$) was modelled by using the following simplified sequential model of nitrification-denitrification processes:

$$\begin{cases} f_1(C_1) = -k_1 C_1 \\ f_2(C_1, C_2) = k_1 C_1 - k_2 C_2 \end{cases} \tag{11.8}$$

where k_1 [T^{-1}] and k_2 [T^{-1}] are the nitrification and denitrification rate coefficients, respectively. To make the sequential model (11.8) more realistic we turn off nitrification by setting $f_1(C_1)=0$ when the oxygen concentration is smaller than the threshold $C_{\lim}$.

Oxygen is transported as a tracer and consumed by nitrification. The associated dynamics can be described by using a transport model similar to equation (11.6) with a first-order decay term describing oxygen consumption:

$$\begin{cases} \dfrac{\partial C}{\partial t} + \dfrac{\partial C}{\partial \tau} = -k_1 C \\ C(\tau, 0) = C_s \\ C(0, t) = C_s \end{cases} \tag{11.9}$$

where C_s is oxygen concentration at saturation. For simplicity in the model in equation (11.9) we assumed that oxygen is at saturation within the downwelling area and at the initial time within the hyporheic zone. In equations (11.6) and (11.9) τ marks the spatial coordinate along the streamline. Since for $t < \tau$ the DO travelling wave is upstream to the position marked by τ (see the following solution) equation (11.9) applies for $t > \tau$, while $C(t, \tau) = C_s$ for $t < \tau$. Under this condition the solution of (11.9) is given by:

$$C(t, \tau) = C_s\left[\exp(-k_1\tau)\, H(t-\tau) + H(\tau - t)\right] \tag{11.10}$$

where H is the Heaviside step function.

According to equation (11.10) in the first part of the streamline, from the entry point in the downwelling area to the position along the streamline where $C = C_{\lim}$, both nitrification and denitrification are active while in the remaining part of the streamline nitrification is switched off since the anoxic environment cannot sustain nitrification. The distance from the streamline origin at which oxygen concentration declines to $C_{\lim}$ can be obtained from equation (11.10) in the following form:

$$\tau_{\lim} = \frac{1}{k_1} \ln\left(\frac{C_s}{C_{\lim}}\right) \tag{11.11}$$

In order to generalize the subsequent analysis we introduce the following dimensionless travel (residence) time:

$$\tau^* = \tau \frac{k_2}{\text{Da}} = \tau \lambda^2 H_{BM} K \tag{11.12}$$

where λ [L^{-1}] is the wavelength of the bedform, H_{BM} [L] is the amplitude of the bedform (Colombini *et al.,* 1987; Ikeda, 1984), and Da is the Damköler number, which is given by the ratio of the characteristic time of flux, $t_{\text{flux}} = (\lambda^2 H_{BM} K)^{-1}$ over time of reaction, $t_{\text{reaction}} = 1/k_2$:

$$\text{Da} = \frac{t_{\text{flux}}}{t_{\text{reaction}}} = \frac{k_2}{\lambda^2 H_{BM} K} \tag{11.13}$$

The dynamics of ammonium and nitrate can be analysed by solving the differential equations obtained by equation (11.6) with specific ammonium ($i = 1$) and nitrate ($i = 2$) concentrations. For $\tau^* > \tau^*_{\text{lim}}$ the corresponding dimensionless solutions assume the following forms

$$\begin{cases} \text{if } t^* \geq \tau^* \begin{cases} C_1^{*ox}(\tau^*, t^*) = \text{RC}\exp(-k_d \text{Da}\tau^*) \\ C_2^{*ox}(\tau^*, t^*) = \exp(-\text{Da}\tau^*) + \text{RC}\dfrac{k_d}{k_d - 1}\left[\exp(-\text{Da}\tau^*) - \exp(-k_d \text{Da}\tau^*)\right] \end{cases} \\ \text{if } t^* < \tau^* \begin{cases} C_1^{*ox}(\tau^*, t^*) = 0 \\ C_2^{*ox}(\tau^*, t^*) = 0 \end{cases} \end{cases} \tag{11.14}$$

While for $\tau^* > \tau^*_{\text{lim}}$ we obtain:

$$\begin{cases} \text{if } t^* \geq \tau^* \begin{cases} C_1^*(\tau^*, t^*) = C_1^{*ox}(\tau^*_{\text{lim}}, t^*) \\ C_2^*(\tau^*, t^*) = C_2^{*ox}(\tau^*_{\text{lim}}, t^*)\exp\left[-\text{Da}(\tau^* - \tau^*_{\text{lim}})\right] \end{cases} \\ \text{if } t^* < \tau^* \begin{cases} C_1^*(\tau^*, t^*) = 0 \\ C_2^*(\tau^*, t^*) = 0 \end{cases} \end{cases} \tag{11.15}$$

where $k_d = k_1/k_2$ is a parameter that measures the reciprocal strength of nitrification and denitrification, and $\text{RC} = C_{1,0}/C_{2,0}$ is the ratio between ammonium and nitrate concentration at the downwelling area. Solutions (11.14) and (11.15) have been obtained with the following initial and boundary conditions: $C_1(\tau,0) = C_2(\tau,0) = 0$, $C_1(0,t) = C_{1,0}$ and $C_2(0,t) = C_{2,0}$. Furthermore, nitrate initial concentration is used to make dimensionless solute concentrations, i.e. $C_i^* = C_i/C_{2,0}$.

After these preparatory steps the mass flux at the upwelling end of the streamline is computed as follows:

$$q^*_{M,i}[\tau^*(\vec{a}^*), t^*] = q^*[\vec{x}^*]C_i^*[\tau^*(\vec{a}^*), t^*] \quad i = 1, 2 \tag{11.16}$$

where $C^*{}_i[\tau^*(\vec{a}^*), t^*]$ is the solute concentration at the exit point $\vec{x}^*$ of the streamline and $q^*[\vec{x}^*]$, is the dimensionless flux at the upwelling area. The total mass discharge at the exit area is obtained by integrating equation (11.16) over the upwelling area A^*

$$Q^*_{M,i}[t^*] = \int_{A^*} q^*\left[\vec{x}^*\right]C_i^*[\tau^*(\vec{a}^*), t^*]dA^* = \int_{A_0^*} q^*[\vec{a}^*]C_i^*\left[\tau^*(\vec{a}^*), t^*\right]dA_0^* \tag{11.17}$$

where continuity equation has been applied between the downwelling and the upwelling ends of the stream tube. The integral in the right hand term of equation (11.17) is solved as follows:

$$Q^*_{M,i}[t^*] = \sum_j q^*[\vec{x}_j^*]\, C^*_{i,j}[\tau_j, t]\, \Delta A_j \tag{11.18}$$

where $q^*[\vec{x}_j^*]$ is the flow at the downwelling end of the streamline.

11.3 RESULTS AND CONCLUSIONS

The river morphology influences nutrient cycling for several combinations of the parameters controlling nitrification-denitrification in our simplified model and RC either larger or smaller than 1. The abatements of ammonium and nitrate with respect to the respective mass flux entering through the downwelling area are evaluated as follows:

$$\Delta Q^*_{M,i} = \frac{\left[Q^*_{M,i}\right]_d - \left[Q^*_{M,i}\right]_u}{\left[Q^*_{M,i}\right]_d} \tag{11.19}$$

where the subscripts d and u denote the initial and the final concentration at the downwelling and upwelling zones, respectively.

Simulations are conducted in a stream, the morphological and dimensional parameters of which are summarized in Table 11.1.

Figures 11.3a and 11.3b show how the abatements of ammonium and nitrate vary with Da for $k_d = 4.9$, i.e. in a situation in which nitrification proceeds at a stronger rate than denitrification. As expected an increase of Da results in a larger abatement of ammonium. However, nitrate dynamics are dramatically different depending on RC. If $RC > 1$ the hyporheic zone acts as a source of nitrate because denitrification consumes only partially the nitrate produced by nitrification. This is supported by the relative high ammonium concentration in the stream ($RC > 1$) and the relatively slow denitrification ($k_2 < k_1$).

When ammonium concentration in the stream is smaller than nitrate concentration ($RC < 1$), denitrification consumes all the nitrate produced by nitrification of ammonium as well as a fraction of the nitrate entering through the downwelling area that

Table 11.1 Parameters characterising the stream utilized in the simulations. In particular β is the aspect ratio of alternate bars, θ in the Shield stress, d_S is the relative submergence, Diam is the mean grain size of the sediments, H^*_{BM} is the dimensionless amplitude of the bedform, B is the half channel width, z_d is the alluvium depth, Y_0 is the mean flow depth, L is the bedform length and λ is the bedform wavelength (see Colombini *et al.* (1987) for more detais).

Test	β	θ	d_s	Diam (m)	H^*_{BM} (−)	B (m)	z_d (m)	Y_0 (m)	L (m)	λ (m^{-1})
1	13	0.08	0.05	0.01	1.4757	2.6	2.6	0.2	32.9	0.497053

increases with Da (Fig. 11.3b). In this case the hyporheic zone acts as a sink of both ammonium and nitrate.

The effect of k_d on the abatement of ammonium and nitrate is shown in Figures 11.4a and 11.4b, respectively. Contrary to the previous case the qualitative behaviour of the curves showing the abatement of the two compounds is the same for RC < 1 and RC > 1. For a given Da, i.e. a given denitrification rate, we observe that a larger k_d leads to a larger abatement of the ammonium load and consequently a larger production of nitrate. For RC < 1 (not shown in Figure) we observe a similar behaviour with the only difference that a larger k_d leads to a smaller abatement of the nitrogen load. Therefore, a larger difference in the two reaction rates caused by an increase of k_1 causes the hyporheic zone to be less effective in removing the nitrate load.

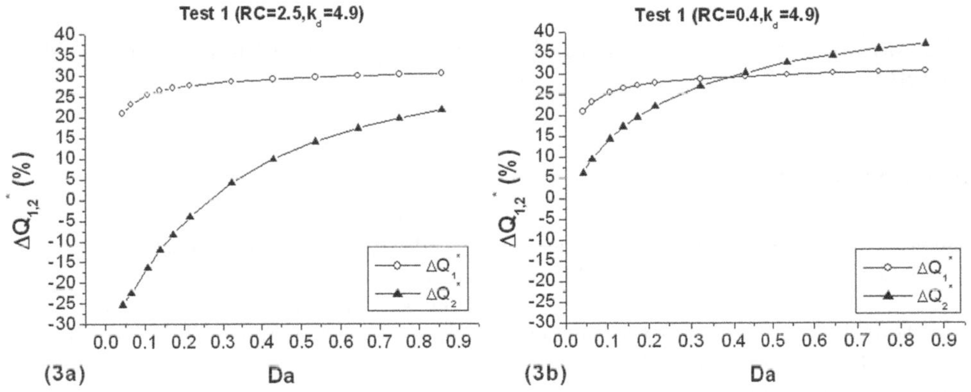

Figure 11.3 Percentage of depletion of ammonia (ΔQ_1^*) and nitrate (ΔQ_2^*) versus the Damköler number for Test 1 and fixed value of k_d in a stream initially rich of ammonium than nitrate (3a) and in a stream initially rich of nitrate than ammonium (3b).

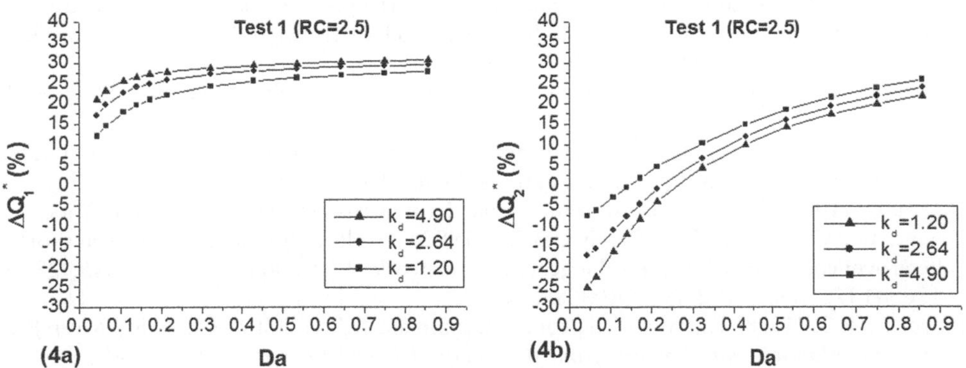

Figure 11.4 Percentage of depletion of ammonia (ΔQ_1^*) and nitrate (ΔQ_2^*) versus the Damköler number for several values of k_d in a stream initially rich with ammonium than nitrate RC = 2.5.

From this preliminary investigation we concluded that while the hyporheic zone acts as a sink of ammonium to an extent that depends on the nitrification rate, it may act as a source or a sink of nitrate depending on the ratio RC between ammonium and nitrate concentration in the stream. For RC > 1 the hyporheic zone acts as a source of nitrate, if the rate at which nitrate is depleted by denitrification is smaller that the nitrification rate, while it acts as a sink of nitrate when RC < 1. The abatement of ammonium increases with k_1 but the increasing rate attenuates at high k_1 because the faster oxygen consumption enlarges the portion of the streamlines under anoxic conditions. For RC > 1 this effect is mirrored in the production of nitrate which is controlled by the amount of ammonium that is transformed to nitrate, while for RC < 1 the abatement of nitrate increases with k_2 because it is not influenced by oxygen concentration.

ACKNOWLEDGEMENTS

The authors gratefully acknowledge the EU project AQUATERRA (contact 505428 – GCOE).

REFERENCES

Alexander, R.B., Smith, R.A. & Schwartz, G.E. (2000) Effect of stream channel size on the delivery of nitrogen to the Gulf of Mexico. *Nature*, 403, 758–761.

Bencala, K.E. & Walters, R.A. (1983) Simulation of solute transport in a mountain pool-and-riffle stream: a transient storage model. *Water Resources Research*, 19 (3), 718–724. doi:10.1029/WR019i003p00718.

Boano, F., Demaria, A., Revelli, R. & Ridolfi, L. (2010), Biogeochemical zonation due to intrameander hyporheic flow. *Water Resources Research*, 46 (2), W02511. doi:10.1029/2008WR007583.

Böhlke, J.K. & Denver, J.M. (1995), Combined use of groundwater dating, chemical, and isotopic analyses to resolve the history and fate of nitrate contamination in two agricultural watersheds, Atlantic Coastal Plain, Maryland. *Water Resources Research*, 31 (9), 2319–2339. doi:10.1029/95WR01584.

Cardenas, M.B., Wilson, J.L. & Zlotnik, V.A. (2004) Impact of heterogeneity, bed forms, and stream curvature on subchannel hyporheic exchange. *Water Resources Research*, 40 (8), W08307, doi:10.1029/2004WR003008.

Colombini M, Seminara G, & Tubino M (1987) Finite-amplitude alternate bars. *J. Fluid Mech.*, 181, 213–232.

Cvetkovic, V. & Dagan, G. (1994) Transport of kinetically sorbing solute by steady random velocity in heterogeneous porous formations. *J Fluid Mech.*, 265, 189–215.

Dagan, G. (1989) *Flow and Transport in Porous Formations*. Springer-Verlag, New York.

Dagan, G., Cvetkovic, V. & Shapiro, A.M. (1992) A solute flux approach to transport in heterogeneous formations: 1. The general framework. *Water Resources Research*, 28 (5), 1369–1376, doi:10.1029/91WR03086.

Edwards, R.T. (1998) The hyporheic zone. In: Naiman, R.J. & Bilby, R.E. (eds.) *River Ecology and Management: Lessons from the Pacific Coastal Ecoregion*. New York, Springer. pp. 347–372.

Elliott, A.H. & Brooks, N.H. (1997) Transfer of nonsorbing solutes to a streambed with bed forms: Theory. *Water Resources Research*, 33 (1), 123–136. doi:10.1029/96WR02784.

Fernald, A.G., Wigington, P.J. & Landers, D.H. (2001) Transient storage and hyporheic flow along the Willamette River, Oregon: Field measurements and model estimates. *Water Resources Research*, 37 (6), 1681–1694. doi:10.1029/2000WR900338.

Haggerty, R., Wondzell, S.M. & Johnson, M.A. (2002) Power-law residence time distribution in the hyporheic zone of a 2nd-order mountain stream. *Geophysical Research Letters*, 29 (13), doi:10.1029/2002GL014743.

Harvey, J. & Bencala, K.E. (1993) The effect of streambed topography on surface-subsurface water exchange in mountain catchments. *Water Resources Research*, 29 (1), 89–98. doi:10.1029/92WR01960.

Ikeda, S. (1984) Prediction of alternate bars wavelength and height. *Journal of Hydraulic Engineering*, 110 (4), 371–386.

Malard, F., Tockner, K., Dole-Oliver M-J. & Ward, J.V. (2002) A landscape perspective of surface-subsurface hydrological exchanges in river corridors. *Freshwater Biology*, 47, 621–640.

Marzadri, A., Tonina, D., Bellin, A., Vignoli, G. & Tubino, M. (2010) Semi-analytical analysis of hyporheic flows induced by alternate bars. *Water Resources Research*, 46, W07531. doi:10.1029/2009WR008285.

Packman, A.I. & Bencala, K.E. (2000) Modelling surface-subsurface hydrological interaction. In: Jones, J.B. & Mulholland, P.J. (eds.)*Streams and Groundwaters*. San Diego, CA, Academic Press. pp. 45–80.

Packman, A.I. & Brooks, N.H. (2001) Hyporheic exchange of solutes and colloids with moving bed forms. *Water Resources Research*, 37 (10), 2591–2605. doi:10.1029/2001WR000477.

Peterson, B. *et al.* (2001) Control of nitrogen export from watershed by headwater streams. *Science*, 292, 86–89.

Rubin, Y. (2003) *Applied Stochastic Hydrogeology*. New York, Oxford University Press

Shapiro, A.M. & Cvetkovic, V. (1988), Stochastic analysis of solute arrival time in heterogeneous porous media. *Water Resources Research*, 24 (10), 1711–1718. doi:10.1029/WR024i010p01711

Sjodin, A.L., Lewis, W.M. Jr. & Saunders III J.F. (1997) Denitrification as a component of the nitrogen budget for a large plain river. *Biogeochem*istry, 39, 327–342.

Strauss, E.A. & Dodds, W.K. (1997). Influence of protozoa and nutrient availability on nitrification rates in subsurface sediments. *Microbial Ecology*, 34, 155–165.

Tonina, D. & Buffington, J.M. (2007) Hyporheic exchange in gravel bed rivers with pool-riffle morphology: Laboratory experiments and three-dimensional modelling. *Water Resources Research*, 43 (1), W01421. doi:10.1029/2005WR004328.

Triska, F.J., Duff, J.H. & Avanzino, R.J. (1993) The role of water exchange between a stream channel and its hyporheic zone in nitrogen cycling at the terrestrial–aquatic interface. *Hydrobiologia*, 251, 167–184.

Wagenschein, D. & Rode, M. (2008) Modelling the impact of river morphology on nitrogen retention – A case study of the Weisse Elster River (Germany). *Ecological Modelling*, 211, 224–232. doi:10.1016/j.ecolmodel.2007.09.009.

Zaramella, M., Packman, A.I. & Marion, A. (2003) Application of the transient storage model to analyze advective hyporheic exchange with deep and shallow sediment beds. *Water Resources Research*, 39 (7). doi:10.1029/2002WR001344.

Chapter 12

Groundwater recharge quantification for the sustainability of ecosystems in plains of Argentina

Mónica P. D'Elia, Ofelia C. Tujchneider, Marta del C. Paris & Marcela A. Perez

Facultad de Ingeniería y Ciencias Hídricas – Universidad Nacional del Litoral. Ciudad Universitaria, Santa Fe, Argentina

ABSTRACT

Groundwater recharge is balanced by discharge from aquifers to springs and rivers under natural conditions. This equilibrium is affected by pumping of groundwater and may cause important losses in discharge areas where wetlands and other riparian ecosystems are developed. Thus, a correct and integrated management of groundwater resources has to take into account the recharge and discharge of aquifer systems and the groundwater user requirements of a region. The objective of this work is to quantify local groundwater recharge and to give a preliminary estimation of the amount of regional groundwater recharge in a central sector of the Santa Fe province (Argentina). The water table fluctuation and chloride mass balance methods were used to estimate local and regional groundwater recharge. The results of this work are important and contribute to a better knowledge of the groundwater system, providing quantitative criteria on which to base the management and protection of groundwater dependent ecosystems.

12.1 INTRODUCTION

Groundwater recharge is balanced by discharge (e.g. springs, rivers, lakes, wetlands, estuaries) under natural conditions. This equilibrium is affected by the pumping of groundwater for human activities but even when this extraction is equal to recharge, it may cause important losses in discharge areas where wetlands and riparian ecosystems are developed.

In flat areas of the centre of Santa Fe province (Argentina), the groundwater system is formed by an unconfined aquifer that is locally recharged from precipitation and a semi-confined aquifer whose recharge is both local (coming from the unconfined aquifer) and regional. Discharge areas are mainly streams, lagoons and wetlands from the unconfined aquifer and the alluvial valleys of the Salado and Paraná rivers for the semi-confined aquifer.

Several important towns are located near these rivers, and human activities are strongly dependent on groundwater. Some terrestrial vegetation and fauna, wetlands and river base flow ecosystems are present in the region, but their groundwater requirements are not completely understood and quantified.

The objective of this work is to quantify local groundwater recharge and to estimate the amount of regional groundwater recharge in a central sector of the Santa Fe province (Argentina) using the water table fluctuation and chloride mass balance methods. The results of this work are important and contribute to a better understanding of the groundwater system, providing quantitative criteria on which to base integrated water management and protection of ecosystems related to groundwater.

12.2 STUDY AREA

12.2.1 Location of the study area

The study area is located in the central sector of the Santa Fe province (Argentina) in the lower part of the Salado River basin, between 31°20′–31°30′S and 61°05′–60°45′W; and 5–20 masl (Fig. 12.1). It is a flat area of 500 km^2 with a slope of about 0.2 to 0.3%, and 1% in areas near to the Salado River. It corresponds to a huge Argentinean plain called 'Llanura Pampeana', where the most important productive activities of the country related to agriculture and livestock are developed.

Esperanza and Rafaela are important cities of the region. Esperanza is situated 30 km to the north west of Santa Fe city – capital of the Santa Fe province – and

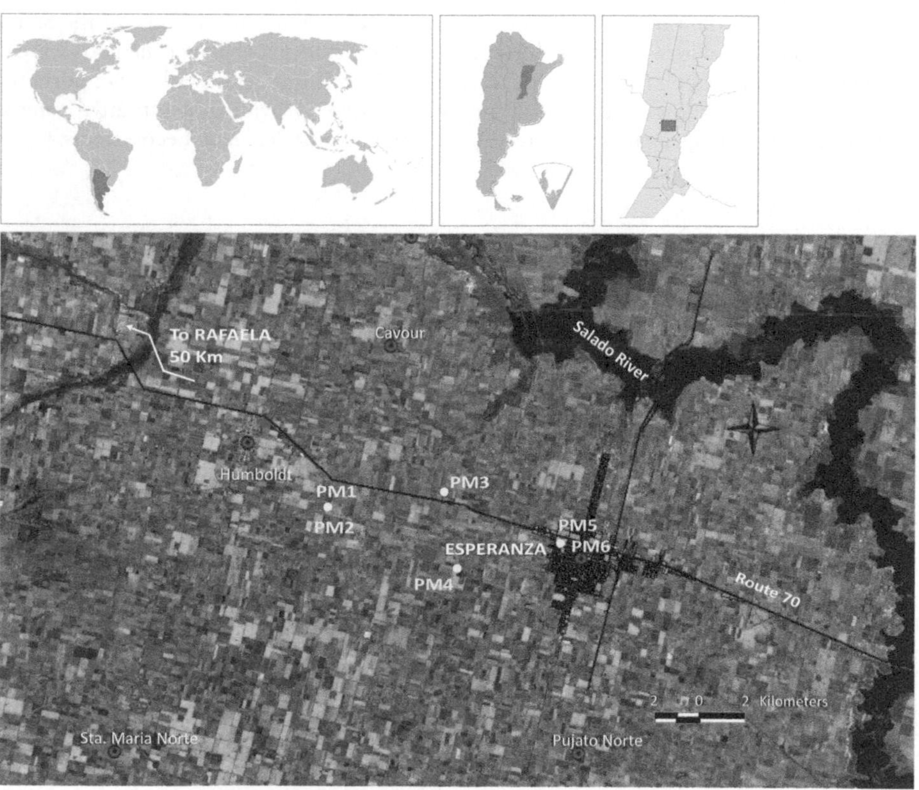

Figure 12.1 Location map of the study area.

it has about 40 000 inhabitants. Rafaela is located 50 km to the west of Esperanza outside the study area and its population is approximately 85 000 inhabitants. Although surface water resources are available in the study area, all human activities depend on groundwater: agriculture, livestock, industry, human consumption, among others.

12.2.2 Geological and hydrogeological settings

The geology and hydrogeology of the region are known from several previous studies (Fili & Tujchneider, 1977; Fili *et al.*, 1999; Tujchneider *et al.*, 2006). The conceptual model of the groundwater system defined by these authors is presented in Figure 12.2.

It is a multilayer aquifer system. The upper part of a Miocene Formation (Paraná Formation), composed by clays and sands of marine origin, is considered to be the hydrogeological basement of the system. Covering these sediments there is an aquifer layer of medium and fine fluvial sands belonging to a Pliocene Formation (Ituzaingó Formation), also called 'Puelches Sands'. The thickness of this layer ranges from 25 to 35 m. This aquifer is semi-confined by clays of about 3 m thickness which correspond to the lower part of a Pleistocene Formation (Pampa Formation). The upper part of this Pampa Formation constitutes an unconfined aquifer composed of silts, loess and clays of aeolian and lacustrine origin and its thickness varies from 15 to 25 m.

The hydraulic parameters of the formations reported by Fili *et al.* (1999) are listed in Table 12.1.

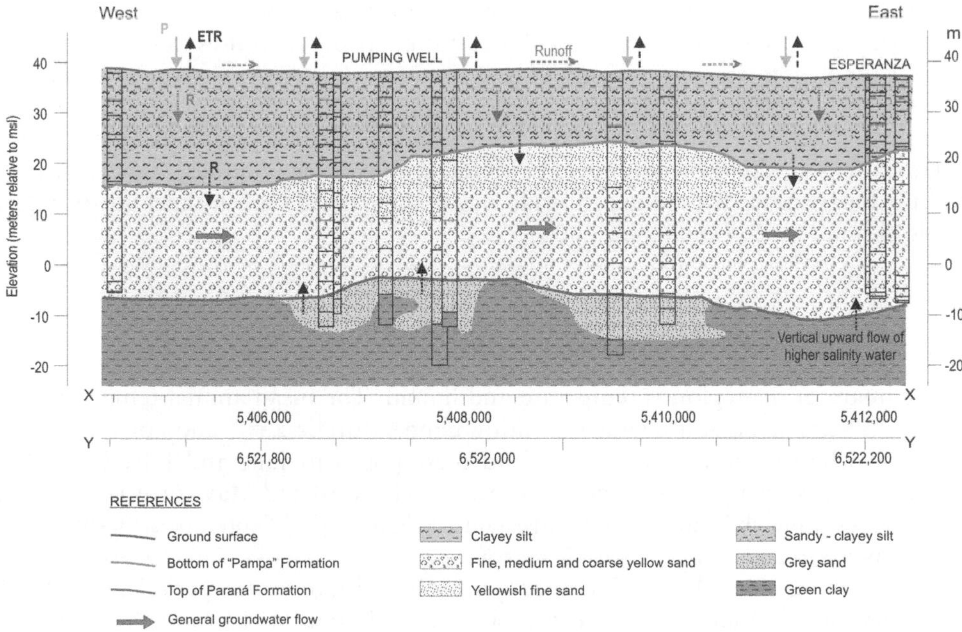

Figure 12.2 Conceptual model of the groundwater system. From Paris (2010).

Table 12.1 Hydraulic parameters of the formations.

Formation	*T* (m^2/day)	*Sy*	K_h (m/day)	K_v (m/day)
Pampa (aquifer)	150	0.05	10	5
Pampa (aquitard)	$1.5 * 10^{-2}$	0.006	$5 * 10^{-3}$	$5 * 10^{-3}$
Ituzaingó	600–950	10^{-4}	30	30

T is transmissivity, *Sy* is specific yield, K_h is horizontal hydraulic conductivity and K_v vertical hydraulic conductivity.

The groundwater flow direction is from west to east. Local and regional groundwater flows were distinguished in previous studies based on the hydrogeological behaviour of the groundwater system and corroborated by isotopic investigations (D'Elia *et al.*, 2006).

The unconfined aquifer is locally recharged from precipitation and discharge areas are mainly small streams and wetlands. It is exploited for water supply to animals in rural areas and for domestic uses in urban areas because of its medium yield and quality.

Recharge of the semi-confined aquifer is both local (coming from the unconfined aquifer) and regional. Vertical upward and downward flows take place through the Pampa Formation clays according to the hydraulic relationships between the unconfined and semi-confined aquifers. Thus, groundwater extraction from the semi-confined aquifer may increase recharge from the unconfined aquifer, which is directly linked to the local hydrological cycle. The alluvial valleys of the Salado and the Paraná Rivers are discharge areas of this aquifer. Due to its good groundwater quality and yield, the semi-confined aquifer is exploited for many different uses in the region.

Groundwater is predominantly of a sodium bicarbonate water type (Tujchneider *et al.*, 1998). In some cases, groundwater of the unconfined aquifer presents important arsenic and fluoride contents (Nicolli *et al.*, 2009). It is important to note that the chemical composition of groundwater in the semi-confined aquifer changed from a sodium bicarbonate to a sodium chloride type in some specific areas under uncontrolled exploitation (Tujchneider *et al.*, 2005).

12.2.3 Climate and soil conditions

The climate of the region is temperate and humid. The mean annual temperature is about 18°C. The highest average monthly temperature is 25°C and corresponds to January, whereas the lowest is 12°C and corresponds to June and July. December, January and February are summer months; March, April and May correspond to the fall season; June, July and August are winter months and September, October and November correspond to the spring season.

The average annual rainfall is approximately 989 mm for the period 1904–2009. The maximum amount of annual rainfall was 1710 mm (1914) and the minimum 423 mm (1906). It is important to note that a gradual increase of the annual precipitation has been taking place from 1960 in the region. Up to that year, there had been an

Table 12.2 Results of water balance.

Period	*P (mm)*	*ETa (mm)*	*EXC (mm)*	*DEF (mm)*
2000–2001	1,060	1,086	0	120
2001–2002	1,142	1,009	48	161
2002–2003	1,619	1,208	374	0
2003–2004	969	1,132	17	91
2004–2005	1,465	1,079	243	143
2005–2006	1,181	1,128	102	43
Average	1,239	1,107	131	93

alternation of dry and wet periods lasting about 3 years each, but then these periods became longer.

The highest average monthly precipitation occurs during summer and autumn (100–140 mm). Winter is the dry season with less than 35 mm precipitation. The subtropical Atlantic Ocean anticyclone originates the principal air masses that produce precipitation in the region during the summer and autumn seasons. Nevertheless, heavy rains take place in winter and in the beginning of spring, as a result of convective storms produced by polar air masses coming from the Atlantic Ocean.

A monthly water balance for the root zone was done in Esperanza for 2000–2006 using the method proposed by Thornthwaite & Mather (1955) (D'Elia *et al.*, 2007). They estimated the potential evapotranspiration and considering the precipitation (*P*) and the soil moisture storage capacity, they calculated the actual evapotranspiration (*Eta*), the water surplus (*EXC*) and the water deficit (*DEF*). The water surplus includes both surface runoff and deep drainage. The latter is considered as potential groundwater recharge.

The results of this balance are listed in annual terms in Table 12.2.

The annual precipitations for the period analyzed were higher than the annual average precipitation of the study area (989 mm), except for 2003–2004. Water excesses were observed in practically the entire period (except for 2000–2001) which suggests that groundwater recharge could have taken place.

The highest amount of precipitation (1619 mm) corresponds to the 2002–2003 hydrological year, in which an extraordinary flood event of the Salado River took place in the region. The highest amount of water surplus (374 mm) and consequently the highest possibility to recharge the unconfined aquifer correspond to this period (D'Elia *et al.*, 2007).

The soil type most widely present in the study area is *Argiudol*. It is a deep and well drained soil principally formed by silts and found in the higher parts of landscape and water divides. In very plain areas, the clay content of the B_{2t} horizon increases and affects the drainage.

12.2.4 Human and ecosystems groundwater requirements

All the human activities developed in the study area are based on groundwater. In the rural areas, groundwater comes from both the unconfined and semiconfined aquifers, but the groundwater extractions have not been quantified yet. The most important activity in these areas is livestock, principally milk production. Soybean, corn and sunflower are the main crops in the region and cover more than 13 000 ha. Industries are related to food, milk, leather and metal production. The water utility of Esperanza and Rafaela cities is Aguas Santafesinas (ASSA). It operates a well-field composed of 30 wells that pump water from the semi-confined aquifer to both cities. The well-field is located in a rural area to the West of Esperanza. ASSA reported that the groundwater extraction was about 13.2 Mm^3/year in 2006. It is important to emphasise that this amount of water does not take into account the groundwater extraction for industry and irrigation uses.

Ecosystems of the study area are well known. Most of these ecosystems belong to different classes of dependency of groundwater (Sinclair, 2001).

The terrestrial vegetation consists of *Celtis tala, Acacia caven, Geoffrorea decoricans, Prosopis nigra, Aspidosperma quebracho-blanco; Sambucus australis, Cestrum parqui, Aloysia gratissima Steinchisma laxa, Nassella hyalina, Setaria verticillata, Blumenbachia insignis, Bowlesia incana,* and *Veronica persica* among others (Pensiero *et al.*, 2005). Some species like *Schoenoplectus*, *Zizaniopsis* and *Typha* are present in lagoons and wetlands. Terrestrial fauna principally includes little mammals like *Lycalopex gymnocercus, Conepatus chinga, Lagostomus maximus*, *Myocastor coypus*, *Galictis cuja*, *Chetophractus villosus*, *Dasypus hybridus*, *Lepus europaeus*; rodents of the *Caviidae* and *Ctenomyidae* families; snakes of the *Colubridae* family, among others. *Prochilodus lineatus, Pseudoplatystoma, Luciopimelodus pati* are species of fish that constitute the aquatic fauna of rivers and lagoons. *Phalacrocorax olivaceus, Ardea alba, Turdus rufiventris* and *Eudromia elegans* are some of the most common species of birds of the study area.

Some flora and fauna species of the study area are shown in Figure 12.3.

12.3 GROUNDWATER RECHARGE ESTIMATION

12.3.1 Available data and methodology

Daily amounts of precipitation of Esperanza for 2000–2006 are available from the *Dirección de Comunicaciones* of the Santa Fe province.

Groundwater levels were measured in six monitoring wells located in the study area, at the outermost limits of the water utility well-field that pumps water from the semi-confined aquifer (Fig. 12.1). PM2, PM3, PM4 and PM6 have a depth of 45 m (semi-confined aquifer) and PM1 and PM5 are 15 m deep (unconfined aquifer). The measurements were done weekly during four years (08/2002 to 08/2006) in the PM5 and PM6 monitoring wells. Additionally, water levels were measured in all the monitoring wells during several field surveys (Table 12.3) and groundwater samples were collected in April 2003, December 2003 and May 2005. The results of the chemical analyses of these groundwater samples, done by the water utility laboratory, are shown in Table 12.4.

In order to measure chloride concentrations, rainfall water samples were monthly collected in Santa Fe station from January 2000 to July 2006. This station is located in the Campus of the Facultad de Ingeniería y Ciencias Hídricas – Universidad Nacional

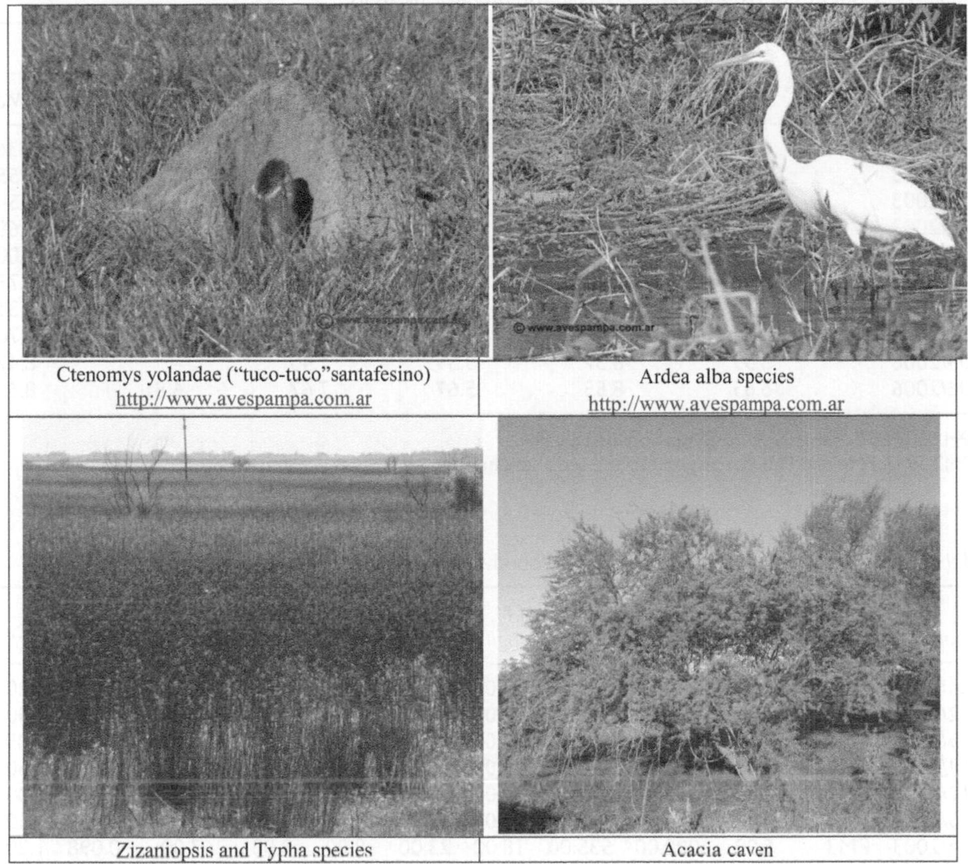

Ctenomys yolandae ("tuco-tuco"santafesino)
http://www.avespampa.com.ar
Ardea alba species
http://www.avespampa.com.ar
Zizaniopsis and Typha species
Acacia caven

Figure 12.3 Flora and fauna species of the study area.

del Litoral (FICH-UNL), Santa Fe, Argentina, about 40 km to the east of the study area. Chloride concentrations of rainfall were determined using the ASTM Tests for Chloride Ion in Water and Wastewater [D 512-67 1974], Method C; in the *Unidad de Vinculación Tecnológica* Laboratory with an analytical error of ±0.05 mg/l (Table 12.5).

The water table fluctuation method was used to estimate the local groundwater recharge of the unconfined aquifer. This method is based on the assumption that the increase of the unconfined aquifer water level is due to recharge water arriving at the water table (Healy & Cook, 2002). Thus, groundwater recharge may be calculated as:

$$R = S_y \frac{dh}{dt} \approx S_y \frac{\Delta h}{\Delta t} \tag{12.1}$$

where S_y is specific yield, h is water level height and t is time. Δh is equal to the difference between the peak of the rise of the water table and the low point of the

Table 12.3 Depth of groundwater levels in the monitoring wells.

	Depth of groundwater level (m)					
Date\Well	*PM1*	*PM5*	*PM2*	*PM3*	*PM4*	*PM6*
09/2002	7.85	9.27	7.92	8.79	7.49	9.36
04/2003	6.87	7.50	7.13	7.08	6.13	7.55
09/2003	5.67	6.43	5.71	6.16	4.63	6.54
10/2003	5.65	6.56	5.76	6.34	4.73	6.68
12/2003	5.95	6.89	5.97	6.76	5.04	7.00
04/2004	6.60	7.62	6.73	7.65	6.03	7.73
09/2004	6.83	8.19	6.88	7.99	6.85	8.31
05/2005	4.54	7.00	4.62	5.90	4.66	7.70
04/2006	5.59	8.37	5.59	7.40	–	8.47
06/2006	5.65	8.55	5.67	7.67	6.81	8.65

PM1 and PM5 correspond to the unconfined aquifer
PM2, PM3, PM4 and PM6 correspond to the semi-confined aquifer

Table 12.4 Results of chemical analysis of groundwater samples.

Date	*Well*	*pH*	*EC* *(μS/cm)*	*TA* *mg/l*	*Cl⁻* *mg/l*	*SO₄⁼* *mg/l*	*Ca⁺⁺* *mg/l*	*Mg⁺⁺* *mg/l*	*Na⁺⁺* *mg/l*	*As* *mg/l*	*F⁻* *mg/l*
04/2003	PM1 (*)	7.90	1190.00	645.00	25.00	46.00	9.00	3.00	305.00	0.078	1.20
12/2003	PM1 (*)	7.90	1230.00	675.00	21.00	43.00	10.00	4.00	317.00	0.060	1.40
05/2005	PM1 (*)	7.80	1080.00	566.00	21.00	23.00	16.00	7.00	231.00	0.073	1.85
04/2003	PM2	7.90	1010.00	545.00	26.00	16.00	15.00	5.00	246.00	0.069	1.50
12/2003	PM2	7.80	1080.00	575.00	30.00	39.00	14.00	6.00	264.00	0.054	2.00
05/2005	PM2	8.00	1250.00	648.00	24.00	41.00	14.00	2.00	296.00	0.089	1.50
04/2003	PM3	8.00	1000.00	535.00	18.00	23.00	8.00	3.00	267.00	0.098	1.20
12/2003	PM3	8.00	1010.00	560.00	18.00	33.00	9.00	3.00	254.00	0.104	1.20
05/2005	PM3	8.00	1020.00	536.00	19.00	30.00	7.00	4.00	237.00	0.115	1.25
04/2003	PM4	7.80	880.00	545.00	20.00	15.00	25.00	10.00	192.00	0.036	1.00
12/2003	PM4	7.70	924.00	505.00	24.00	25.00	24.00	11.00	186.00	0.031	1.00
05/2005	PM4	7.70	918.00	477.00	20.00	20.00	25.00	11.00	170.00	0.041	1.10
04/2003	PM5 (*)	7.60	600.00	355.00	6.00	34.00	66.00	28.00	21.00	0.025	0.60
12/2003	PM5 (*)	7.30	629.00	370.00	3.00	7.00	62.00	28.00	21.00	<0.010	0.65
05/2005	PM5 (*)	7.40	635.00	350.00	4.00	7.00	63.00	29.00	19.00	0.010	0.75
04/2003	PM6	7.40	1130.00	585.00	25.00	20.00	30.00	13.00	256.00	<0.010	0.80
12/2003	PM6	7.50	1310.00	595.00	33.00	23.00	39.00	19.00	254.00	0.020	0.90
05/2005	PM6	7.50	1350.00	556.00	51.00	25.00	43.00	22.00	231.00	0.030	0.80

EC: Electrical Conductivity; TA: Total Alkalinity; Cl^-: Chloride; $SO_4^=$: Sulfate; NO_3^-: Nitrate; Ca^{++}: Calcium; Mg^{++}: Magnesium; Na^{++}: Sodium; As: Arsenic; F^-: Fluoride
PM1 and PM5 correspond to the unconfined aquifer
PM2, PM3, PM4 and PM6 correspond to the semi-confined aquifer

extrapolated antecedent recession curve at the time of the peak. To estimate Δh a decreasing exponential equation was considered as a recession curve:

$$h = h_0 e^{-\alpha t} \tag{12.2}$$

Table 12.5 Chloride concentrations of rainfall.

	Chloride concentrations of rainfall (mg/l)						
Month\Year	*2000*	*2001*	*2002*	*2003*	*2004*	*2005*	*2006*
January	3.88	0.19	0.53	0.48	0.41	0.16	0.15
February	0.79	0.21	1.65	ND	0.65	0.03	0.21
March	1.36	0.51	0.54	0.58	0.49	0.19	0.54
April	1.05	0.92	0.33	ND	0.23	0.22	1.45
May	0.61	1.71	0.02	0.36	3.54	ND	ND
June	ND	0.26	ND	0.38	0.72	0.22	1.15
July	1.13	ND	0.44	ND	1.61	ND	4.52
August	0.57	0.52	1.46	0.48	ND	ND	ND
September	0.56	0.50	0.79	1.11	2.09	ND	ND
October	0.48	0.59	0.08	0.50	0.77	0.31	ND
November	0.09	0.24	0.08	0.80	0.43	0.04	ND
December	0.57	0.48	0.17	0.24	0.33	0.04	ND

ND: No data are available

where h is the recession level, h_0 is the initial level, t is time and α is a coefficient depending on hydrological and geometric characteristics of groundwater reservoir. For this work, this coefficient is assumed to be:

$$\alpha = \pi^2 \frac{T}{4} S_y L^2 \tag{12.3}$$

where T is transmissivity, S_y is specific yield (effective porosity in unconfined aquifers) and L is the distance between the point of measurement and the discharge level in the groundwater flow direction (Custodio & Llamas, 1983). A specific yield of 0.05 and a transmissivity of 150 m^2/d – estimated for the 'Pampa Formation in previous studies – were considered. A distance of about 17 000 m between the PM5 monitoring and the Salado River – estimated as the discharge level of groundwater in the flow direction – was also taken into account. A weekly time step was used to evaluate groundwater recharge and then the results were monthly integrated.

The Curve Number (CN) method of the U.S. Soil Conservation Service (U.S. Soil Conservation Service, 1964) was used to estimate the surface runoff. This method considers:

$$\frac{(P - Ia) - Q}{S} = \frac{Q}{(P - Ia)} \tag{12.4}$$

where P is rainfall, Ia is the initial abstraction, Q is surface runoff and S is potential maximum retention. That is to say the ratio of actual retention in a basin and potential maximum retention equals the ratio of runoff and potential runoff. If considering $Ia = 0.2S$, as practical experience has shown, it is possible to write:

$$Q = \frac{(P - 0.2S)^2}{(P + 0.8S)} \tag{12.5}$$

An empirical relationship between the maximum retention (in inches) and the runoff curve number (CN) was found:

$$CN = \frac{1000}{10 + S} \tag{12.6}$$

Potential maximum retention will be null when CN equals 100 (impermeable areas) and it will be infinity when CN equals 0 (no surface runoff occurs). Hence, an estimation of CN has to be made in order to calculate the runoff. CN is defined as a function of the basin characteristics such as hydrological soil condition, land use and the antecedent moisture condition when rainfall occurs. According to their ability to transmit water, four hydrological soil groups are classified: Group A has a high water transmission rate, whereas this rate in Group B is moderate, in Group C low and in Group D very low. On the other hand, taking into account the five-day total antecedent rainfall for the dormant and growing seasons, three antecedent soil moisture condition classes are distinguished.

According to its ability to transmit water, the principal soil of the study area was classified as B – moderate rate of water transmission. Considering this characteristic of the soil together with the land use cover, the low slope of the area and the antecedent moisture condition of the soil – in relation to the daily precipitation – CN values and consequently daily surface runoff were estimated. These results were then integrated in monthly and annual terms and the average annual surface runoff was calculated.

A chloride mass balance estimate was performed to estimate both the local recharge to the unconfined aquifer and the local and regional recharge to the semi-confined aquifer.

Recharge to the unconfined aquifer was calculated using the chloride mass balance with a zero-dimensional mixing cell model (Cook & Herczeg, 1998). It is considered that the mass into the system (precipitation, P) times the chloride concentration in P (C_p) is balanced by the mass out the system (drainage, D) times the chloride concentration in drainage water in the unsaturated zone (Scanlon *et al.*, 2002). Drainage is assumed to be recharge to aquifer (R) and surface runoff (SR), the chloride concentration in recharge is C_r and the chloride concentration in surface runoff is C_{sr}. Thus:

$$P \times C_p = R \times C_r + SR \times C_{sr} \tag{12.7}$$

And recharge to aquifer may be calculated as:

$$R = \frac{P \times C_p - SR \times C_{sr}}{C_r} \tag{12.8}$$

An analysis of the chloride concentration in rainfall was done and the average chloride concentration of rainfall (C_p) was calculated for the period analysed. The average chloride concentration of runoff (C_{sr}) was assumed to be the same as C_p and the average of the groundwater chloride concentration was estimated both for the unconfined and semi-confined aquifers as the average of the chloride concentrations determined to groundwater samples collected.

Finally, a one-dimensional mixing cell model was used to preliminarily estimate the regional recharge to the semi-confined aquifer between two sections along the groundwater flow (in accordance with the PM1-PM2 monitoring wells at the West and PM5-PM6 monitoring wells at the East). A homogeneous and isotropic media is assumed. Thus:

$$R_2 \times C_2 = R_1 \times C_1 + R_0 \times C_0 \tag{12.9}$$

$$R_2 = R_1 + R_0 \tag{12.10}$$

where R_1 is the incoming regional flow through the upstream section and R_2 is the outgoing regional flow through the downstream section in the groundwater flow direction. R_0 is the downward vertical flow through the aquitard layer. C_1 and C_2 are the chloride concentrations of groundwater in the two sections and C_0 is the chloride concentration of groundwater coming from the aquitard layer (equal to C_r obtained for the unconfined aquifer layer).

12.3.2 Results

The analysis of the groundwater levels in the study area indicates that the variability of the water table is similar to the semi-confined piezometric level. Both levels ranged from 4 m to more than 9 m of depth. It is also observed that there were two cycles of about two years in which groundwater levels increased approximately 3 m during the first (2002–2004), and 2 during the second cycle (2004–2006). Figure 12.4 shows monthly

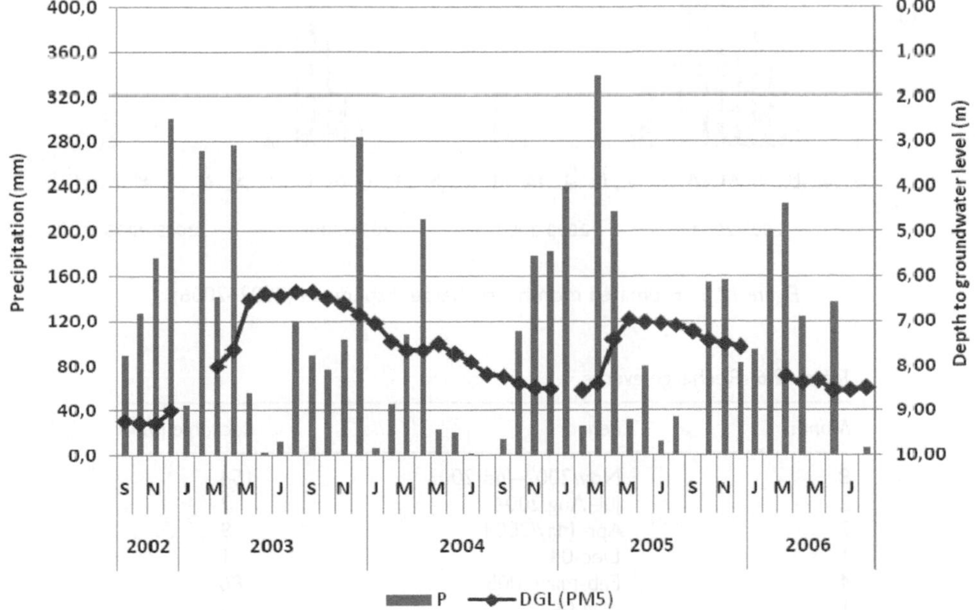

Figure 12.4 Depth to groundwater level in PM5 and monthly precipitation. Esperanza (2002–2006).

groundwater level fluctuations of the PM5 monitoring well in relation to monthly rainfall of Esperanza. Monthly groundwater recharge resulting from the water table fluctuation method is shown in Figure 12.5.

The most important events of recharge took place during summer and autumn. The amounts of recharge of each event and the annual recharge amounts are presented in Table 12.6 and Table 12.7 respectively.

For the period 2002–2006, the annual recharge to the unconfined aquifer ranged from 5 to 160 mm with a mean value of 65 mm. It is important to note that an extreme humid period and an extraordinary flood of the Salado River happened during October 2002 to April 2003.

The daily surface runoff estimated by the CN method was integrated in monthly and annual terms, and the average annual surface runoff resulted in 170 mm.

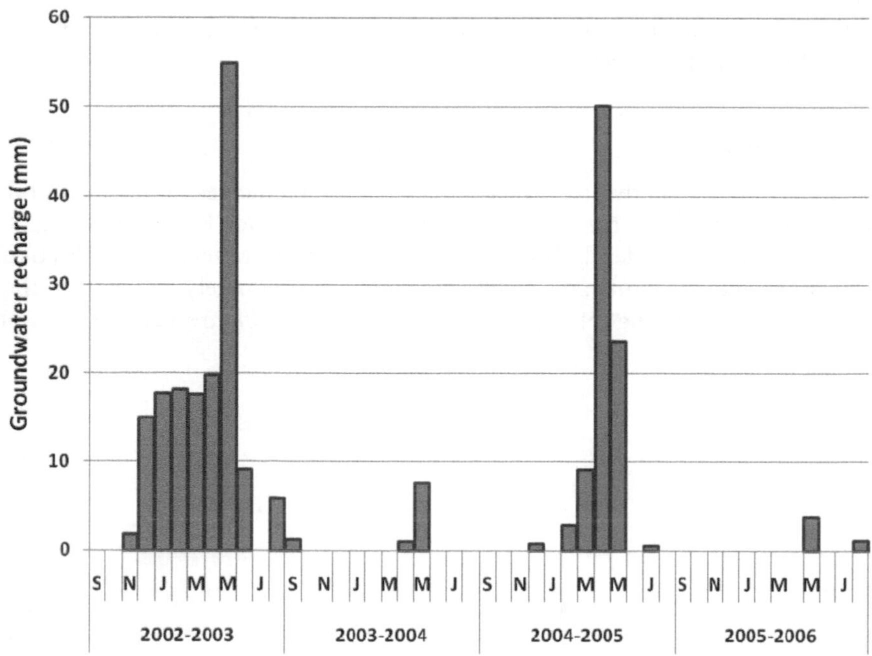

Figure 12.5 Estimated monthly recharge. Esperanza (2002–2006).

Table 12.6 Recharge events.

Month	*Period*	*Recharge (mm)*
8	Nov/2002–Jun/2003	154
2	Jul–Aug/2004	7
2	Apr–May/2004	9
1	Dec-04	1
4	Feb-May/2005	86
1	May-06	4
2	Aug/2006	1

As was indicated in previous studies, groundwater is predominantly of a sodium bicarbonate water type, (except for PM5 in which groundwater is of a calcium bicarbonate type). Chloride contents in groundwater varied from 3 to 51 mg/l.

The annual mean distribution of rainfall and the monthly average chloride concentrations in precipitation are presented in Figure 12.6. It is observed that the lower the precipitation the higher the chloride concentrations in precipitation.

The mean monthly chloride concentration in rainfall (C_p) for the period 2000–2006 was 0.73 mg/l.

The average chloride concentration of the unconfined aquifer ($C_r = C_0$) was 13.3 mg/l. In the two investigated sections of the semi-confined aquifer, the average of the groundwater chloride concentration were $C_1 = 27$ mg/l and $C_2 = 36$ mg/l. Average annual precipitation was 1309 mm and average annual surface runoff 170 mm for the 2002–2006 period. The amount of water recharging the lower aquifer through the aquitard layer (R_0) was estimated to be 18 mm, taking into account the thickness and the specific yield of this layer. The application of the chloride mass balance shows that

Table 12.7 Annual recharge.

Time	*Annual rainfall (mm)*	*Annual recharge (mm)*	*Annual recharge (Mm³)*	*% Pa*
2002–2003	1619	160	80.0	9.9
2003–2004	969	10	5.0	1.0
2004–2005	1465	87	43.5	5.9
2005–2006	1181	5	2.5	0.4
Average	1309	65	32.5	4.3

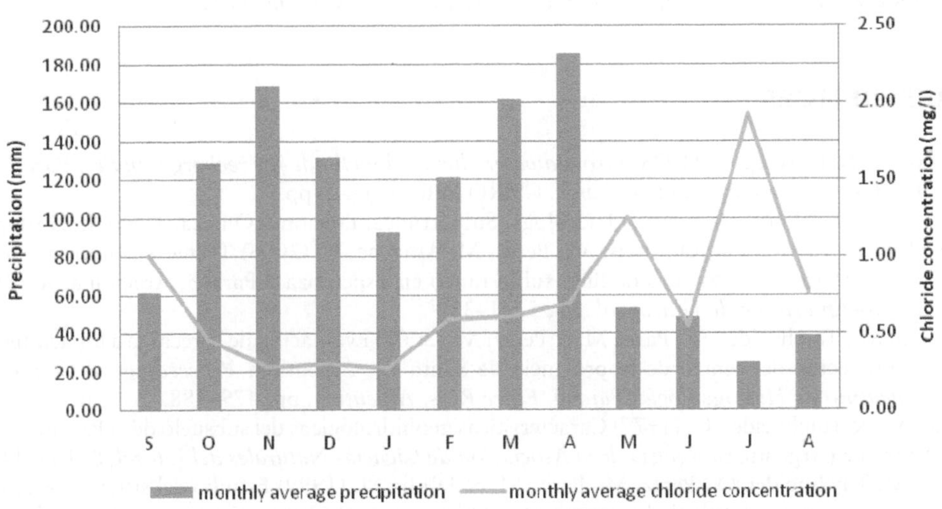

Figure 12.6 Annual mean distribution of rainfall and monthly average chloride concentrations in precipitation. Santa Fe (2000–2006).

recharge to the unconfined aquifer R was 63 mm/year. Besides, in relation to the semi-confined aquifer the incoming regional flow (regional recharge) R_1 was 41 mm/year and the outgoing regional flow R_2 was 59 mm/year.

12.4 CONCLUDING REMARKS

This is the first study that analyses and quantifies groundwater recharge in the investigated region.

Very similar groundwater recharge estimates for the unconfined aquifer were obtained using two widely applied approaches: the water table fluctuation method (65 mm/year) and a chloride mass balance method (63 mm/year). Recharge to the semi-confined aquifer was also estimated by the last method, both at local (18 mm/year) and regional scales (41 mm/year).

The most important events of recharge of the unconfined aquifer took place during summer and autumn (in accordance with the rainy seasons) and the amount of these events varied from 1 to more than 150 mm (2.5 to 77 Mm^3), due to the alternation of dry and wet periods and the Salado River flood.

The groundwater system was affected by the extraordinary flood event of the Salado River in 2003 and caused great economical losses in productive activities and a disturbance of the ecosystems of the area. Losses in productive activities added up to millions of dollars, although the disturbance of the ecosystems was not quantified. Dry periods also affect groundwater systems and ecosystems.

The results of this work contribute to the knowledge of the groundwater system and provide quantitative criteria on which to base management and protection of ecosystems related to groundwater. But, other factors such as urban development, agricultural land drainage, soil erosion and groundwater quality deterioration must be taken into account in the interest of the ecosystem sustainability and more efforts are needed to make this compatible with the other groundwater uses.

REFERENCES

Cook, P. & Herczeg, L. (1998) *Groundwater chemical methods for recharge studies. Part 2 of the Basics of recharge and discharge.* CSIRO Publishing. 24pp.

Custodio, E. Llamas, M. (1983) Hidrología Subterránea. Ediciones Omega. Tomo, I. 2350pp.

D'Elia M., Tujchneider, O., Paris, M., Perez, M. Aravena, R. (2006) Técnicas isotópicas en la caracterización de sistemas de flujo subterráneo en Esperanza y Paraná, Argentina. *Revista Latino-Americana de Hidrogeología*, 5, 31–38.

D'Elia M., Tujchneider, O., Paris, M. & Perez, M. (2007) Evaluación de la recarga a los acuíferos en un sector del centro de la provincia de Santa Fe, Argentina. *Proceedings V Congreso Argentino de Hidrogeología. Paraná, Entre Ríos, Argentina*. pp. 479–488.

Fili, M. & Tujchneider, O. (1977) Características geohidrológicas del subsuelo de la Provincia de Santa Fe (Argentina). *Revista de la Asociación de Ciencias Naturales del Litoral*, 8, 105–113.

Fili, M.,Tujchneider, O., Perez, M., Paris, M. & D'Elia M. (1999) Estudio del sistema de aguas subterráneas en el área de Esperanza-Humboldt y zona de influencia. Servicio Especializado de Asistencia Técnica. Convenio Aguas Provinciales de Santa Fe y Universidad Nacional del Litoral. Informe Final.

Healy, R. & Cook, P. (2002) Using groundwater levels to estimate recharge. *Hydrogeology Journal*, 10 (1), 91–109.

Nicolli, H., Tujchneider, O., Paris, M., Blanco, M. & Barros, A. (2009) Movilidad del arsénico y oligoelementos asociados en aguas subterráneas del centro – norte de la provincia de Santa Fe, Argentina. VI Congreso Argentino de Hidrogeología. Seminario: Presencia de flúor y arsénico en aguas subterráneas. Santa Rosa, La Pampa. Argentina. Proceedings. pp. 81–90.

Paris, M. (2010) Métodos estadísticos multivariados aplicados en Hidrologia Subterránea. Ph.D. Thesis, Doctorado en Ciencias Geológicas-Facultad de Ciencias Exáctas, Fisico-Químicas y Naturales, Universidad Nacional de Río Cuarto. 173 pp. Unpublished.

Pensiero, J.F., Gutiérrez, H.F., Luchetti, A.M., Exner, E., Kern, V., Brinch, E., Oakley, L., Prado, D. & Lewis, J.P (2005) Flora vasclar de la provincia de Santa Fe. Claves para el reconocimiento de las familias y géneros. *Catálogo sistemático de las especies*. 1st Edition. Universidad Nacional del Litoral.

Scanlon, B., Healy, R. & Cook, P. (2002) Choosing appropriate techniques to quantifying groundwater recharge. *Hydrogeology Journal*, 10 (1), 18–39.

Sinclair, K.M. (2001) *Environmental Water Requirements to Maintain Groundwater Dependent Ecosystems*. Environmental Flows Initiative Technical Report. Report number 2. 122pp.

Thornthwaite, C. & Mather, J. (1955) The water balance. *Publications in Climatology*. Vol. VIII Number 1. Drexel Institute of Technology. Laboratory of Climatology. Centerton, New Jersey. 104pp.

Tujchneider, O., Paris, M., Fili, M., D'Elía M. & Perez, M. (1998) Protección de aguas subterráneas. Caso de estudio: ciudad de Esperanza (República Argentina). Primera fase: Diagnóstico del sistema. IV Congreso Latinoamericano de Hidrología Subterránea., 2, 805–821. Montevideo, Uruguay.

Tujchneider, O., Perez, M., Paris, M., & D'Elia, M. (2005) Deterioro de fuentes de agua subterránea por ascenso de agua salada. IV Congreso Argentino de Hidrogeología y II Seminario Hispano – Latinoamericano sobre temas actuales de la Hidrología Subterránea. Río Cuarto, Córdoba, Argentina. Tomo II. pp. 217–226

Tujchneider, O., Paris, M., Perez, M. & D'Elia, M. (2006) Singularidad constitutiva de sistemas geohidrológicos de llanura y la gestión de los recursos hídricos subterráneos. *Revista Latino-Americana de Hidrogeología*, 5, 117–121.

U.S. Soil Conservation Service (2006) *National Engineering Handbook (NEH), Section, 4, Hydrology. Part I: Watershed Planning*. Washington, DC, US Department of Agriculture.

Chapter 13

Nutrient sources for green macroalgae in the Ria Formosa lagoon – assessing the role of groundwater

Tibor Y. Stigter[1], Amélia Carvalho Dill[2], Erik-jan Malta[3] & Rui Santos[3]

[1]Geo-Systems Centre/CVRM – Instituto Superior Técnico, Lisbon, Portugal
[2]Geo-Systems Centre/CVRM – University of the Algarve, Faro, Portugal
[3]ALGAE – Marine Plant Ecology Research Group, CCMAR, University of the Algarve, Faro, Portugal

ABSTRACT

The role of nitrate-contaminated groundwater discharge in the appearance of algal blooms in the Ria Formosa lagoon, south Portugal, is assessed by setting up a regional water balance for its drainage basin and determining annual N loads on groundwater. Of the estimated 80.8 million m^3 (hm^3) of natural recharge, 34% is consumed for irrigation, mainly in the western sector near Faro, largely reducing coastal groundwater discharge (CGD) in that area. In the east, since 2001 irrigation is performed with surface water, which has lead to an average annual increase in CGD by 24 hm^3, 3.5 hm^3 of which is artificial recharge. It is not known what part of CGD enters the lagoon, but estimates point towards 40–50%. Geophysical surveys are carried out to study groundwater seepage along geological faults. Total N load on the drainage basin of the lagoon is calculated in the order of 570 ton/yr, but due to restricted outflow near Faro and deep circulation, 300 ton/yr is estimated to enter the lagoon. Groundwater discharge may be important for triggering winter algae blooms on the mudflats, with little access to ocean water, as well as for the summer algae, which seemingly depend on the Ria Formosa as a nutrient source.

13.1 INTRODUCTION

Human activities, particularly those related to agricultural practices, have highly increased the nutrient load on groundwater and caused its contamination in many areas of the Algarve, in the south of Portugal. Currently, the highest nitrate concentrations are observed in the drainage basin of the Ria Formosa (Fig. 13.1), a mesotidal lagoon that is recognised both on a national and an international level as an extremely valuable and sensitive ecosystem. Every winter, blooms of green ulvoid algae develop in the intertidal zone, while in summer ulvoid blooms develop as well on the sandy beaches of the adjacent coastal zone. Winter species in the Ria Formosa are mainly *U. prolifera* forming thick mats on the mud flats, whereas summer species are mainly free-floating *U. rigida* and *U. rotundata*. A further increase of the nutrient load on the lagoon may cause denser populations of algal blooms and the development of new

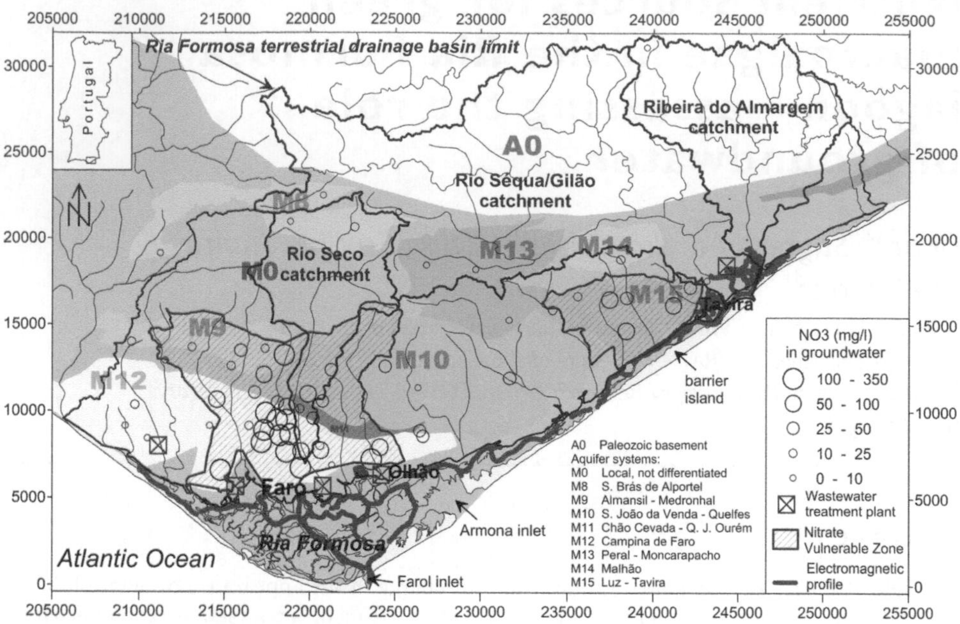

Figure 13.1 Location of the Ria Formosa lagoon, its drainage basin and the three major stream catchments, defined aquifer systems, nitrate concentrations in groundwater, the designated Nitrate Vulnerable Zones, waste water treatment plants and electromagnetic profiles.

bloom sites, with potentially serious impacts on biodiversity, seagrass meadows, shellfish and fish populations and tourism development. Groundwater flow may form an important contribution to the nutrient load of the lagoon, as monitoring in the area provides evidence that well-defined nitrate contaminant plumes are moving towards the lagoon (Stigter *et al.*, 2007; Stigter, 2012).

In 2004 a scientific research project was initiated that seeks to identify the species-specific nitrogen (N) metabolism of the blooms both within and outside the lagoon and to relate them with the N mass balance between the lagoon and the adjacent terrestrial and coastal zones. A fundamental task in this project was the determination and quantification of the main sources of nutrients in the Ria Formosa, particularly N, as this element is considered to limit primary production in most shallow ecosystems and thus to be one of the driving forces in ulvoid growth rates (Howarth and Marino, 2006). Figure 13.2 shows all the known sources and sinks of N for the lagoon. The input of N (I1+I2) from coastal groundwater discharge (CGD) is difficult to assess, as several discharge mechanisms are known to exist: (G1) 'diffuse' outflow along the fresh/saltwater interface near the coast; (G2) 'preferential' outflow along geological faults that form water conduits and (G3) deep groundwater circulation that may extend beyond the limits of the Ria Formosa, leading to submarine groundwater discharge (SGD). In a study for the northern end of the Wadden Sea in Denmark, Andersen *et al.* (2007) concluded that the most significant freshwater discharge occurs in distinct zones near the high tide line.

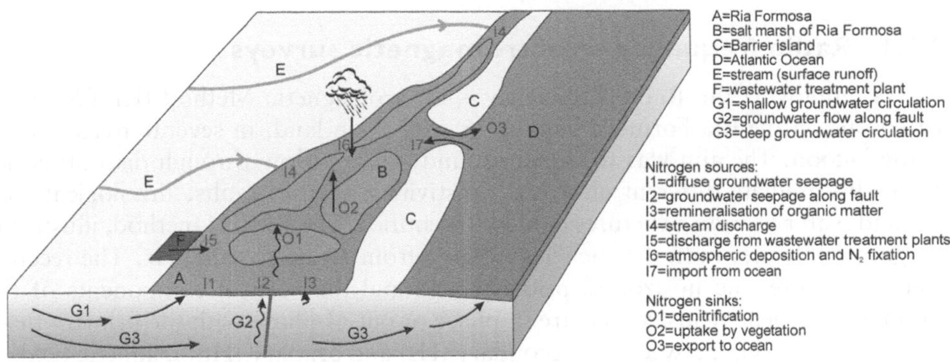

Figure 13.2 Schematic diagram (unscaled) of sinks and sources for nitrogen in the Ria Formosa.

13.2 STUDY AREA

The drainage basin of the Ria Formosa lagoon, located in the south of Portugal, is characterised by a Mediterranean climate with hot dry summers and mild rainy winters. Mean annual air temperature recorded in Faro is 17.3°C (Silva, 1988), whereas mean yearly precipitation in the basin is 664 mm (Nicolau, 2002). The total terrestrial drainage basin area is 741 km^2, with the two most important sub-basins of the rivers Rio Séqua/Gilão and Ribeira do Almargem (Fig. 13.1) which together cover 43% of the area and account for approximately 80% of total runoff. These two rivers have an intermittent behaviour, whereas all other streams are ephemeral. The hydrogeology is characterised by a Palaeozoic age basement of schists and graywackes that crop out in the north, with extremely low permeability (Fig. 13.1). The main aquifers are formed of karstified limestones and dolomites (with high secondary porosity) and fine- to course-grade sands.

Land use is characterised by intensive irrigated citrus and horticulture in two areas of the basin that have been designated Nitrate Vulnerable Zones (Fig. 13.1), in compliance with the Nitrates Directive (91/676/EEC). Nitrate concentrations in groundwater of these areas are well above the guideline value for drinking water (50 mg-NO_3/l) and locally can exceed 300 mg/l, mainly as a result of leaching of mineral fertilisers, irrigation and, to a lesser extent, losses from leaky septic tanks.

The Ria Formosa is a mesotidal lagoon, extending for approximately 55 km along the south coast of Portugal. Five sand barrier islands and two peninsulas separate the lagoon from the Atlantic Ocean and six inlets provide the water exchange with the ocean. The average water depth is less than 2 m; tidal height varies between 3.7 m (maximum spring tides) and 0.4 m (minimum neap tides). This results in the flushing of a large part of the water volume, thereby imposing an intense exchange of dissolved nutrients and particulate matter between the lagoon and the adjacent coastal waters. The Ria Formosa receives secondarily treated sewage inputs from three major cities bordering the lagoon, Faro, Olhão and Tavira, representing an important source of dissolved inorganic nutrients and organic matter.

13.3 METHODOLOGY

13.3.1 Radio frequency– electromagnetic surveys

Surveys employing the Radio Frequency – Electromagnetic Method (RF-EM) were carried out on the Ria Formosa lagoon, as well as on land, in several areas bordering the lagoon. The aim was to detect groundwater outflow through discontinuities and to discriminate different electrical resistivities, so that faults, lithological contacts and water-bearing structures could be identified. The RF-EM method, illustrated in Figure 13.3, uses radio frequencies ranging from 12 up to 300 kHz. The receiver antenna captures the horizontal primary field and the vertical components of the secondary magnetic fields, which are in-phase or out-of-phase with the primary field.

The relationship between the secondary (Hs) and primary (Hp) magnetic fields is studied as a percentage-expressed Hs/Hp ratio. The investigation depth (P) is a function of the resistivity (Rho) of the strata and the radio frequency (F, in Hertz) used:

$$P = 503\sqrt{\frac{Rho_{ap}}{F}} \tag{13.1}$$

The equipment has been designed for fast and extensive mapping of geological contacts by combining a Data logger, which registers every two seconds and a Global Positioning System. The field data are georeferenced, transformed into 3D data profiles and coupled with all the available information, by means of GIS format software. The direction of the profiles should be as much as possible perpendicular to the structure strike (Fig. 13.3). This is often difficult to achieve (Carvalho Dill *et al.*, 2009), due to the fact that geological contacts, faults, dykes and veins, are often deformed, folded or faulted, hindering the quantitative analysis of field data.

This method was used for the first time on a coastal lagoon. Previous work onshore on the peninsula of Tróia (south of Setúbal) had already shown, despite the attenuation effect of the salt water, that valuable information could be obtained with electromagnetic methods (Carvalho Dill *et al.*, 2009). Structures like fractures and faults are

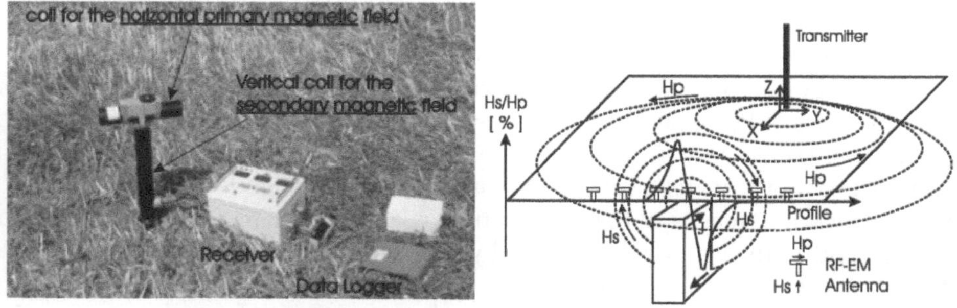

Figure 13.3 (*left*) Equipment used for the Radio Frequency-Electromagnetic survey; (*right*) principle of the RF-EM method, adapted from Turberg and Müller (1992).

revealed by the presence of freshwater, which circulates along them, creating significant contrasts with salt water. 104 km of profiles were performed on land (28 km) with a motorised vehicle and within the lagoon (76 km) using the manual antenna mounted on a boat. The direction of the profiles was conditioned by the navigability of the channels, which was tide-dependent. The antenna was put as far as possible from the outboard engine, in order to minimise interference, as the equipment could not be isolated. Electrical interference from the motor was detected occasionally, particularly whenever the velocity increased. Nevertheless it was possible to attained quite good measuring conditions.

13.3.2 Quantification of coastal groundwater and nitrogen discharge

In order to assess the nitrogen (N) transport towards the lagoon, the annual groundwater discharge from land needed to be quantified. A water balance approach was used, and for this purpose the groundwater catchment was assumed to have the same boundaries as the surface water drainage basin. In reality this is often not the case (e.g. Krause and Bronstert, 2005), due to geological heterogeneities and deep groundwater circulation. However, since the aim is to estimate total groundwater outflow from land, only a small error is involved, because the most significant recharge areas are located within the surface water drainage basin (see Fig. 13.1). Assuming there are no changes in storage, the average annual water balance of the Ria Formosa drainage basin can be considered to be:

$$G_o = P - ET - S_o - I_{ground,eff} + IRF_{surf} \tag{13.2}$$

where G_o is the groundwater outflow, P is precipitation, ET is evapotranspiration, S_o surface water outflow, $I_{ground,eff}$ effective groundwater irrigation, i.e. irrigation minus return flow and IRF_{surf} return flow from surface water irrigation in the eastern part of the basin, where it constitutes an additional recharge component since 2001 (Stigter *et al.*, 2006b). Average annual precipitation was obtained from Nicolau (2002), who used a kriging interpolator with external drift, using elevation as auxiliary variable, to map the spatial distribution of rainfall with a resolution of 1 km^2. Recharge ($P - ET - S_o$) was determined based on estimated infiltration rates for each outcropping lithology. These values are based on a review of recharge estimations performed in the Ria Formosa basin using: (i) the Kessler method in areas of carbonate rock outcrops; (ii) the semi-empirical formulae of Thornthwaite, Coutagne and Turc that calculate real evapotranspiration in areas of sedimentary deposits (Silva, 1984; Silva, 1988) and (iii) the chloride mass balance approach (Stigter *et al.*, 1998; De Bruin, 1999).

Average annual irrigation water requirements correspond to 800 mm/yr in areas of citriculture and horticulture (two crop cycles per year), based on data provided by the Instituto de Desenvolvimento Rural e Hidráulica (IDRHa) for the region. Irrigation return flow is estimated to be 15% for drip irrigation systems such as those used in the area (Beltrão, 1985; Keller and Bliesner, 2000). Keller and Bliesner (2000) refer that 'the unavoidable excess depth of applied water is at least 10% on all parts of an area that is sufficiently irrigated to meet evapotranspiration demands'. In addition, it is a common and recommended practice to irrigate in excess of crop water requirements, especially in arid and semi-arid environments, in order to control soil water salinity and avoid salt accumulation. The excess irrigation constitutes additional recharge in surface-water irrigated areas (Stigter *et al.*, 2006b).

The N load on groundwater mainly originates from fertilisation and domestic effluents (septic tanks). Rainfall has revealed to constitute a relevant additional source, through dry and wet deposition of ammonia and nitrate, the presence of which in the atmosphere is mainly caused by the referred anthropogenic factors. The mean annual N budget (N_g) for groundwater of the Ria Formosa drainage basin can be written as:

$$N_g = R_n \times N_{Rn} + R_{IRF} \times N_{IRF} + R_{effl} \times N_{effl} - G_o \times N_{Go} - N_{sink} \quad (13.3)$$

where, on a mean annual basis, Rn, R_{IRF} and R_{effl} are natural recharge, recharge from irrigation return flow (IRF) and effluents, respectively, whereas N_{Rn}, N_{IRF} and N_{effl} are leached N contents from natural recharge, IRF and effluents, respectively. G_o is groundwater outflow, part of which enters the lagoon and N_{Go} is the N concentration in groundwater outflow. N_{sink} refers to all existing N sinks in groundwater, mainly reduction to N_2 (denitrification).

N_{Rn}, N_{IRF} and N_{effl} are difficult to quantify separately, in part because natural recharge from rainfall leaches out all existing N sources, from rainwater, soil, fertilisers and domestic effluents. Some simplifications are required in order to successfully determine the groundwater N budget. Soil N is considered not to change on a mean annual basis, whereas denitrification is considered not to occur, as has been discussed by Stigter *et al.* (2006c). Losses from fertilisation mainly occur in the two designated nitrate vulnerable zones (Fig. 13.1), where action programmes have been implemented to attempt to reduce these losses (Stigter *et al.*, 2007; Stigter, 2012). Annual N requirements are 200–300 kg/ha for the vegetable crops and citrus trees grown in the area and for this study average N losses to the groundwater are estimated to be in the order of 15–20%, based on literature findings. Alva *et al.* (2006) indicate a 10–15% leaching of applied fertiliser-N for nitrogen best management practices in citrus orchards, using fertigation. Similar practices are gaining importance in the studied region, but losses are currently expected to be higher and will have been considerably higher in the past. Other studies indicate much higher leaching rates for citrus culture (e.g. Dasberg *et al.*, 1984; Paz and Ramos, 2004; Boman and Battikhi, 2006) and horticulture (e.g. He *et al.*, 2006), as high as 50%.

Leaching of domestic effluents from septic tanks is an important point-source of N in areas not connected to the sewerage network, where roughly 20% of the population lives (CCDR-Alg, 2007). This is clearly revealed by the microbiological contamination observed in many groundwater wells, which is mostly of domestic origin, as livestock farming is practically non-existent. The total N load from septic tanks was estimated on the basis of average population density in each of the municipalities located within the limits of the Ria Formosa basin, water use per capita, average N concentrations in wastewater ($\approx$70 mg/l) and N removal efficiency of the "septic tank + soil" system ($\approx$25%, Costa *et al.*, 2002).

Mean annual outflow volumes of groundwater and dissolved N were compared to those of stream discharge, wastewater treatment plants (WWTPs) and direct rainfall on the lagoon (respectively I4, I5 and I6 in Fig. 13.2), for the hydrological year 2005/2006, when rainfall was equal to the yearly average. The quantification of discharges from streams and WWTPs is presented by Stigter *et al.* (2006a). For atmospheric deposition rain water was also sampled on a few occasions, but since nutrient concentrations in rainwater depend highly on the intensity, duration and frequency of rain events, an estimate of average nutrient content, based on collected samples, past rainwater samples and minimum observed nutrient increases in surface water was used. Total annual nutrient input from precipitation was then simply calculated as the product of rain volume and concentration.

To estimate the exchange of nutrients between the Ria Formosa and the ocean (I7 and O3 in Fig. 13.2), sampling campaigns were set up during spring tide and neap tide once every two months during one year (March 2006 to March 2007). Samples were taken every 2 h over a 24 h tidal cycle from the Farol inlet (see location in Fig. 13.1), responsible for 45–50% of the total water exchange (Pacheco *et al.*, 2010). During the other months, samples were taken at daily high and low tide during spring tide and neap tide from the Farol as well as the Armona inlet, together representing more than 90% of the total water exchange, to detect possible spatial differences. Water temperature, pH and salinity were recorded. Samples were filtered over glass fibre filters to determine particulate organic matter, particulate nutrients (C, N and P) and Chlorophyll-*a* (phytoplankton). Filtered water was analysed for DOC and inorganic nutrients.

13.4 RESULTS AND DISCUSSION

13.4.1 Geophysical study of the lagoon

Good measuring conditions were achieved and in some areas it was even possible to perform parallel profiles on the land in the adjacent shallow waters of the Ria Formosa. This fact not only confirmed that the anomalies detected on land are also detectable in shallow salt waters, which validates the use of this method in these environments, but also gave important information about the existence of geological structures, their orientation and their role in the genesis of such coastal environments.

Figure 13.4 illustrates a typical fault system anomaly followed by a change in lithology towards the sea: a geological formation with higher resistivity and which is covered by more recent sediments and water. The higher resistivity is reflected by a lowering of the Hp/Hs ratio (blue line) – 'lower step' – in this prototype device. The black line drawn on top indicates the running average of the Hs/Hp. The upper EC graphic (red line) shows that this anomaly is accompanied by a decrease in water salinity, possibly due to groundwater outflow associated with the fault system. The anomaly was detected along the River Gilão (Tavira) but was also identified towards the southwest (hatched area), parallel to the channel (NE/SW), suggesting the structural control of this lagoon.

Another example is shown in Figure 13.5 corresponding to a lateral changing in lithology (step-fault?) that was detected south east of Faro (Fig. 13.1). Its direction seems roughly parallel to one of the directions of the meandered channel, possibly causing its form. This can lead to the hypothesis that tectonic features have much more to do with the morphology of the Ria Formosa than previously considered.

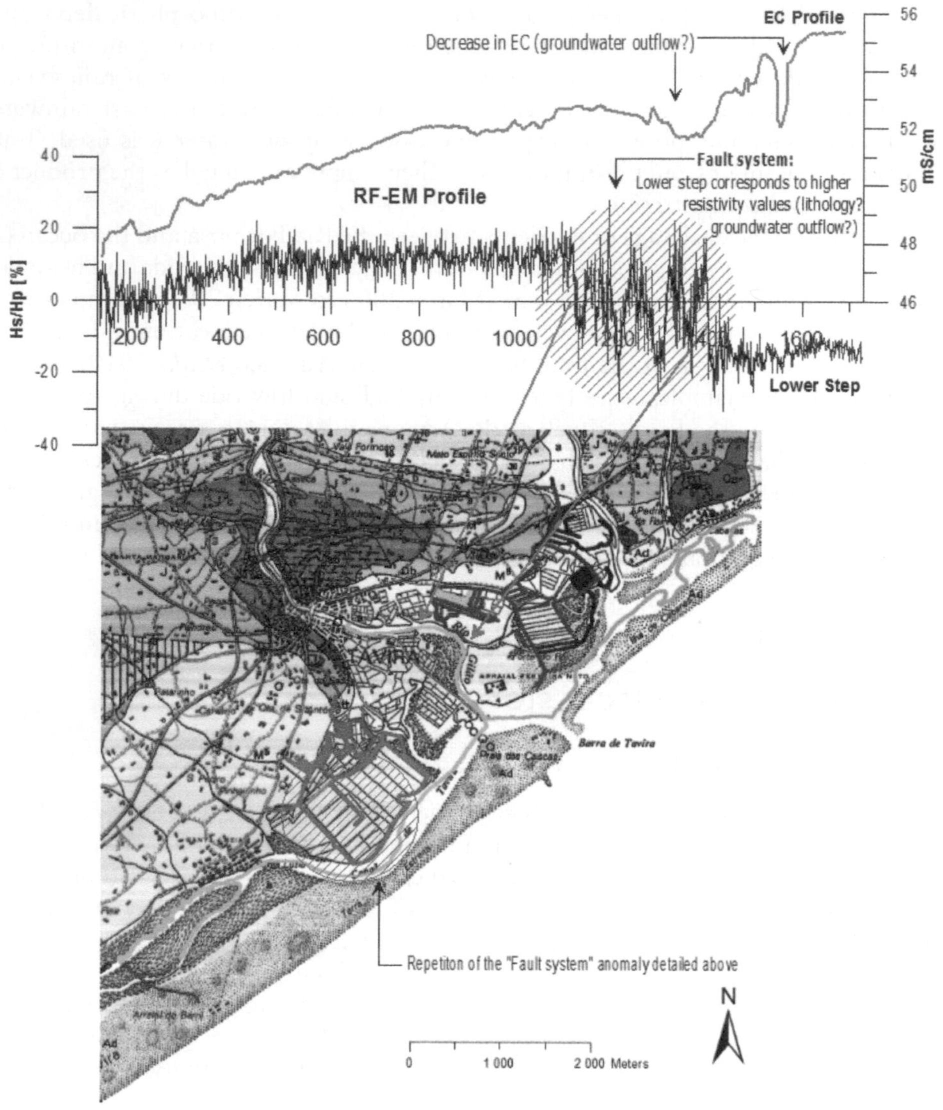

Figure 13.4 RF and EC profiles showing a major anomaly corresponding to a fault system detected along the river Gilão (Tavira) and also southwestwards (hatched area). The RF profiles are plotted on the Geological Map 53-B, 1:50 000 (copyright LNEG). Lithology: J^2c, J^3 and J^4: Jurassic marly limestone and marl; M^5 Miocene sandy limestone; Qa Quaternary fine silt and sand; Qb Quaternary sand and gravel; Ad Holocene dune sand.

13.4.2 Groundwater and nitrogen discharge into the lagoon

Table 13.1 characterises the hydrogeological units of the Ria Formosa drainage basin (see Fig. 13.1 for location) and quantifies some of the water balance components on

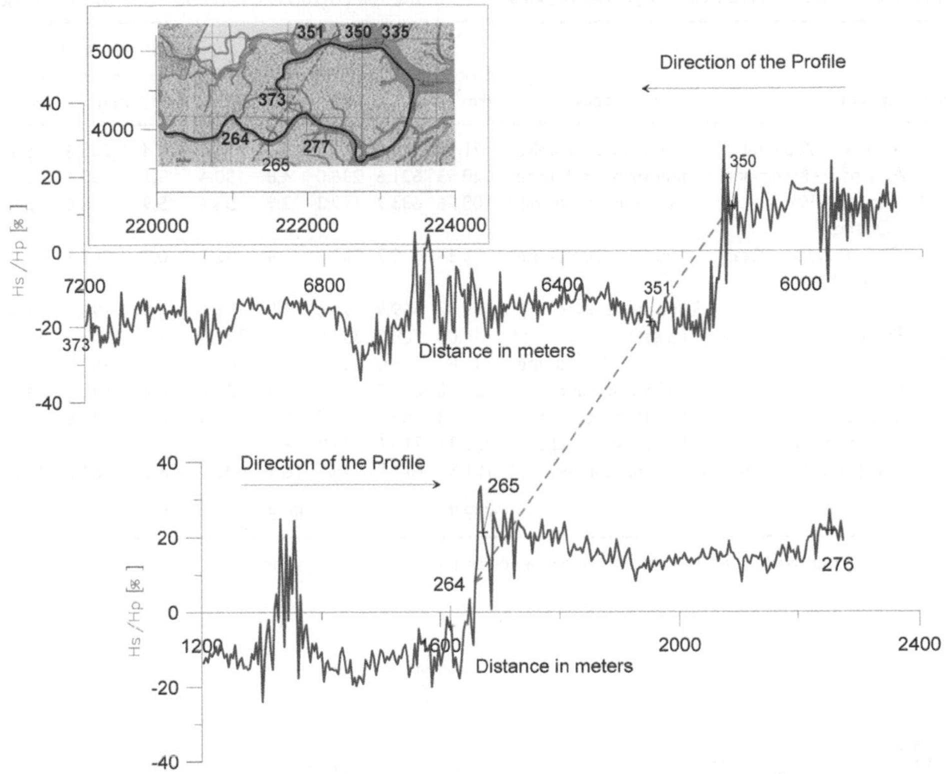

Figure 13.5 Parallel profiles showing the lateral changing of lithology due to a step-fault system, with increasing resistivity towards the east; note the vertical scale (Hs/Hp [%]) ranging from −40 to +40%; also shown is the location of the profile in the lagoon on topographic map (© IGeoE, Portuguese National Grid coordenate system).

an average yearly basis, namely natural recharge ($R_n = P - ET - S_o$) and groundwater outflow (G_o) before and after the implementation of the surface water irrigation district in the eastern part of the basin. The spatial distribution of these two parameters (R_n and G_o) is shown in Figure 13.6. Though P is higher in the north, recharge is very low, due to the low infiltration capacity in the Paleozoic basement. Average S_o is 30% of rainfall, whereas only 5% is considered to infiltrate. The remaining 65% is lost via ET. In the aquifer systems, the infiltration capacities are higher, reaching 50% in the highly karstified limestones (Stigter *et al.*, 2009). Systems M8, M9, M11 and M13 receive the highest recharge per m^2, but due to their large area, besides M13, M0 and M10 receive the highest volumes. Total calculated natural recharge is 80.8 hm^3($\times$ 10^6 m^3). For the aquifer systems the values are comparable to those reported by Almeida *et al.* (2000), though the total is 15% higher.

On an average annual basis, 34% of natural recharge is consumed for irrigation, mainly in the aquifer systems M10 and M12, creating a large deficit in the latter. This deficit is compensated by groundwater flowing in from the north, preventing a

Table 13.1 Characterisation of hydrogeological units of the Ria Formosa basin and their water balance.

Aquifer system		Main lithology	Area (km^2)	P (mm)	Recharge mm	Recharge hm^3	Go (present) mm	Go (present) hm^3	Go (past) mm	Go (past) hm^3
M8	S. Brás de Alportel	limestone, dolomite	21.15	829.6	255.6	5.4	255.6	5.4	255.6	5.4
M9	Almansil – Medronhal	limestone, dolomite	20.53	621.6	236.0	4.8	150.4	3.1	150.4	3.1
M10	S. João da Venda – Quelfes	sand, limestone, marl	108.56	633.2	119.3	12.9	54.4	5.9	−8.6	−0.9
M11	Chão Cevada – Qta. João Ourém	limestone, dolomite	5.34	619.7	259.7	1.4	42.3	0.2	42.3	0.2
M12	Campina de Faro	limestone, sand	79.37	569.9	94.6	7.5	−96.1	−7.6	−96.1	−7.6
M13	Peral – Moncarapacho	limestone	44.06	708.2	330.5	14.6	331.0	14.6	327.2	14.4
M14	Malhão	limestone, dolomite	11.83	638.0	168.2	2.0	169.1	2.0	162.9	1.9
M15	Luz – Tavira	limestone, sand	27.72	605.0	172.9	4.8	229.1	6.4	−146.0	−4.0
M16	S. Bartolomeu	limestone, dolomite	3.28	585.5	114.7	0.4	127.6	0.4	41.6	0.1
A0	Paloezoic basement	Shales, greywackes	226.31	716.1	35.8	8.1	35.8	8.1	35.8	8.1
M0	Local, not differentiated	clay, marl, sand	191.55	673.5	98.7	18.9	95.7	18.3	64.2	12.3
Total			739.7			80.8		56.8		33.0

G_o (past) = groundwater outflow before the existence of surface water irrigation

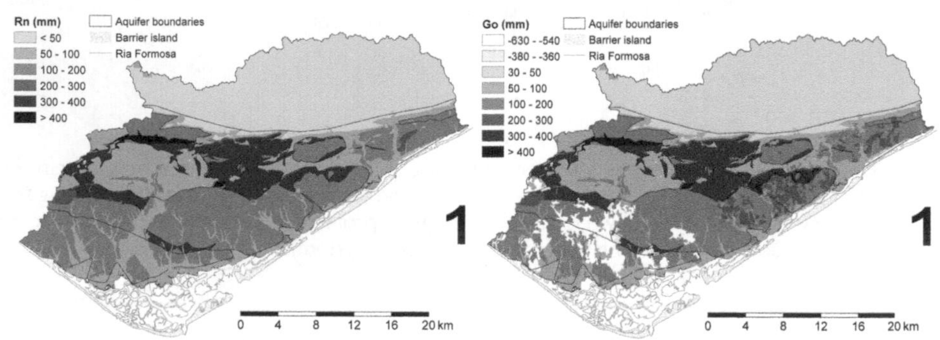

Figure 13.6 Mean annual groundwater recharge (*Rn, left*) and outflow (G_o, *right*), based on spatial distribution of rainfall, determined infiltration/precipitation ratios, as well as irrigation and return flow volumes.

decrease of water levels. Indeed, hydraulic heads of the M12 aquifer only show decreasing trends in consecutive years of below-average rainfall. Groundwater discharge along the coastline of this area is limited and the long residence times promote groundwater salinisation and nitrate contamination (Stigter *et al.*, 2006c). Before the integration of the eastern area in the surface water irrigation district in 2001, 60% of groundwater recharge was consumed by agriculture. The shift to surface water irrigation provides 3.5 hm^3 of artificial recharge per year, and together with the ceased groundwater extractions in this area leads to an overall increase of groundwater outflow from the Ria Formosa basin of 24 hm^3.

Table 13.2 Quantification of water and nitrogen discharge from land for each outflow component.

Discharge	*Mechanisms*	*Water (hm^3/yr)*	*Ntot (ton/yr)*	*Period*	*Method*
G1/I1+ G2/I2+ G3	Groundwater	37	570[a]	Throughout the year	Estimated from average N losses from fertilisation, septic tanks and precipitation, considering that input = output from land
I4	Streams	40	27	Only in winter	Calculated (see Stigter *et al.*, 2006a)
I5	WWTPs	10	417	Throughout the year	Calculated (see Stigter *et al.*, 2006a)
I6	Direct rainfall on the lagoon	50	12	Only in winter	Calculated, based on N analyses and spatial distribution of rainfall (Nicolau, 2002)

[a] 300 ton/yr is estimated to enter the lagoon, due to restricted outflow near Faro and deep circulation

It is not known precisely how the present groundwater outflow (*Go*) of 57 hm^3 is distributed among each of the referred outflow components of Figure 13.2. Deep circulation below the lagoon (*G3*) is particularly important in the east and in a GIS environment was estimated to be 50–60% of *Go*, by considering the contribution of recharge north of aquifer systems M10 and M15 and east of M8.

A value of 570 ton-N/yr was obtained on the basin. Once again, it is not known in detail what fraction enters the lagoon (I1 + I2), but considering the restricted outflow in the area of Faro and combining the location of the N sources with that of the areas where recharge contributes to groundwater discharge into the lagoon (G1), it is estimated that 300 ton-N/yr enters the lagoon. Table 13.2 compares this value with other N sources and shows the importance of groundwater. Based on local seepage measurements, Leote *et al.* (2008) also identify coastal groundwater discharge in the Ria Formosa as a major source of N to the lagoon's internal nutrient mass balance. However, the authors do not consider the large variations between the eastern and western sectors.

The WWTPs provide an important point source, but may have a limited range, as was indicated in earlier studies (Cabaço *et al.*, 2008), in the order 500/1000 m from the discharge point. Most likely, nutrients here are rapidly assimilated in biological processes, such as uptake by vegetation and bacterial denitrification. Curiously, direct rainfall accounts for the largest freshwater input, but N concentrations are much lower. Stream discharge is calculated for the hydrogeological year of 2005/2006. Rainfall in this year was very near the average annual values, so that stream discharge can be compared to the other components, except for the catchment of Rio Seco. Here runoff was low, because the preceding period of long-lasting drought had drastically increased the infiltration capacity of the soil.

The effect of freshwater discharge can be determined by analysing salinity. A very clear signal of stream discharge could be observed during the spring tide 24 h campaign of 7 and 8 November 2006. The days preceding this campaign were characterised by heavy rainfall, causing significant stream discharge. Rain stopped shortly after the first samples were taken, so the effect of direct atmospheric deposition on the Ria is

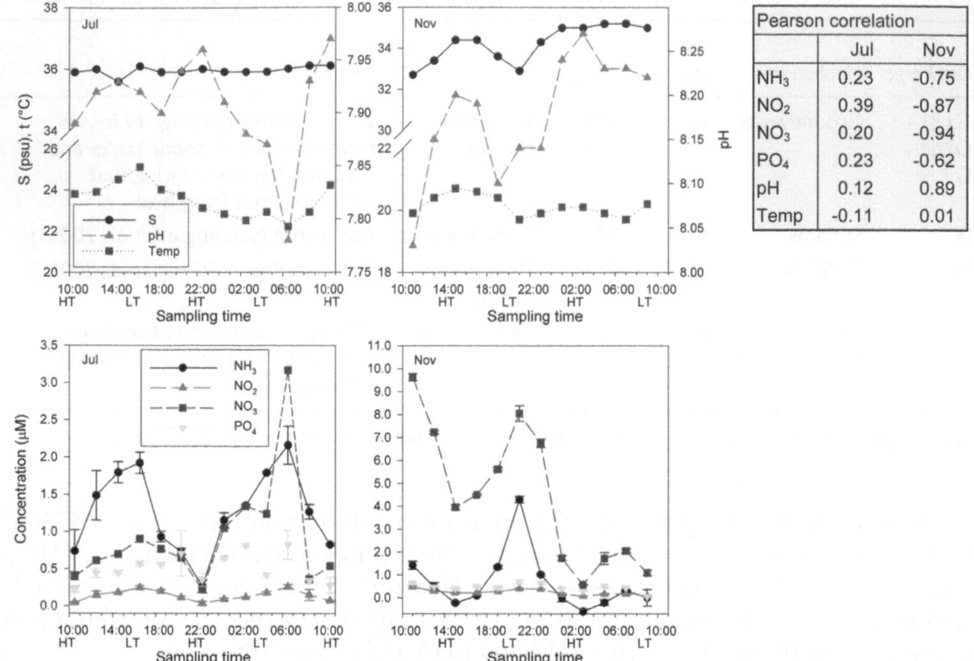

Pearson correlation		
	Jul	Nov
NH_3	0.23	-0.75
NO_2	0.39	-0.87
NO_3	0.20	-0.94
PO_4	0.23	-0.62
pH	0.12	0.89
Temp	-0.11	0.01

Figure 13.7 Exchange of the Ria Formosa lagoon with the Atlantic Ocean, results of samples taken every 2 h during a 24 h tidal cycle in the main exchange channel (Barra do Farol) in July and November 2006 (note different scales).

assumed to be low. The clear tidal dynamics of nutrient concentrations (see Fig. 13.7) show a strong negative correlation with salinity, which is most noticeable for nitrate, indicating a dominating terrestrial input and river outflow (in rainfall ammonium is the dominant N compound). Another interesting aspect is that with incoming tide, nutrient concentrations are still high, whereas salinity is still lower than the typical ocean salinity. This means that mixing of outgoing water with ocean water is not complete and that some of the outgoing water returns with the next incoming tide, so that the nutrients will again be available for plant uptake in the Ria Formosa. Finally, the last period to low tide (outgoing water) shows the short-living effect of river outflow both in nitrate and in salinity, corroborating the discharge observations of the most important streams in this region of the Ria Formosa.

A complete presentation of the exchange data with the ocean is provided by Malta *et al.* (*in prep*). In general, the results based on inorganic nutrients seem to confirm earlier studies (Newton and Mudge, 2005) in that, on an annual basis, there is a net export of nutrients to the ocean. It is clear however that there is large heterogeneity in time. Most export takes place in winter, whereas in summer export is limited. Occasionally, import from the ocean may occur (see July night sample in Fig. 13.7) probably due to upwelling of nutrient rich water from the deep sea.

13.5 FINAL CONSIDERATIONS

The role of groundwater as a nutrient source for the algal blooms continues to be a matter of investigation. It is known that winter blooms reach their peak biomass in December–January after which they decline. The algae probably survive the summer heat and high light as tiny life stages (germlings) in the sediment of tidal mud flats (Malta & Santos, 2008). The role of sediment fluxes, i.e. groundwater and remineralisation, can be particularly important for the winter algae, present on the mudflats and with little access to nutrients in ocean water. Algae sampled earlier in the year generally have higher nutrient contents than algal samples in late spring and summer, even when growth rates are high as is the case for the winter blooms, which could also be related to a higher N input. Groundwater discharge in the summer will be much lower, especially in the western part of the drainage basin, where irrigation consumes a large fraction of annual recharge. In the east this is not the case now that irrigation is performed with surface water, causing a water and N flux towards the lagoon and the ocean throughout the entire year, with potential consequences for the appearance of algal blooms. Interestingly, major summer blooms spots were located in the eastern part of the lagoon and along the eastern beaches (Malta *et al.*, 2007). As these blooms start to develop in early spring, the contribution of rainfall and stream outflow to the nutrient budget is already minimal and oceanic concentrations are low, hence it is likely the groundwater outflow and/or remineralisation are the major nutrient sources for the algae in this period.

The lower nutrient content in the algae in the summer is also due to the improved conditions (light, temperature) that stimulate production and cause dilution of internal cell nutrient quota by high growth rates. Apart from the seasonal trend, there are indications that algae growing in the Ria Formosa have higher nutrient contents than algae grown in the sea, independent of the season. From the aerial and underwater surveys carried out in 2006, it became clear that the main algal concentrations in summer can be found in and around the tidal inlets connecting the lagoon with the ocean, suggesting that the Ria Formosa is the main source for these blooms. Moreover, results from trawls carried out by boat in March 2008 showed the presence of high amounts of summer bloom algae on the bottom of the channels, especially around the Armona outlet and the central – eastern sector of the lagoon, supporting this hypothesis. There may be some production in the ocean as well, but considering the low production rates at oceanic nutrient concentrations found in the experiments, this might be just enough to maintain the biomass (Malta *et al.*, 2007).

Algal biomass is also highly variable between years; in the summers of 2005 and 2006 blooms were hardly observed, whereas 2004, 2007 and 2008 showed large accumulations on the beaches of the eastern Algarve. This could be related to climate factors, such as water temperature and rainfall, and surface and groundwater discharge (Malta *et al.*, 2007). Rainfall in 2005 was extremely low, but in 2006 reached average values, so that groundwater storage could be replenished. The winters of 2004 and 2007 also had below-average rainfall (65%), so apparently no direct link can be established, although both years were preceded by relatively wet years. Also in this case further study is required.

ACKNOWLEDGEMENTS

The present study is performed in the scope of the research projects POCI/MAR/58427/2004 and PPCDT/MAR/58427/2004 and the authors gratefully acknowledge the *Fundação para a Ciência e a Tecnologia* for funding the projects.

REFERENCES

Almeida, C., Mendonça, J.J.L., Jesus, M.R. & Gomes, A.J. (2000) *Aquifer Systems of Continental Portugal* [In Portuguese]. Report INAG, Lisbon. http://snirh.pt/index.php?idMain=4&idItem=3&idISubtem=link1.

Andersen, M.S., Baron, L., Gudbjerg, J., Gregersen, J., Chapellier, D., Jakobsen, R. & Postma, D. (2007) Discharge of nitrate-containing groundwater into a coastal marine environment. *Journal of Hydrology*, 336, 98–114.

Alva, A.K., Paramasivamb, S., Fares, A., Obreza, T.A. & Schumann, A.W. (2006) Nitrogen best management practice for citrus trees. II. Nitrogen fate, transport, and components of N budget. *Scientia Horticulturae*, 109, 223–233.

Beltrão, J. (1985) *Localised Irrigation* [In Portuguese]. Report Universidade do Algarve, Faro, Portugal. 31 pp.

Boman, B.J. & Battikhi, A.M. (2006) Growth, evapotranspiration, and nitrogen leaching from young lysimeter-grown orange trees. *Journal of Irrigation and Drainage Engineering – ASCE*, 133, 350–358.

Cabaço, S., Machás, R., Vieira, V. & Santos, R. (2008) Impacts of urban wastewater discharge on seagrass meadows (*Zostera noltii*). *Estuarine, Coastal and Shelf Science*, 78 (1), 1–13.

Carvalho Dill, A., Turberg, P., Müller, I. & Parriaux, A. (2009) The combined use of radio-frequency electromagnetics (RF-EM) and radiomagnetotellurics (RMT) methods in non ideal field conditions for delineating hydrogeological boundaries and for environmental problems. *Environmental Geology*, 56 (6), 1071–1091.

CCDR-Alg (2007) Population served by wastewater drainage and treatment facilities. http://www.ccdr-alg.pt/sids/indweb/indicador.asp?idl=7&idt=2. Cited 29 June 2007.

Costa, J.E., Heufelder, G., Foss, S., Millham, N.P. & Howes, B. (2002) Nitrogen removal efficiencies of three alternative septic technologies and a conventional septic system. *Environment Cape Cod*, 5 (1), 15–24.

De Bruin, J. (1999) Report on groundwater research in Luz de Tavira, Algarve Portugal: Water balance, hierarchical clustering chemical data, Slingram electromagnetic survey. M.Sc. Thesis, Vrije Universiteit, Amsterdam, the Netherlands, 43 pp.

He, F.F., Chen, Q., Jiang, R.F., Chen, X.P. & Zhang, F.S. (2006) Yield and nitrogen balance of greenhouse tomato (*Lycopersicum esculentum Mill.*) with conventional and site-specific nitrogen management in Northern China. *Nutrient Cycling in Agroecosystems*, 77, 1–14.

Howarth, R.W. & Marino, R. (2006) Nitrogen as the limiting nutrient for eutrophication in coastal marine ecosystems: evolving views over three decades. *Limnology & Oceanography*, 51, 364–376.

Keller, J. & Bliesner, R.D. (2000) *Sprinkle and trickle irrigation*. Caldwell, NJ, USA, Blackburn Press. 652 pp.

Krause, S. & Bronstert, A. (2005) An advanced approach for catchment delineation and water balance modelling within wetlands and floodplains. *Advances in Geosciences*, 5, 1–5.

Leote, C., Ibánhez, S.P. & Rocha, C. (2007) Submarine Groundwater Discharge as a nitrogen source to the Ria Formosa studied with seepage meters. *Biogeochemistry*, 88, 185–194.

Malta E-j., Tavares, D. & Santos, R. (2007) Ulva blooms along the Algarve beaches:inferring bloom potential from photosynthetic production. *4th European Phycological Congress, Oviedo, Spain, 23–28 July 2007.*

Malta E-j. & Santos, R. (2008) The effects of light and desiccation on photosynthesis and recovery potential of intertidal Ulva prolifera (Ulvales, Chlorophyta). *8th GAP workshop, Group for Aquatic Primary Productivity, Eilat (Israel), March 2008.*

Newton, A. & Mudge, S.M. (2005) Lagoon-sea exchanges, nutrient dynamics and water quality management of the Ria Formosa (Portugal). *Estuarine, Coastal and Shelf Science*, 62, 405–414.

Nicolau, R. (2002) *Modelling and Mapping of Spatial Distribution of Precipitation – An Application to Continental Portugal* [In Portuguese]. Ph.D. thesis, Universidade Nova de Lisboa, Portugal. 356 pp.

Paz, J.M. & de Ramos, C. (2004) Simulation of nitrate leaching for different nitrogen fertilization rates in a region of Valencia (Spain) using a GIS–GLEAMS system. Agriculture, Ecosystems & Environment, 103 (1), 59–73.

Silva, M.J.B.L. (1988) *Hidrogeologia do Miocénico do Algarve [Hydrogeology of the Miocene of the Algarve].* Ph.D. Thesis, Universidade de Lisboa, Lisbon, Portugal. 496 pp.

Silva, M.O. (1984) *Hidrogeologia do Algarve Oriental [Hydrogeology of the Eastern Algarve].* Ph.D. Thesis, Universidade de Lisboa, Lisbon, Portugal. 260 pp.

Stigter, T.Y. (2012) Restoration of groundwater quality to sustain coastal ecosystems productivity. In: Wolanski, E. & McLusky D.S. (eds.) *Treatise on Estuarine and Coastal Science.* Vol. 10, pp. 245–262. Waltham: Academic Press.

Stigter, T.Y., Van Ooijen, S.P.J., Post, V.E.A., Appelo, C.A.J. & Carvalho Dill, A.M.M. (1998) A hydrogeological and hydrochemical explanation of the groundwater composition under irrigated land in a Mediterranean environment, Algarve, Portugal. *Journal of Hydrology*, 208, 262–279.

Stigter, T., Carvalho Dill, A., Malta, E. & Santos, R. (2006a) Quantification of nitrogen and phosphorous nutrient discharge to the Ria Formosa via surface runoff [In Portuguese]. *Paper Presented at the 5th Iberian Congress on Water Management and Planning, Faro, 4–8 December 2006.* 12 pp.

Stigter, T.Y., Carvalho Dill, A.M.M., Ribeiro, L. & Reis, E. (2006b) Impact of the shift from groundwater to surface water irrigation on aquifer dynamics and hydrochemistry in a semi-arid region in the south of Portugal. *Agricultural Water Management*, 85 (1–2), 121–132.

Stigter, T.Y., Ribeiro, L. & Carvalho Dill, A.M.M. (2006c) Evaluation of an intrinsic and a specific vulnerability assessment method in comparison with groundwater salinisation and nitrate contamination levels in two agricultural regions in the south of Portugal. *Hydrogeology Journal*, 14 (1–2), 79–99.

Stigter, T.Y., Carvalho Dill, A.M.M., Ribeiro, L. & Reis, E. (2007) Groundwater status in the two nitrate vulnerable zones of the Algarve – concerns for the adjacent wetland and agroecosystems. *XXXV IAH Congress, Groundwater and Ecosystems, Lisbon, Portugal, 17–21 September 2007.*

Stigter, T.Y., Monteiro, J.P., Nunes, L.M., Vieira, J., Cunha, M.C., Ribeiro, L., Nascimento, J. & Lucas, H. (2009) Screening of sustainable groundwater sources for integration into a regional drought-prone water supply system. *Hydrology and Earth System Sciences*, 13, 1–15.

Turberg, P. & Muller, I. (1992) La méthode inductive VLF-EM pour la prospection hydrogéologique en continu du milieu fissuré. *Annales Scientifiques de l'Universite de Besançon,Memoire Hors Serie*, 11, 207–214.

Chapter 14

Relationships between wetlands and the Doñana coastal aquifer (SW Spain)

Marisol Manzano[1], Emilio Custodio[2], Edurne Lozano[3] & Horacio Higueras[1]

[1]*Technical University of Cartagena, Cartagena, Spain*
[2]*Technical University of Catalonia and International Centre for Groundwater Hydrology, Barcelona, Spain*
[3]*GEOAQUA S.C., Zaragoza, Spain*

ABSTRACT

The Doñana area holds important fluvial marshes and hundreds of small-to-medium size wetlands. They are mostly groundwater dependent, and their hydrology is controlled by the particular location within the groundwater regional flow scheme. Intensive groundwater pumping for irrigation since the 1980s has dramatically changed the groundwater flow regime. Accumulated inter-annual lowering in the piezometric levels of the preferentially exploited deep aquifer layers induced a progressive water-table drawdown, resulting in a decrease of natural water discharge through seepage and phreatic evapotranspiration. Wetlands water sources and hydroperiods were modified. Hydrogeochemical, environmental isotopes and modelling studies validated the conceptual model of wetlands and gave insight about how groundwater pumping and vegetation management modify wetlands hydrology. One major insight of modelling is that after a change has been produced, the system as a whole needs about 30 years to reach midway between the initial and the final steady state, and thus springs, seepages and river flows tend to decrease further if groundwater development persists.

14.1 INTRODUCTION

The Doñana aquifer system is situated in the south west Atlantic coast of Spain, between the Guadalquivir River and the Tinto River, not far from the Portuguese border. The surface area is around 3000 km^2 and hosts the Guadalquivir River marshes and two protected natural areas of international relevance, accounting some 1100 km^2. A large part was declared Reserve of the Biosphere in 1981, RAMSAR site in 1982, and Natural World Heritage Site in 1994.

In this largely uninhabited area human settlements started in the 1930s and 1940s by reclamation of part of the wetland for rice cultivation, and later on by introducing pine-tree forest, which was a failure in sandy and in shallow water-table areas. Then extensive eucalyptus tree plantations for timber were successfully introduced. Currently most of the Doñana core area is uninhabited, except for people in charge of the protection, developers of natural resources, and visitors. However, flourishing human communities developed in the surroundings, at least since 5000 years BP. More recently, in the late 1970s, large areas for irrigation with local groundwater were developed outside the present protected zones. This was the result of a development plan from the late 1960s, implemented from studies sponsored by the United Nations Food and Agriculture Organization and the Spanish Government (FAO, 1972).

Tourism is also an important economic activity that started in the late 1970s. It is mainly beach–based and concentrated both seasonally (spring and summer) and spatially (Matalascañas and Mazagón coastal resorts; capacity of ca. 300 000 people). New developments were halted to protect the area.

The above mentioned eucalyptus trees replaced native trees and shrub vegetation, which includes some phreatophyte species in shallow water table areas. The result is an increased phreatic evapotranspiration, which is an added term to groundwater use.

The climate is Mediterranean sub-humid with Atlantic influence: dry summers and wet winters. Mean rainfall, concentrated between October and March, is 500–600 mm/year, depending on the area, but has very high inter-annual variability, from 250 mm to 1100 mm/year. Mean yearly temperature is around 17 °C near the coast and 18 °C in the centre of the area. The mean number of annual sun-hours is close to 3000.

The two large rivers that surround the aquifer, the Tinto River to the north west and the Guadalquivir River to the east, do not contribute water to it. They are drainage boundaries with a poor connection to groundwater. Surface water inside the territory is limited to a couple of small groundwater-fed permanent rivers and some seasonal brooks, some permanent lagoons, and several hundreds of temporal lagoons in small depressions where the water table crops out.

Intensive groundwater exploitation since the early 1980s, mostly for irrigation and to supply urban and tourist areas, but also for some environmental uses, led to conspicuous groundwater level drawdown, especially in the deep aquifer layers. The corresponding water table lowering has decreased aquifer natural discharge through springs, streams and phreatic evapotranspiration, inducing slow but progressive vegetation changes at the numerous phreatophyte areas, and modifying the hydrological behaviour of wetlands (Manzano *et al.*, 2005; Coleto, 2003; Lozano, 2004; Lozano *et al.*, 2005).

Since 1985, many hydrogeological studies have been carried out focussing in different sectors of the aquifer. Hydrodynamic studies and groundwater flow modelling at regional and local scales were performed to understand aquifer functioning (Custodio & Palancar, 1995; UPC, 1999; Trick & Custodio, 2004) and its relationship with wetlands (Lozano *et al.*, 2005), both under natural conditions and after the water table artificial lowering. Chemical and environmental isotope information is abundant and has been the subject of several studies (Baonza *et al.*, 1984; Poncela *et al.*, 1992; Iglesias, 1999; Manzano *et al.*, 2001; Lozano *et al.*, 2005; Manzano *et al.*, 2007a; Higueras *et al.*, 2011). Hydrogeochemical modelling has been carried out to establish a conceptual model to explain groundwater chemistry origin and evolution. However, in spite of the numerous studies, due to the large dimensions and complexity of the area some aspects remain poorly known.

14.2 GEOLOGY AND HYDROGEOLOGY

The aquifer system consists of detrital, unconsolidated Plio–Quaternary sediments overlapping impervious Miocene marine marls (Fig. 14.1). The Pliocene materials are mostly impervious (marls, silts and sandy silts). The Quaternary sediments consist

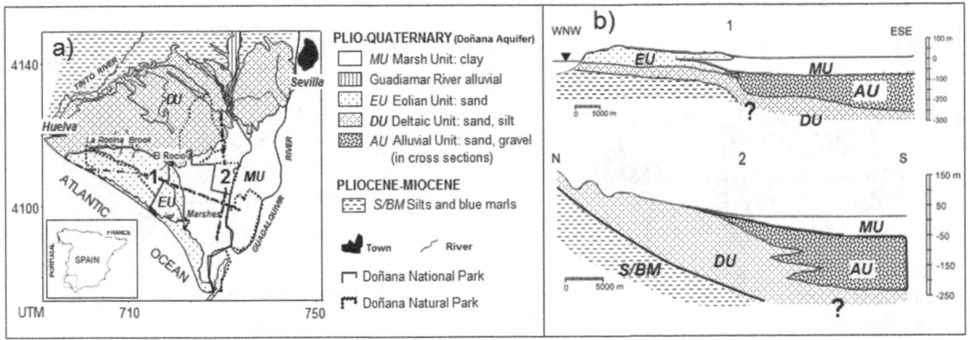

Figure 14.1 Location and geology of the Doñana aquifer system. The wetlands are situated on top of an extended eolian mantle which forms the upper layer of the formation called Eolian Unit (EU).

of deltaic and alluvial silts, sands and gravels to the north and the littoral area, and alluvial and eolian sands to the west. They are dominantly formed by amorphous silica grains, with minor contents of K- and Na-feldspars, illite, chlorite and kaolinite. Some carbonates may be present either as detrital grains or shell remains, except in the upper sand layers of the western sector, from where they have been already dissolved by the acidic rain.

The Quaternary layers thicken from north to south and from west to south east. To the south east, the coarse sediments are covered by a thick (50–80 m) sequence of estuarine and marshy clays separated from the ocean by a recent sand spit. The aquifer has a surface area of some 3000 km^2 and variable thickness (20 m inland to >150 m at the coast line). At regional scale the upper part of the aquifer shows two lithologic domains: a sandy one to the north and west of the marshes with extensive areas blanketed by eolian sands, which roughly behaves as an unconfined aquifer, and a clayey domain in the marsh area (1800 km^2 of surface area) under which a large sector of the aquifer is confined.

Direct rainfall recharge occurs in the sandy areas, especially where the upper layer is of eolian sand. There is also recharge through excess irrigation water in the agricultural zones (Fig. 14.2a), but it derives from local groundwater. Groundwater flows mostly to the south and south east and part of it discharges by upward flows along the contact between the eolian sands and the marshy clays, originating a very ecologically rich ecotone (Fig. 14.2b), while other part flows to the aquifer sector confined under the marshes. There is also discharge to a main brook called La Rocina (Fig. 14.2b), to the ocean, and, under natural flow conditions, to numerous small phreatic wetlands (ponds and small water courses) situated on top of the sands, and through phreatic evapotranspiration. Some discharge may also occur as upward flow through the Quaternary marsh clays. The south east sector of the confined layers contains almost stagnant, partly evaporated connate marine water, which have not been flushed out due to the low hydraulic head prevailing since the late Holocene sea level stabilisation some 6 ka BP (Zazo *et al.*, 1996; Manzano *et al.*, 2001).

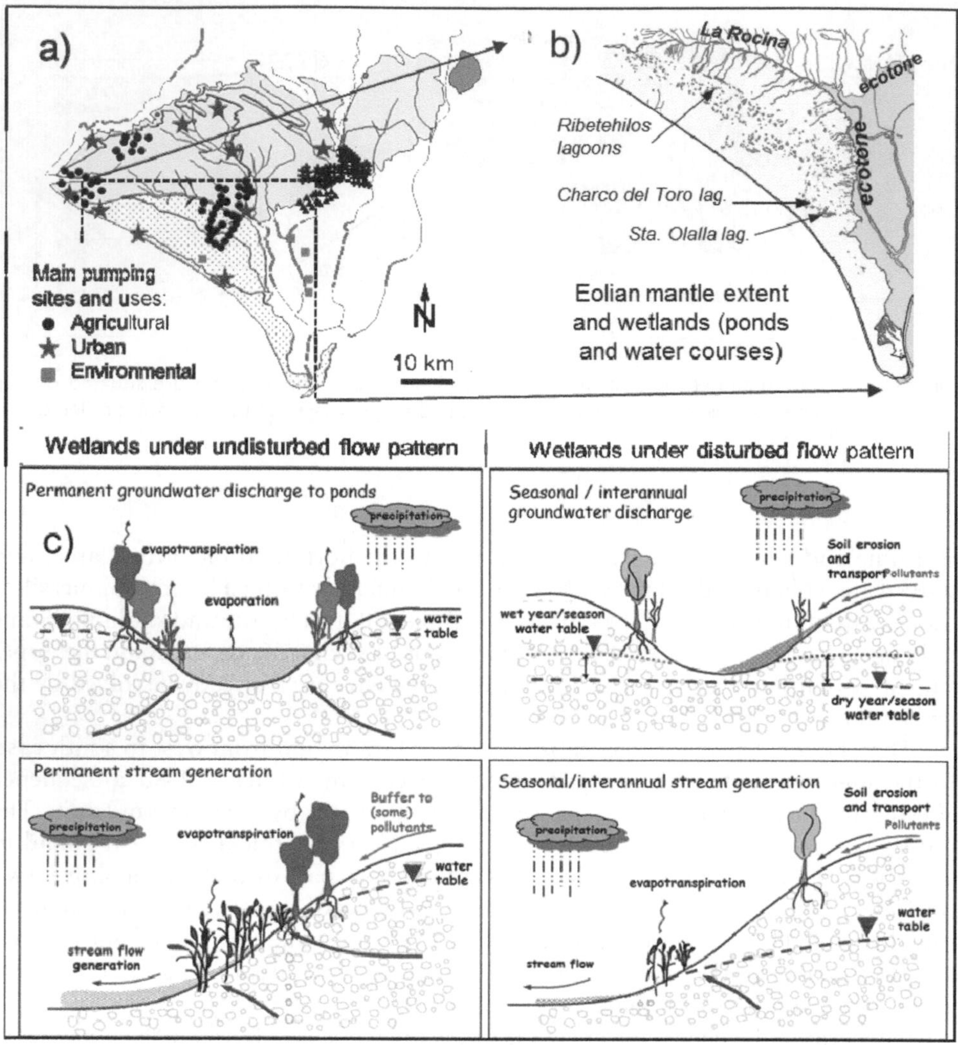

Figure 14.2 a) Location of the main wells and groundwater uses. b) Eolian mantle extent and groundwater supported wetlands. c) Main features of the major types of wetlands in the eolian mantle, ponds and small water courses, under undisturbed (left) and disturbed (right) groundwater flow conditions.

But the natural flow pattern is strongly modified in many areas because of the intensive groundwater exploitation since the early 1980s. Agricultural wells are concentrated near the ecotones to the north east and north west of the marshes (Figs. 14.2a and 14.2b). Intensive groundwater abstraction induced a deep layer head drawdown and consequently a water-table drawdown through vertical leakage. Nowadays a large proportion of recharge water is pumped out from the deep layers of the unconfined area close to the marshes. Thus, pumping has partially and progressively depleted natural discharge to springs, streams, wetlands, phreatophyte areas and seepage, inducing slow but progressive vegetation changes and modifying the hydrological behaviour of some wetlands (Fig. 14.2c) (Trick & Custodio, 2004; UPC, 1999; Manzano *et al.*, 2002; Coleto, 2003).

14.3 GROUNDWATER CHEMISTRY AND ENVIRONMENTAL ISOTOPE CONTRIBUTION TO THE UNDERSTANDING OF THE AQUIFER FLOW SYSTEM

Until 1992 the monitoring network consisted of multi-screened agricultural wells and a few shallow wells, measured by the Geological Institute of Spain and the Agriculture Development Institution. Then, groundwater samples were a mixture of waters from different depths and variable transit times. Also head values were an average value along the wells penetration. For such a thick aquifer, groundwater flow has downward and upward vertical components, enhanced by groundwater development. The information obtained did not allow getting insight into vertical gradients, chemical variations with depth or travel times. Late in the 1980s, electrical conductivity and temperature downhole logs and saline tracing tests demonstrated the existence of vertical flows (Custodio *et al.*, 1996). This guided the design and drilling along the 1990s of a dedicated monitoring network of nested boreholes, first by the Public Works Geological Service of Spain and afterwards by the Guadalquivir River Basin Water Authority. The new multi-piezometer emplacements allowed clarifying the 3-D groundwater flow pattern, to establish flow paths and to perform depth-oriented sampling, which provided information on transport processes and transit time along groundwater flow paths.

14.3.1 Conceptual model for groundwater chemistry

The conceptual model to explain groundwater chemical background, its origin and modifications, is the result of studies integrating statistics, chemical trends, time tracers and geochemical modelling. Groundwater derives from rain infiltration. Rainwater is slightly acidic (pH = 4.5–6.5) and of the sodium-chloride type due to the airborne marine influence. As rainwater infiltrates its chemical composition is modified by evapotranspiration, which causes solute concentration (small in the eolian sands areas, more intense in the fluvio-marine sand areas, and sometimes locally intense in the lagoon areas), by dissolution of soil CO_2, equilibrium with silica, dissolution of Na- and K-feldspars, and dissolution of $CaCO_3$ remnants from shells, when they are present. Phreatic groundwater has a low mineralisation ($<0.2\ mS\,cm^{-1}$ at 25°C), varying from the sodium–chloride type, in the areas formed by almost pure silica sands, to the sodium–calcium–chloride–carbonate type, in the areas where there is some calcite (Iglesias, 1999; Manzano *et al.*, 2001).

Following the downward flow through the sandy unconfined area no other additional reactions seem to take place. Groundwater mineralisation reaches maximum values of $1\ mS\,cm^{-1}$ (25°C) close to the unconfined–confined boundary, and also in some of the shallow water-table areas surrounding the lagoons. Calcite dissolution is a space-dependent process, as this mineral is present only in some areas and depths.

In the upper part of the unconfined areas groundwater composition is affected by different human activities: in the cultivated areas, agriculture-derived substances (SO_4^{2-}, NO_3^-, Br^-, pesticides) are currently present down to depths of 25–35 m; in the areas where agriculture has never been active, the greater concentrations of some elements (mainly metals and sulphate) in the upper layers of the saturated zone with respect to the deeper ones seem to reflect the atmospheric supply of agrochemical-derived pollutants to the aquifer, as shown by $\delta^{34}S$ and $\delta^{15}N$ contents (Higueras *et al.*, 2011).

As groundwater flows to the confined area under the marshes it becomes increasingly brackish because of mixing with old marine water trapped in the sand, gravel and clay layers. A broad mixing zone develops from NW to SE under the marshes, but its geometry is poorly known due mostly to the difficulties to operate inside the mashes, and also to get samples representative of single flow lines in a multilayered aquifer probably with density driven flows (Konikow & Rodríguez-Arévalo, 1993). Groundwater salinity ranges between 1 and 80 mS cm^{-1} and is mostly of the Na–Ca–Cl–HCO_3 type. In the confined area, groundwater composition changes along flow paths mostly due to mixing with modified old marine water, equilibrium with calcite, and Na/Ca–Mg cation exchange (only close to the northern ecotone, where groundwater abstractions induce the movement of the fresh–saline waters front), with sulphate reduction and probable C incorporation after the mineralisation of sedimentary organic matter.

14.3.2 Conceptual model for groundwater isotopic composition

Local rain water isotopic values fit reasonably the Mean World Meteoric Water Line, though some precipitations seem to have a Mediterranean origin (deuterium excess ≈13–15‰; Fig. 14.3. left). Shallow groundwater in the water-table areas point to the following signature for averaged local rain: $\delta^{18}O = -4.7$ to -5.5‰ SMOW; $\delta^2H = -28$

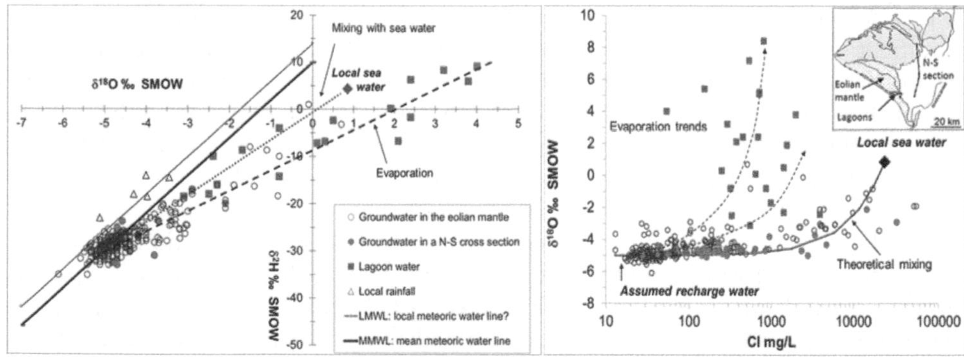

Figure 14.3 Left: Isotopic signature of ground and surface (lagoon) waters. Evaporation in the lagoons and from the very shallow water table is clearly observed, while mixing with sea water is not so clear. Right: The existence of both evaporation and mixing are clearly displayed. Also, two mixing processes seem to take place: for samples from the N-S cross section, mixing with marine water existing in the confined sector under the marshes; for lagoon water (and very shallow groundwater samples nearby the lagoons) the salinisation seem to occur through the incorporation of airborne marine salts.

to −35‰ SMOW. The spatial variations of $\delta^{18}O$ and $\delta^{2}H$ points to the occurrence of two main modifying physical processes:

1 Evaporation from lagoons and from places where the water table is very shallow (Fig. 14.3. left).
2 Mixing of fresh and marine-like saline groundwater. This process seem to occur through two different mechanisms: a) for samples along a north to south cross-section, deep mixing of fresh and saline groundwater which is early Holocene marine water confined under the marshes; b) for samples from the western eolian mantle, incorporation of airborne marine salts to surface water in lagoons and to very shallow groundwater around the lagoons. Figure 14.3. right shows that both processes are present.

14.4 RELATIONSHIP BETWEEN WETLANDS AND GROUNDWATER

Doñana is characterised by an extraordinary abundance of wetlands. The geomorphological origin and hydrology of these wetlands varies greatly, but most of them are dependent on groundwater (Custodio, 2000; Manzano, 2001): springs, small watercourses, small ponds, lagoons, riparian forests, meadows, and patches of phreatophyte vegetation. A determining factor of wetlands hydrology is their location with respect to the regional flow pattern in the aquifer. This influences the mechanisms of lagoon water inflow and depletion, the hydroperiod (flooding frequency and duration), and the water mineralisation and ionic type (Manzano *et al.*, 2002). Much of the extraordinary biodiversity that makes Doñana world famous is the integrated result of combining the different hydrological types of wetlands together with their location, water salinity and chemical composition. A systematic classification work has been carried out within the framework of research studies, which is the basis of the Andalusian Government Wetland Management Programme (PAH, 2002).

The two most common wetland types on the sandy areas, especially on the western eolian mantle, are small erosive or eolian depressions located at the foot of stabilised dune fronts, and small watercourses flowing to the marshes (Figs. 14.2b and c). Both types of wetlands are mostly hypogenic (groundwater is the main water source), seasonal (they hold water only in the wet season), and their location is related to local or medium-scale groundwater flow paths. The watercourses generate where the water-table intersects the land surface, and they develop by scouring the sandy surface. Depressions at the foot of dune fronts get their water from dune seepage, but in most of the cases they are just flow-through wetlands located along very shallow phreatic flow paths. In very wet periods some of those shallow flows may become surface flows and connect several lagoons which usually appear isolated, but in reality are located on the same flow paths that become active in heavy rainfall moments. Thus, wetlands related to the eolian mantle are located in the aquifer recharge area, and they get their water from local groundwater.

Water chemistry in the hypogenic wetlands of the recharge area reflects mostly groundwater composition. At regional scale the main change in groundwater composition is the mineralisation degree, which varies on a spatial basis, but the ionic type

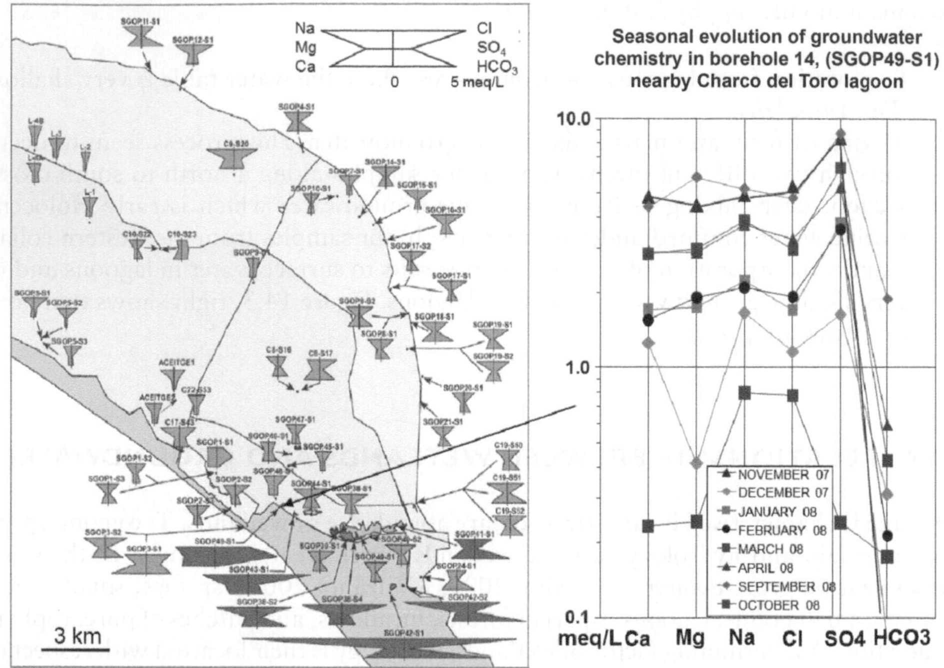

Figure 14.4 Left: Characteristic chemical composition of groundwater samples from the eolian mantle during the 1990s and 2000s (clear grey diagrams). Relevant temporal changes appear only in shallow groundwaters nearby the peri-dunal lagoons (dark grey diagrams), which change from sodium-chloride to sodium-sulphate facies on a seasonal pattern. Right: Example of seasonal chemical changes in shallow groundwater, assumed to be due to sulphur reduction and oxidation cycles (see text for explanation). Modified after Lozano *et al.* (2005).

is highly stable in time (Fig. 14.4. left). In general, wetlands waters are of the sodium–chloride type, or intermediate between sodium-chloride and calcium–bicarbonate type. However, some of the lagoons show temporary changes in mineralisation and ionic type, which depend mostly on changes in the balance between water inputs and outputs and on the chemical reactions occurring within the lagoons, in which the biota plays a crucial role. These reactions may considerably change water salinity, pH, and ion composition on a temporal basis.

The studies performed to improve the knowledge about chemical processes include hydrochemical and isotopic analysis ($\delta^{18}O_{H2O}$, $\delta^{2}H_{H2O}$, $\delta^{34}S_{SO4}$ and $\delta^{18}O_{SO4}$) and modelling. The results indicate that regional or local groundwater flow systems develop around some lagoons depending on the season of the year (dry or wet), driven by changes in the hydraulic gradient between the lagoons and the water-table in the surroundings (Delgado *et al.*, 2001; Lozano *et al.*, 2005; Manzano *et al.*, 2007b). Gradient direction changes imply that the major chemical modifications occurring in the wetland water bodies are transferred to the nearby shallow groundwater. For example, in some lagoons strong evaporation in the dry season causes salt precipitation (sodium chloride, calcium sulphate and may be some calcium carbonate), often accompanied by

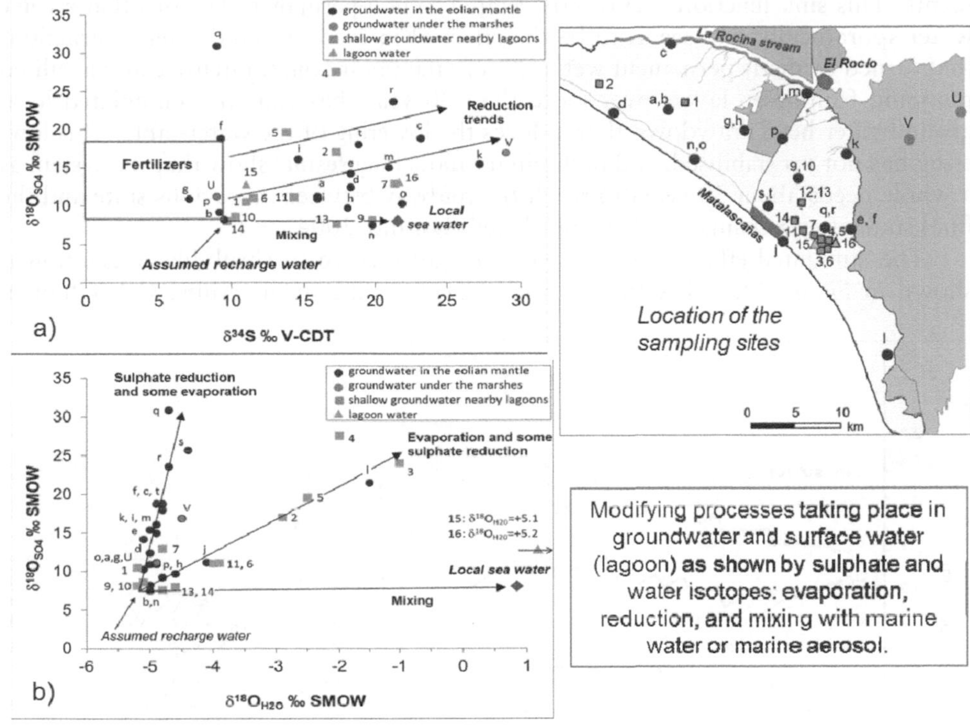

Figure 14.5 a) Reduction processes in groundwater of the eolian mantle and in two samples from the confined area (samples U and V) are clearly observed. Two more processes seem to take place: pollution with fertilizers (some samples), and apparent mixing with marine water, though the affected samples are not from areas with marine water in the aquifer. b) The existence of reduction processes is confirmed by comparing the values of $\delta^{18}O_{H2O}$ and $\delta^{18}O_{SO4}$, and also the strong concentration effect of evaporation and evapotranspiration. However, the mixing with marine water is discarded (for the samples shown in the figure), thus supporting the hypothesis of atmospheric supply or marine salts.

sulphate reduction and sulphur accumulation in the lagoon bottom sediments. In the next wet season, a large part -but not all- of these salts are re-dissolved and transferred to the nearby surrounding phreatic water along with sulphur, after being oxidised to sulphate when exposed to oxygen action (Fig. 14.4).

The occurrence of sulphate reduction processes in the aquifer has been observed with sulphur isotopes (Higueras *et al.*, 2011). Temporal chemical changes in shallow groundwater nearby some lagoons are driven not only by reduction process: evaporation (and evapotranspiration) and mixing with marine salts, probably as airborne supply, are also significant processes (Figs. 14.5a and b). This modifies groundwater salinity close to the water-table and, eventually, its ionic composition. The process is seasonal around some lagoons (e.g. Santa Olalla, samples # 15 and 16 in Fig. 14.5), but in other cases (e.g. Charco del Toro, sample # 14 in Fig. 14.5, see also Fig. 14.4) sulphate concentration in groundwater nearby the ponds is increasing in the last years, pointing to enhanced oxidation after desiccation of the S accumulated in lake sediments. This sink function is currently related to non-permanent lagoons that receive water sporadically, and seems to be the effect of intensive groundwater abstractions that turned formerly permanent wetlands into temporary ones. Intensive and localised pumping from deep layers over more than 30 years has caused accumulated local groundwater head drawdown that induces the lowering of the water-table. This lowering has not yet stabilised, and preliminary modelling results show that the system as a whole needs about 30 years to reach the midway between a previous state and the final state after a change has been produced (Lozano, 2004).

The simulated effect of land cover changes relative to shrub-like vegetation is shown in Figure 14.6. The figure caption explains the main results. Other human

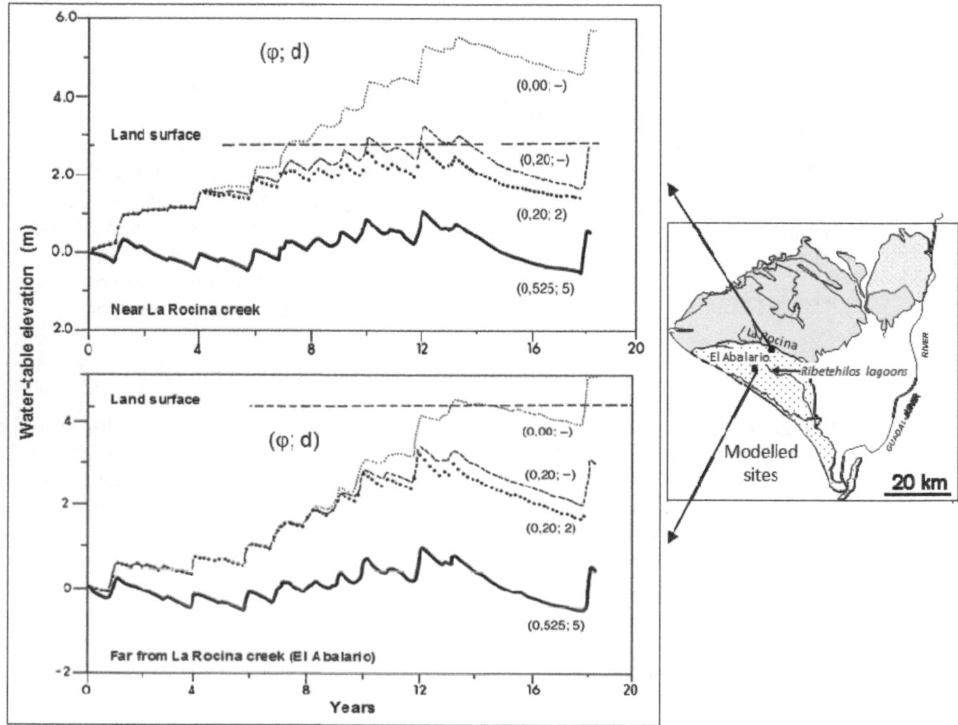

Figure 14.6 Modelled water-table evolution in El Abalario area after different vegetation-management alternatives. Pumping volumes maintain as in 1998. φ is the maximum phreatic evaporation (m/year) and **d** is the maximum depth available for phreatic evaporation (m). The simulated scenarios are: $\varphi = 0.525$ m/y and d = 5 m is the actual calibrated situation in 1998; $\varphi = 0.2$ m/y and d = 2 m is the forecasted evolution if eucalyptus were substituted by native vegetation with **d** around 2 m; $\varphi = 0.2$ m/y and d = "-" is the evolution if eucalyptus were substituted by a native vegetation able to adapt to any "d" (d is variable), and $\varphi = 0.0$ m/y and *d* = "-" is the evolution if eucalyptus were eliminated but not substituted by any other vegetation. The figure suggest that: 1) eucalyptus elimination without reforesting would lead to a phreatic level rise above soil surface in less than 8 years close to La Rocina creek, and in some 18 years far to it; 2) eucalyptus substitution by a vegetation able to reach whichever water-table depth would rise the phreatic level some 2 m close to La Rocina, and between 2 and 3 m far to it; 3) the impact of eucalyptus trees introduced some 50 years ago in El Abalario area was a water-table lowering between 0.5 m and a few metres. Modified after Custodio (2000).

activities disturbing the aquifer natural regime and that of the related wetlands are groundwater recharge changes and sediment relocation due to deforestation, the introduction about fifty years ago of eucalyptus trees – with greater phreatic water consumption than native Mediterranean vegetal species – and their eradication between 1995 and 2002, and groundwater pollution by local urban sewage, agricultural wastes (nutrients, pesticides, diffused over large areas), airborne contaminants deposition, and heavy metals and volatile fertilizers (NH_3).

14.5 CONCLUDING REMARKS

The integrated hydrodynamic, hydrochemical and isotopic studies carried out along the years in Doñana, combined with groundwater monitoring improvement, have greatly contributed to the knowledge of the groundwater flow pattern, transit and residence times, and the role of groundwater in the wetlands. Most wetlands in the Doñana region are groundwater-dependent, and those located on the sandy recharge area are very sensitive to groundwater flow modifications.

Intensive groundwater withdrawal during more than three decades localised in natural aquifer discharge areas near the marshes, has led to an evolving transient period. Accumulated drawdown of the more permeable deep aquifer piezometric levels induced water-table lowering. Seepage to wetlands and to phreatophyte areas located in the aquifer sandy recharge area has decreased which results in severe hydrological damage to most wetlands by reducing the inundation and/or the soil saturation frequency and lasting. Water table lowering also hinders phreatophyte roots to reach the water table level. Consequently, many small phreatic lagoons around the cultivated areas which were permanent 20–30 years ago, are nowadays seasonal or even sporadic, holding water only in very wet years. Also the formerly permanent ravines discharging to La Rocina creek or to the marshes along La Vera (western ecotone) are nowadays seasonal, some of them fed mostly with excess irrigation drainage water from the nearby crop fields to the west.

The present hydrodynamic situation is a transient one evolving towards a new equilibrium with the current relation between aquifer recharge pattern and discharge. Preliminary modelling indicates that after some change has been produced the system needs about 30 years to reach midway between a previous state and the final state. This means that to restore the wetlands natural functioning, the water balance terms of the aquifer have to be restored and enough time has to be allowed for the transient evolution to fade out. This could be partially attained either by reducing total abstraction or by well relocation.

Modelling also helped to evaluate the impact of eucalyptus tree introduction some 50 years ago in broader areas of the eolian mantle, more intensively in El Abalario central area. It produced a water–table lowering between 0.5 m and a few metres, which is a very significant value from the point of view of natural shrub-like vegetation and of the small, seasonal ponds depending on local, phreatic water discharges. Those eucalyptuses were eliminated between 1995 and 2000, and the natural vegetation was allowed to regenerate by itself. This, together with a couple of wet years (1998–99), favoured a water-table uprising around 0.5–1 m. Also, some of the originally seasonal lagoons that were converted into sporadic (e.g. the Ribatehilos ponds, NE of El Abalario) are again holding water every wet season. This behaviour clearly shows the relevant role of land and vegetation cover management on wetland functioning, even on a mid-term scale.

Hydrogeochemical and environmental isotope studies show that although groundwater composition dominates wetlands chemistry at regional scale, reactions taking place within the lagoons have relevant influence on the composition of the nearby shallow groundwater: changes in water salinity, ion composition and isotopic concentration originated in the lakes can be found in groundwater nearby the wetlands, and they are persistent. The largest, permanent lagoons (Dulce, Santa Olalla) are net evaporation surfaces that become progressively salty in dry periods, but the saline balance seem to be maintained by episodic flooding in wet periods, during which salts are exported via occasional surface outflows. However, in other wetlands (e.g. Charco del Toro) the balance of salts has been disturbed after the modification of the wetland hydroperiods derived from the water-table lowering due to intensive groundwater abstraction, and the hydrochemical processes taking place within the wetlands are changing. Thus, groundwater management is having a serious impact not only on groundwater discharge to wetlands by seepage and evapotranspiration, but also on wetland and groundwater chemistry.

ACKNOWLEDGEMENTS

The different works have been founded by the Spanish Government (Interministerial Commission for Science and Technology, Projects P1387.0842, AMB-92.636; AMB-95.0372; HID.99-205; REN 01.1293, and CGL2009-12910-CO3-03, the Ministry of Environment –Project Doñana 2005–, the European Commission (Projects ENV4.CT95.0156 and EVK1-1999.0032P/2002.00527) and the International Atomic Energy Agency, CRP project "Isotopic Techniques for Assessment of Hydrological Processes in Wetlands" (2007–2011).

REFERENCES

Baonza, E., Plata, A. & Silgado, A. (1984) *Hidrología isotópica de las aguas subterráneas del Parque Nacional de Doñana y zona de influencia [Isotopic Hydrology of Groundwater in the Doñana National Park and Influence Zone]. Monograph.* Centro de Estudios y Experimentación de Obras Públicas, Madrid, C7, 1–139.

Coleto, I. (2003) *Funciones hidrológicas y biogeoquímicas de las formaciones palustres hipogénicas de los mantos eólicos de El Abalario-Doñana (Huelva) [Hydrological and biogeochemical functions of hypogenic palustrine formations in the aeolian mantle of El Abalario-Doñana (Huelva)].* Ph.D. Thesis, Universidad Autónoma de Madrid, Spain.

Custodio, E., Manzano, M. & Iglesias, M. (1996) Análisis térmico preliminar de los acuíferos de Doñana. *IV Simposio sobre el Agua en Andalucía (SIAGA-96). Almería II*, 57–87.

Custodio, E. (2000) Groundwater-dependent wetlands. *Acta Chimica Hungarica*, 43 (2), 173–202.

Custodio, E. & Palancar, M. (1995) Las aguas subterráneas en Doñana [Groundwater in Doñana]. *Revista de Obras Públicas*, 142 (3340), 31–53.

Delgado, F., Lozano, E., Manzano, M. & Custodio, E. (2001) Use of environmental isotopes and chemical tracers to characterize the relationships between phreatic and saline fresh water lakes and the aquifer. *Proceedings of the 3rd International Conference on Future Groundwater Resources at Risk, IAH, Lisbon, Portugal.*

FAO (1972) Proyecto de utilización de aguas subterráneas para el desarrollo agrícola de la cuenca del Guadalquivir. Anteproyecto de transformación en regadío de la zona Almonte–Marismas (margen derecha). [Project for using groundwater for the agricultural development in the Guadalquivir River basin. Pre-project on transformation into irrigated-land of the Almonte-Marismas zone (river right margin)]. United Nations Programme for Development, Food and Agriculture Organization. Technical Report I, AGL: SF/SPA 16. Roma, 2 volumes.

Higueras, H., Manzano, M., Custodio, E., Juárez, I., Puig, R. & Aravena, R. (2011) Isotopic assessment of the impact of agriculture on the hydrology of the aquifer and wetlands at the Doñana Ramsar site, SW Spain. *Geophysical Research Abstracts*, 13, 6789.

Iglesias, M. (1999) *Caracterización hidrogeoquímica del flujo del agua subterránea en El Abalario, Doñana, Huelva [Hydrogeochemical characterisation of groundwater flow in El Abalario, Doñana, Huelva].* Ph.D. Thesis, Technical University of Catalonia, Barcelona, Spain.

Konikow, L.F. & Rodríguez-Arévalo, J. (1993) Advection and diffusion in a variable-salinity confining layer. *Water Resources Research*, 29 (8), 2747–2761.

Lozano, E. (2004) *Las aguas subterráneas en los Cotos de Doñana y su influencia en las lagunas [Groundwater in the Cotos area of Doñana and its influence on lagoons].* Ph.D. Thesis, Technical University of Catalonia, Barcelona, Spain.

Lozano, E., Delgado, F., Manzano, M., Custodio, E. & Coleto, C. (2005) Hydrochemical characterisation of ground and surface waters in "the Cotos" area, Doñana National Park, southwestern Spain. In: Bocanegra, E.M., Hernández, M.A. & Usunoff, E. (eds.) *Groundwater and Human Development.* International Association of Hydrogeologists, Selected Papers on Hydrogeology 6, Leiden, Balkema.

Manzano, M. (2001) *Clasificación de los humedales de Doñana atendiendo a su funcionamiento hidrológico [The Doñana wetlands typification according to their hydrology]. Hidrogeología y Recursos Hidráulicos*, 24, 57–75.

Manzano, M., Custodio, E., Loosli, H.H., Cabrera, M.C., Riera, X. & Custodio, J. (2001) Palaeowater in coastal aquifers of Spain. In: Edmunds, W.M. & Milne, C.J. (eds.) *Palaeowaters in Coastal Europe: Evolution of Groundwater since the late Pleistocene. Geological Society of London, Special Publication*, 189, 107–138.

Manzano, M., Custodio, E., Mediavilla, C. & Montes, C. (2002) Metodología de tipificación hidrológica de los humedales españoles con vistas a su valoración funcional y a su gestión. Aplicación a los humedales de Doñana [A methodology for the hydrological classification of the Spanish wetlands as a basis for their functional evaluation and management. Application to the Doñana wetlands]. *Boletín Geológico y Minero*, 113, 313–330.

Manzano, M., Custodio, E., Mediavilla, C. & Montes, C. (2005) Effects of localised intensive aquifer exploitation on the Doñana wetlands (SW Spain). In: Sahuquillo, A., Capilla, J., Martínez-Cortina, J.L. & Sánchez-Vila, J. (eds.) *Groundwater Intensive Use. Selected Papers on Hydrogeology* 7. Leiden, Balkema.

Manzano, M., Custodio, E., Iglesias, M. & Lozano, E. (2007a) Groundwater baseline composition and geochemical controls in the Doñana aquifer system (SW Spain). In: Edmunds, W.M. & Shand, P. (eds.) *The Natural Baseline Quality of Groundwater.* Oxford, Blackwell Publishing.

Manzano, M., Custodio, E. & Higueras, H. (2007b) Groundwater and its functioning at the Doñana RAMSAR site wetlands (SW Spain): Role of environmental isotopes to define the flow system. *International Symposium in Advances in Isotope Hydrology and its Role in Sustainable Water Resources Management.* International Atomic Energy Agency, Vienna. pp. 149–160.

PAH (2002) Plan Andaluz de Humedales [Andalusian Wetlands Plan]. Consejería de Medio Ambiente, Junta de Andalucía. Sevilla. http://www.juntadeandalucia.es/medioambiente/site/web/menuitem.a5664a214f73c3df81d8899661525ea0/?vgnextoid=ce0d731f73277010VgnVCM1000000624e50aRCRD&vgnextchannel= 3259b19c7acf2010VgnVCM1000001625e50aRCRD&lr=lang_es. Viewed 28 February 2011.

Poncela, R., Manzano, M. & Custodio, E. (1992) Medidas anómalas de tritio en el área de Doñana [Anomalous tritium values in the Doñana area]. *Hidrogeología y Recursos Hidráulicos*, 17, 351–365.

Trick, Th. & Custodio, E. (2004) Hydrodynamic characteristics of the western Doñana Region (area of El Abalario), Huelva, Spain. *Hydrogeology,Journal*, 12, 321–335.

UPC (1999) *Regional groundwater flow model in the Almonte-Marismas aquifer. Groundwater-Hydrology Group of the Technical University of Catalonia and Geological Institute of Spain, Madrid*. Unpublished report, in Spanish.

Zazo, C., Goy, J.L., Lario, J. & Silva, P.G. (1996) Littoral zone and rapid climatic changes during the last 20,000 years: the Iberian study case. *Z. Geomorph. N.F.; Berlin–Stuttgart, Suppl.*, 102, 119–134.

Chapter 15

Groundwater dependent ecosystems associated with basalt aquifers of the Alstonville Plateau, New South Wales, Australia

Richard T. Green[1], *Ross S. Brodie*[2] & *R. Michael Williams*[3]

[1]*NSW Office of Water, Department of Primary Industries, Grafton NSW, Australia*
[2]*Geoscience Australia (GA), Canberra ACT, Australia*
[3]*NSW Office of Water, Department of Primary Industries, Sydney NSW, Australia*

ABSTRACT

Water reforms in Australia recognise the environment as a legitimate user of a groundwater resource. This presented a challenge for scientists to determine the environmental dependency on groundwater from a Tertiary basalt aquifer in northern New South Wales (NSW) as part of an aquifer management plan. The Alstonville Plateau fractured rock basalt aquifer is characterised by a high degree of surface water connectivity and a shallow dynamic watertable. Three fundamental types of groundwater dependent ecosystems (GDEs) were recognised as existing on the plateau, being (1) groundwater-fed wetlands such as seepages and springs; (2) riparian and aquatic ecosystems (including the hyporheic zone) within or adjacent to streams fed by groundwater baseflow and (3) terrestrial vegetation communities that have seasonal groundwater dependency, such as rainforest remnants that access the shallow groundwater. Hydrogeological assessment of the basalt aquifer was combined with detailed mapping of springs and seepages and associated floral and faunal dependencies. This GDE mapping was incorporated into the gazetted water sharing plan for the aquifer (DLWC, 2003) and formed the basis for management rules defined for buffer zones around high-priority ecosystems.

15.1 INTRODUCTION

The Alstonville Plateau fractured rock basalt aquifer covers an area of about 391 km^2 in the north-east of the State of New South Wales (NSW), Australia. It is located around the town of Alstonville situated inland from the coastal regional centre of Ballina, about 700 km north of Sydney (Fig. 15.1). The Lismore Basalt volcanic flows and interbedded sediments form the plateau that rises up to 200 m above the coastal plain. Due to its latitude and elevated position close to the coast, the Alstonville Plateau experiences relatively high annual rainfall with an average of 1300 mm/yr (Brodie & Green, 2002). Average monthly rainfall statistics highlight the relatively dry late-winter to spring period with September typically the driest month.

The plateau was originally covered by rainforest, which was initially cut for hardwoods, followed by clearing for dairying by the early 1900s. From the 1970s, horticulture intensified and more recently the regional population has rapidly grown. This means that the greatest demand for groundwater on the North Coast of NSW

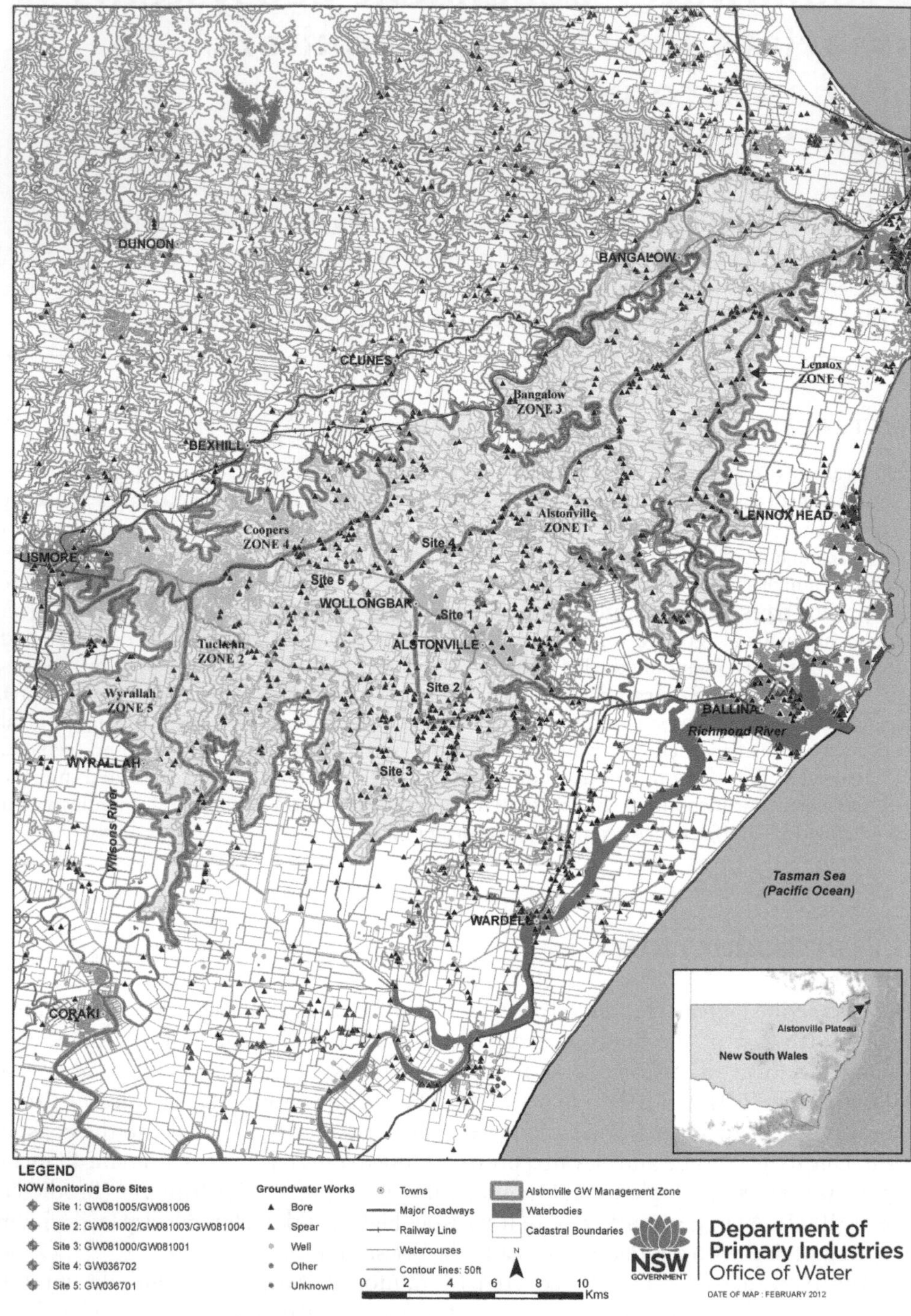

Figure 15.1 Location of the Alstonville Plateau basalt groundwater management area, NSW.

is in the area around Alstonville. The low salinity groundwater is generally used for horticulture, town water supply, domestic and stock. Yields generally range from 1 to 15 l/s, with maximum yields of 38 l/s. Groundwater is extracted from shallow wells, excavations, natural springs and bores. Since the 1980s the trend has been to construct deeper bores to obtain more reliable groundwater supplies. There are about 800 licensed bores on the Alstonville Plateau, with 150 being medium to high-yielding bores used mostly for the irrigation of stone fruit orchards, pasture, and to a lesser extent, vegetable crops, as well as for municipal supply. Interference effects caused by over extraction of groundwater have been reported between competing water users. The pattern of development defined for the "Alstonville Groundwater Sources" across the plateau is shown in Figure 15.1.

This paper details the assessments undertaken to determine the fundamental types of groundwater dependent ecosystems (GDEs) that exist on the Alstonville plateau and also describes how they were incorporated into the groundwater sharing plan for this aquifer.

15.2 ALSTONVILLE PLATEAU GROUNDWATER MANAGEMENT

In 1992 the Commonwealth, state and territory governments of Australia agreed to the *National Strategy for Ecologically Sustainable Development* (Commonwealth of Australia 1992). Following this, the Council of Australian Governments (COAG) released a *National Water Reform Framework* agreement outlining the reforms to the management of the nation's water resources (COAG, 1994). COAG then commissioned reform recommendation reports specifically on groundwater and water for ecosystems (SLWRMC, 1996a, 1996b). In essence these recommend that the State agencies responsible for groundwater management ensure groundwater management plans include identification of the sustainable yield and the appropriate levels of allocation and use of aquifers.

The NSW Office of Water (Department of Primary Industries) administers the NSW Water Management Act (DLWC, 2000), which is the primary mechanism for managing groundwater on the Alstonville Plateau. The basalt aquifer was classified as high risk in terms of over allocation and potential for contamination (DLWC, 1998a), so was defined as a groundwater management area (GWMA804) enabling development of a water sharing plan (Fig. 15.1).

Only where water sharing plans are in place does the *Water Management Act 2000* (DLWC) apply and replace the *Water Act 1912* (NSW, 1912). Under both Acts, the rights to the control, use and flow of all water in aquifers are vested in the State. The Acts provide the direction for the statutory process to allow individuals and other entities access to the resource by way of licences and approvals. Water sharing plans establish the rules for sharing water between the environment and water users and between competing consumptive water users. The water sharing plans, once finalised and gazetted under the Water Management Act 2000, have legal effect for 10 years.

The *Environmental Planning and Assessment Act 1979* (DUAP, 1979), is the primary mechanism for managing land uses that may threaten groundwater. The protection of groundwater from contamination is primarily governed by the *Protection of the Environment Operations Act* 1997 (EPA, 1997). The legislation is supported by the

NSW State Groundwater Policy, comprising a Framework Document (DLWC, 1997), a *Groundwater Quality Protection Policy* (DLWC 1998b), a *Groundwater Dependent Ecosystems Policy* (DLWC, 2002), and a *draft Groundwater Quantity Management Policy* (DLWC, in draft). The policies establish management principles and practice for achieving the State's goal of managing the State's groundwater resources so that they can sustain environmental, social and economic uses (DLWC, 1997).

The Water Sharing Plan for the Alstonville Plateau Basalt Groundwater Sources (DLWC, 2003) commenced on 1 July 2004 and applies to 30 June 2014. It contains a number of objectives aimed at sustainable groundwater extraction, protection of groundwater quality and maintenance of groundwater dependent ecosystems (GDEs), as follows:

(i) ensure the on-going maintenance and enhancement of groundwater quantity and quality across these groundwater sources;
(ii) provide sustainable access to water for town water supplies;
(iii) maintain the groundwater contribution to surface waters for the protection of water dependent ecosystems and extractions in downstream sub-catchments;
(iv) provide sustainable access to groundwater for irrigation and commercial purposes;
(v) preserve and enhance ecosystems that depend on groundwater in these groundwater sources;
(vi) minimise the risk of contamination of these groundwater sources, and;
(vii) manage extraction in order to maintain the beneficial use category of these groundwater sources.

The State government convened a water management committee to develop a draft groundwater sharing plan (DLWC, 2002). In 2002, the Northern Rivers Water Management Committee upon receipt of information that there was a noticeable decline in groundwater levels over the previous decade made the decision to adopt the precautionary principle by reserving 80% of the average annual recharge for the environment. The sustainable yield (or extraction limit) is that proportion of the long-term average annual recharge, which can be extracted each year without causing unacceptable impacts on the environment or other users.

On the plateau, groundwater divides are assumed to coincide with the boundaries of surface water sub-catchments. They are defined as groundwater sources as the shallow aquifers are recharged by rainfall, which supports the local streams. The six water sources are also known as zones one to six with the names of Alstonville, Tuckean, Bangalow, Coopers, Wyrallah and Lennox (Fig. 15.1).

The primary form of recharge for the basalts is from rainfall infiltration. A recharge rate of 8% of rainfall was selected as an average to be calculated across the surface of the Alstonville aquifer. Recharge was calculated based on assessed rainfall, area and proportion of rainfall infiltrating the aquifer:

$$\text{Average Annual Recharge (AAR)} = \text{Aquifer Area} * \text{Proportion Infiltration} * \text{Average Annual Rainfall}$$

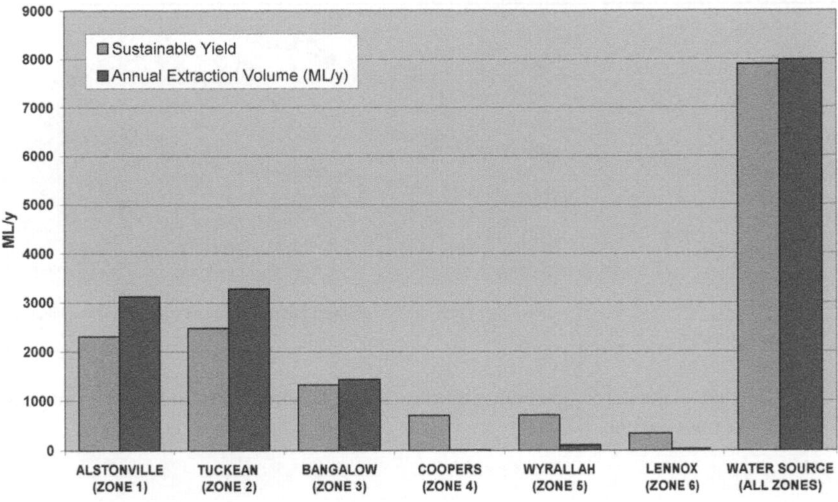

Figure 15.2 Alstonville Plateau basalt groundwater sources – sustainable yield volume compared to annual groundwater extraction allocation (includes access licenses plus estimated stock and domestic usage from 2003).

The AAR for the Alstonville groundwater management area was calculated to be 44 472 Ml/yr), with 80% of this annual recharge allocated to the environment and 20% for extractive use. The sustainable yield (or long-term average extraction limit – LTAEL) for the system is 8895 Ml/yr with licensed allocation estimated to be at capacity. An embargo was enacted on the issuing of new access licenses as current allocations were nearing or exceeding extraction limits for many management zones. The aquifer sustainable yield and estimated annual extraction volumes are shown in Figure 15.2, based on the 2003 water sharing plan document (DLWC, 2003).

15.3 HYDROGEOLOGICAL AND ECOLOGICAL ASSESSMENT

A collaborative project between the Department of Land and Water Conservation (DLWC) and the Australian Government agency of the Bureau of Rural Sciences (BRS) was instigated to provide the baseline information to support the development of the water sharing plan. This involved investigating aquifer characteristics as well as defining the ecosystem dependencies on the groundwater resource on the plateau, including:

(i) Developing a conceptual model of the plateau hydrogeology;
(ii) Assessing the groundwater dynamics of shallow and deep aquifers;
(iii) Investigating groundwater discharge to the plateau streams;
(iv) Constructing a numerical groundwater flow model;
(v) Identifying and mapping groundwater dependent ecosystems.

These investigations are described below.

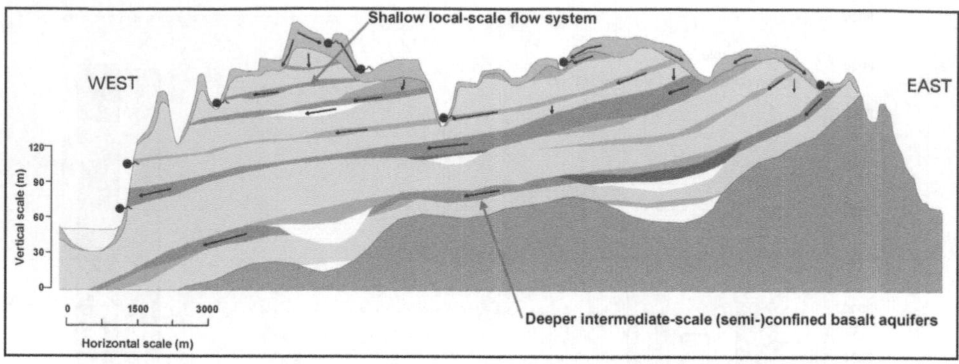

Figure 15.3 Conceptual model for groundwater flow in the Alstonville groundwater management area (Brodie and Green, 2002).

15.3.1 Hydrogeological conceptual model

The Alstonville Plateau is a dissected remnant of basaltic flows and interbedded sediments of the Lismore Basalt, overlying Triassic-Cretaceous sediments of the Clarence-Morton Basin. These flows are the most southerly extension of the Tertiary Lamington Volcanics that are associated with the Tweed Shield Volcano (20–23 Ma). The flows are interpreted to be continental basaltic successions of Cas and Wright (1995), displayed by flood and valley-fill lavas that were extruded across the landscape. Multiple vents or fissures on the shield flanks, rather than just the central vent explain the large regional extent of the Lismore Basalt. These flows rarely exceed 20 m in thickness and they show a broad range of textures including massive extremely fine-grained, coarser even-textured, porphyritic, vesicular, amygdaloidal or laminar variants (McElroy, 1962). Also aerial deposition of scoria and tuff occurred. Macro structures range from intense fracturing, very close random jointing, columnar jointing to large coherent masses (Chesnut & Swane, 1976). Also Drury (1982) noted deposition of calcite, analcite, common opal, prehnite, chalcedony, chabosite, sulphides and labradorite in fractures and vesicles.

A series of hydrogeological cross sections were compiled for the Alstonville Plateau, using the available borehole logs (Brodie & Green, 2002). A conceptual model for the plateau hydrogeology was subsequently developed and shown in Figure 15.3. The Lismore Basalt consists of a series of basalt flows with a combined maximum thickness estimated at 225 m, which generally dip shallowly to the north-west. Aquifer horizons consist of medium to highly fractured or vesicular basalt, old soil profiles or sediments. These developed during a prolonged hiatus between volcanic activity, before the deposition of further lavas. At least eight of these marker horizons have been identified by deep drilling (Ross *et al.*, 1989). Interflow fluvial systems tend to be incisive and erosional, hence sand and gravel deposits are neither thick nor extensive and tend to be preserved as deep leads. In addition some lacustrine and swamp environments formed from the damming of valleys by lavas, resulting in discrete diatomite layers. Following the cessation of Tertiary volcanism, erosion has removed a significant proportion of the lava pile, indicated by the undulating topography and incised streams. A significant weathering profile has also developed, forming the basalt soils valued for irrigation on the plateau (Brodie and Green, 2002).

In summary, the two main groundwater systems in the basalt plateau sequence are:

(i) A shallow unconfined aquifer in the upper mantle of soil and weathered or highly fractured basalt. Depending on the depth of fracturing, the unconfined aquifer can exceed 40 m in thickness. The basalt soils are poor sealing clays, so readily allow rainfall infiltration (Green *et al.*, 2001). Local-scale groundwater movement is topographically driven, discharging as hill-slope springs, valley seepages and baseflow to the plateau streams. Isotope analysis indicates that the shallow groundwater is modern (Williams, 1998; Budd *et al.*, 2000).

(ii) Deeper intermediate-scale aquifers in the interlayered and fractured horizons within the volcanic sequence. These aquifers become semi-confined to confined to the north-west. The deeper aquifers discharge at lower levels in the landscape, typically as springs along the base of the plateau escarpment. Isotope studies confirm that the deeper groundwater is over 50 years old (Budd *et al.*, 2000; 2002).

Relatively thick sequences of massive poorly fractured basalt can restrict vertical movement of groundwater to depth. However, vertical columnar jointing or fracturing can provide a degree of vertical interconnection in certain areas.

15.3.2 Groundwater dynamics

Monitoring bores have been constructed at various depths to assess the dynamics of the shallow and deep aquifers (Green, 2006). Analysis of groundwater levels for the period 1999 to 2006 confirms that the shallow aquifers (<50 m depth) are rapidly recharged by rainfall events but are also the first aquifers to be impacted by droughts (Fig. 15.4). The fluctuation in shallow groundwater levels typically varies annually by 1 to 4 m between the wetter summer and dryer winter seasons (Fig. 15.1. Sites: 1 to 5). It is apparent from the groundwater hydrographs that the recharge rates in the shallow aquifers are high. Brodie & Green (2002) found that there is a characteristic rapid rise in the shallow watertable immediately following rainfall events exceeding 100 mm/week, followed by a gradual watertable decline until the next significant rainfall event replenishes the aquifer. The groundwater level recession is due to discharge to local plateau streams, as well as evapotranspiration and groundwater extraction.

When these aquifers are not fully replenished during the wet season due to a prolonged drought, the result is that the watertable will drop to unseasonally low levels. This is reflected in reduced baseflow to streams and may cause the temporary loss or reduction in size of some springs and wetlands. During these drought periods the defined groundwater dependent ecosystems will be most at risk from significant shallow groundwater extraction.

Analysis of groundwater monitoring for the period 1999 to 2006 shows that the deep aquifers (>50 m depth) are not as responsive to significant rainfall events as the shallow system. The DLWC monitoring bore hydrographs are shown in Figure 15.5 for the deep aquifer with bores depths ranging from 58 to 168 m. In the

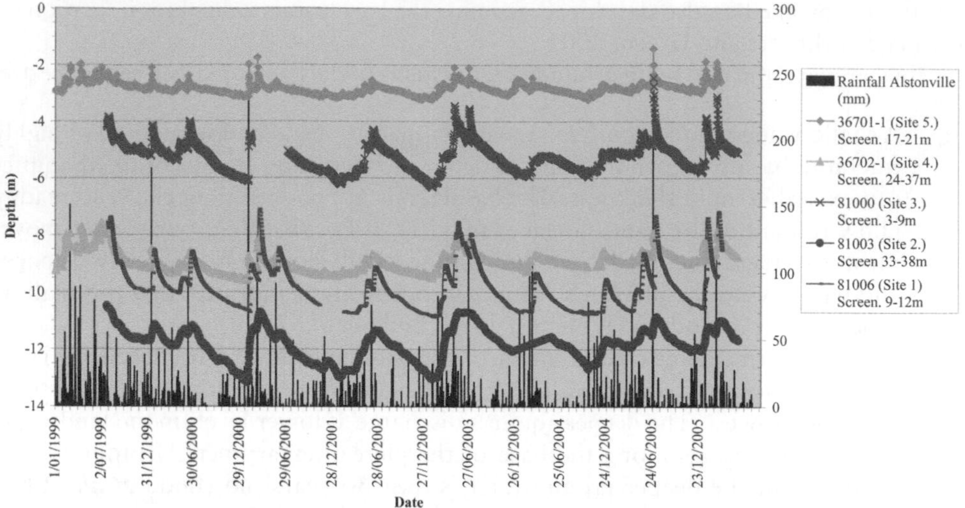

Figure 15.4 Comparison of shallow groundwater level monitoring and rainfall 1999–2006.

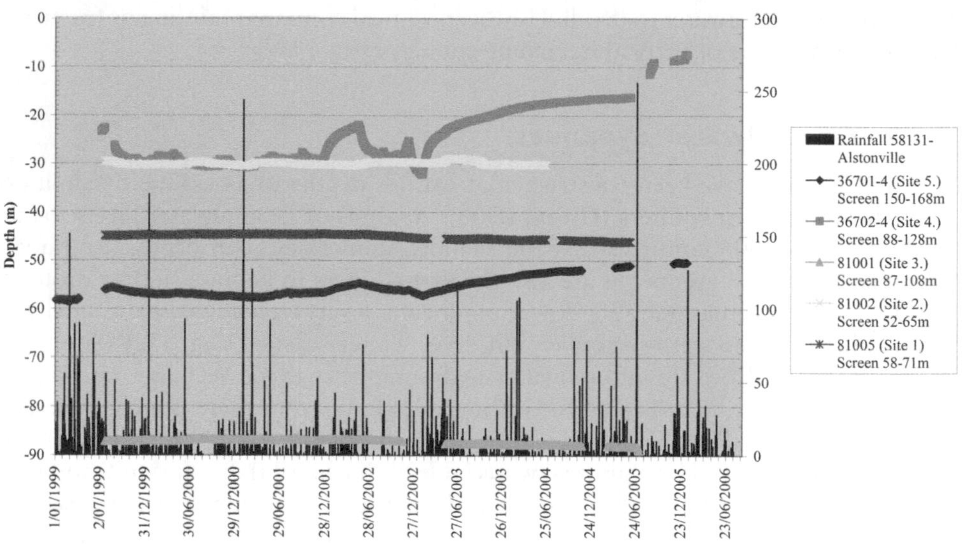

Figure 15.5 Comparison of deep groundwater level monitoring and rainfall 1999–2006.

south-eastern area of the plateau (Fig. 15.1. Sites: 1, 2 & 3), deep groundwater levels were relatively stable over the monitoring period. Data from deep monitoring bores including 81005, 81002 and 81001 show that from mid 2002 until mid 2005 there was a gradual decline in groundwater levels of up to 1.6 m in this area (Fig. 15.5). This slight decline started in the 2002 drought year, and continued through to 2006.

In contrast, the deep aquifers in the central part of the plateau west of Alstonville (Fig. 15.1. Sites: 4 & 5) have been impacted by over-extraction. The water levels in monitoring bores 36701-4 and 36702-4 in this area were exceptionally low due to significant extraction, until a recovery started in early March 2003, coincident with the end of the drought (Fig. 15.5). Recharge occurred in this area of the deep aquifer for the following three and a half years as shown in the bore hydrographs. Aquifer levels rose by 8 to 25 m across the aquifer, in the area previously affected by over-extraction. This rise in groundwater levels is due to the significant reduction in pumping, brought about by above average rains that broke the drought in early 2003.

Test pumping of bores screened in the deep aquifer has shown no impact on nearby shallow monitoring bores. A recently installed shallow monitoring bore (<10 m) located next to the town water supply bore at Lumley Park in Alstonville showed no drawdown during deep groundwater extraction (>60 m). This is important as the extraction bore is located near springs that feed nearby Maguire's Creek. Hence, this deep extraction has no effect on shallow aquifers which help to sustain these local ecosystems during dry times.

15.3.3 Groundwater discharge to plateau streams

The major streams on the plateau are perennial and gain groundwater from the basalt aquifers. The shallow unconfined aquifer has a high connectivity with local plateau streams, whereas the deeper aquifers generally discharge at the plateau base. Historical stream flow records show that even when there is no rainfall (and hence no runoff), streamflows progressively and significantly increase downstream. This was indicated by discrete streamflow measurements of major plateau streams during June and August 2002. The June stream gauging was undertaken after a significant rainfall event, while the August flows reflected a snapshot of the drier period of the 2002 drought. As an example, Gum Creek was shown to increase in flow by 100% from the top of the plateau to the base, in both surveys. It should be noted that tributaries into Gum Creek are minor and the majority of the increase in stream flow is from shallow groundwater discharge. The larger Marom Creek showed stream flow at the base of the plateau about 14 to 19 times the flow in the upper reaches, which could not be accounted for by tributary inflows.

Fluctuations in shallow groundwater levels also reflect changes in the discharge to local streams. The most recent drought year of 2002 showed shallow groundwater levels dropping to historically low levels and stream discharges from the plateau decreasing considerably.

15.3.4 Groundwater flow model

A MODFLOW groundwater model was constructed by DLWC for assessing the impacts of groundwater extraction from the deep aquifer system. Bilge (2003) developed the 2-layer model using both shallow and deep groundwater monitoring and the hydrogeological cross sections compiled by Brodie & Green (2002). For modelling purposes, the plateau hydrogeology was conceptually divided into the upper shallow unconfined layer and a lower layer. The regional groundwater flow in the Alstonville Plateau is to the north-west. The lower aquifer is generally confined to semi-confined, however along the north-west and south-west boundary it discharges at shallower depths, which free drain at the plateau base. The variability of the shallow aquifer system precluded

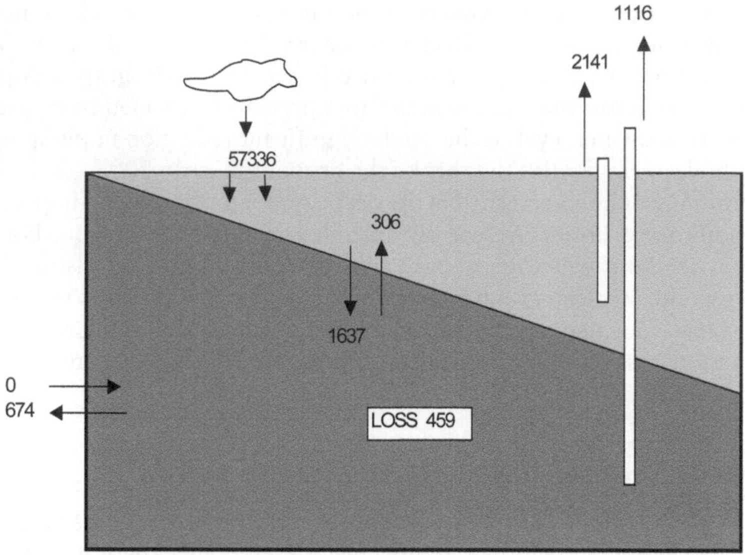

Figure 15.6 Water budget of the Alstonville Plateau groundwater model for July 1987 to June 2001, units are Ml/yr or 10^3 m^3/y, (Bilge, 2003).

it from being explicitly modelled. It is conceptualised as both a source of groundwater supplying local streams and some leakage to the deeper aquifers. The water balance was derived by Bilge (2003) for the calibration period spanning 14 years from July 1987 to June 2001. The estimated average annual groundwater budget for the model area is shown in Figure 15.6. The Modflow model was used to calibrate the yearly loss or gain in storage which is dependent on the amount of yearly extraction and net recharge from rainfall leakage to the deep aquifer. The storage loss in the model area for the deep aquifer amounted to 459 megalitres a year (Ml/yr). The model reflected the deep aquifer groundwater level results and showed that recharge to the deep aquifer is limited and that usage led to a drop in the level of the deep aquifer system. For the model to become operational, Bilge (2003) recommended that additional activities such as further construction of deep and shallow monitoring bores, metering and collection of water use data and additional groundwater level monitoring be undertaken.

15.3.5 Mapping of groundwater dependent ecosystems

An assessment of groundwater dependent ecosystems (GDEs) was undertaken for the Alstonville Plateau. Initially, an expert panel approach was taken with the convening of representatives from state agencies, universities, local councils and community groups. The workshop defined three key GDE types in the form of:

(i) *Wetlands*, fringing vegetation and aquatic communities permanently or temporarily underwater or waterlogged by groundwater, such as seepages and springs;

(ii) *River Base Flow Systems*, riparian and aquatic ecosystems (including the hyporheic zone) within or adjacent to streams fed by groundwater flow, and;

(iii) *Terrestrial Vegetation*, communities that have seasonal groundwater dependency, such as rainforest remnants that access the shallow groundwater.

Wetland seepages on the plateau can be dominated by sedges or bulrush that provide important habitat, or by melaleuca and used as an over-winter food source for migratory birds. Groundwater discharge is also important in sustaining major plateau streams during extended dry times, which is critical for aquatic ecosystems. It is considered that the maintenance of baseflow and stream pools by groundwater discharge is the main reason for a relatively high platypus population in plateau streams. There is also the potential for the shallow watertable to be accessed by native vegetation. Deep rooted rainforest trees such as black bean (*Castanospermum australe*) can have a seasonal or opportunistic dependency on shallow groundwater (Graham, 2001). This is of particular concern as only isolated remnants of subtropical rainforest totalling 140 hectares remain on the plateau.

To support the GDE assessment, over 1500 springs and about 150 seepage zones were mapped across the plateau (Brodie *et al.*, 2002). Aerial photography from the 1940s was used, as most of the plateau was then cleared dairy pasture with limited horticultural development or infestation by exotic species such as camphor laurel or lantana. Mid-slope springs were identified as dark patches interpreted as enhanced pasture growth, and in-stream springs were identified by the sharp transition from dry gully to flowing stream. Groundwater seepage areas could be identified as broader swampy reaches of the plateau streams. This enabled reaches of the plateau streams to be categorised into:

(i) *Intermittent*, with no or limited groundwater input and identified as dry watercourses;

(ii) *Perennial*, with low to moderate groundwater input and identified as watercourses containing water but limited development of adjacent swampy seepage areas; and

(iii) *Seepage*, with high groundwater input and identified as watercourses bordered by swampy seepage areas.

This seepage and spring mapping was combined with other datasets such as detailed vegetation mapping (Graham, 2001) and sightings of threatened flora and fauna as well as platypus sightings (Rohweder, 1992) to produce a map showing significant GDEs. An example of the GDE mapping near Alstonville is shown in Figure 15.7. This mapping was used as the basis for defining buffer zones around these ecosystems which were incorporated into the gazetted water sharing plan. For example, groundwater extraction exceeding 20 Ml/yr from new or replacement bores is excluded within 100 m of the high-priority ecosystems. The buffer distances were set so that there was adequate protection of the significant plateau ecosystems but also that there were adequate zones where groundwater extraction could be directed to. The water sharing plan for this aquifer was implemented in 2004, with a significant proportion of the average annual recharge reserved for the GDEs.

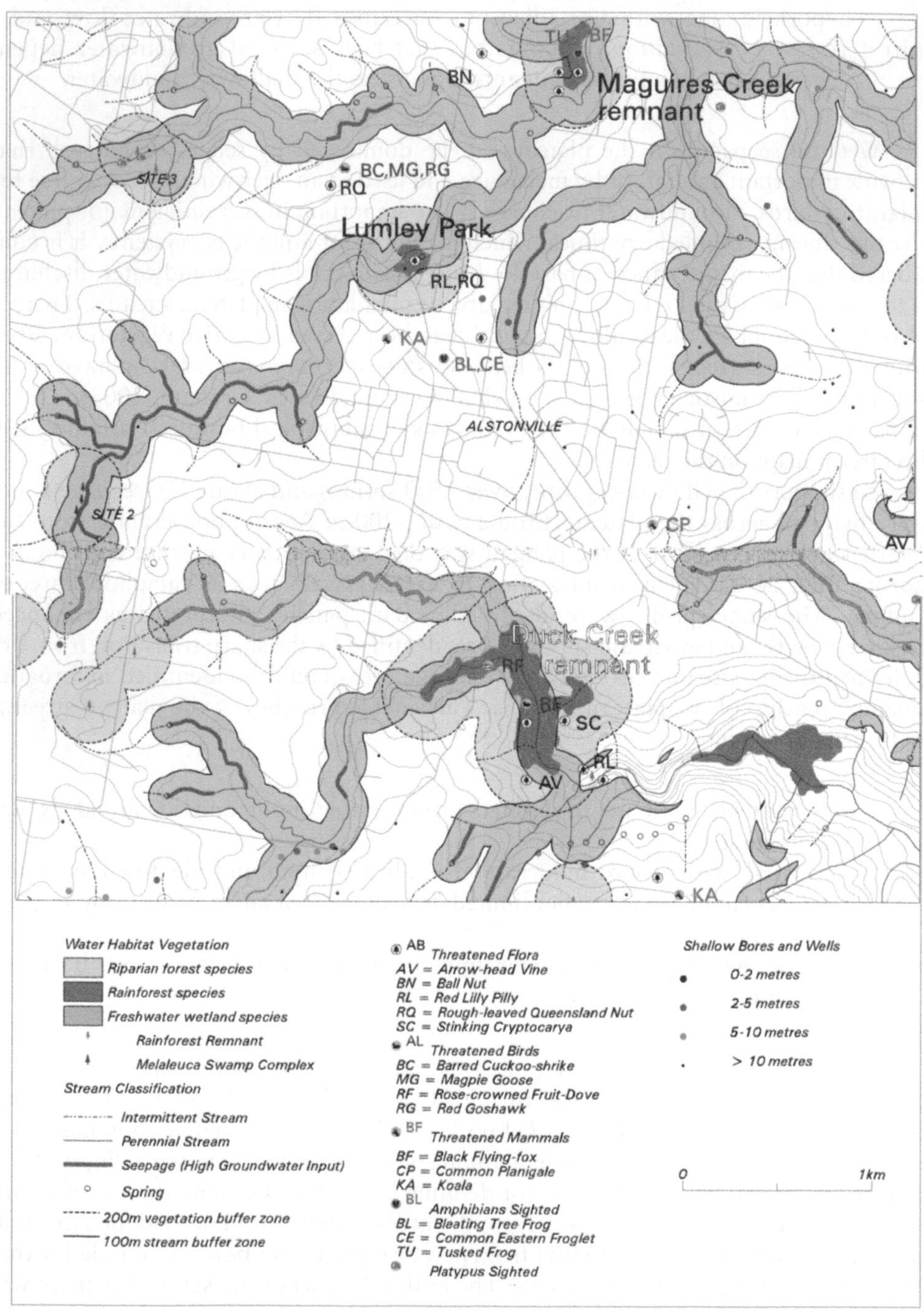

Figure 15.7 Groundwater dependent ecosystem mapping near Alstonville (Brodie *et al.*, 2002).

15.4 CONCLUSIONS

Hydrogeological investigations have shown that the complex multi-layered basalt aquifer of the Alstonville Plateau has a high degree of connectivity with the surface water system. As plateau streams rely on groundwater baseflow particularly during dry periods, they have been defined as groundwater dependent ecosystems and included in the legislation developed for the new water sharing plan. Detailed understanding of hydrogeological processes and ecosystem dependencies is crucial for managing such aquifers so that there are clear rules and guidelines for those stakeholders using groundwater. It will also assist government in assessing the potential extraction impacts on GDE's for groundwater trading and thus allow an informed decision-making process, in order to maintain important ecosystems into the future.

As part of the increased management required under the Water Sharing Plan, the NSW Office of Water completed installation of additional monitoring bores on the plateau in 2006. This was undertaken in order to enhance the groundwater network so that improved science would assist in the effective sustainable management of the aquifer.

REFERENCES

Bilge, H. (2003) *Alstonville Plateau Groundwater Model – Model Development and Calibration.* Sydney, NSW Department of Sustainable Natural Resources.

Brodie, R.S. & Green, R.T. (2002) *A Hydrogeological Assessment of the Fractured Basalt Aquifers of the Alstonville Plateau, NSW.* Canberra, Bureau of Rural Sciences.

Brodie, R.S., Green, R.T. & Graham, M. (2002) Mapping groundwater dependent ecosystems: a case study in fractured basalt aquifers of the Alstonville Plateau, NSW. *Proceedings, Darwin IAH Conference.*

Budd, K., Plazinska, A. & Brodie, R.S. (2000) *A Groundwater Quality Assessment of the Fractured Basalt Aquifers on the Alstonville Plateau, NSW.* Canberra, Bureau of Rural Sciences.

Budd, K., Brodie, R.S. & Green, R.T. (2002) Beneficial use of groundwater: A case study of groundwater quality protection in fractured basalt aquifers of the Alstonville Plateau, NSW. *Proceedings, Darwin IAH Conference.*

Cas, R.A.F. & Wright, J.V. (1995) *Volcanic Successions: Modern and Ancient.* London, Chapman & Hall. 528 pp.

Chesnut, W.S. & Swane, I. (1976) *Geological Factors Influencing Urban Development in the Lismore Region.* Geological Survey of NSW. Report GS1976/349 18pp.

Commonwealth of Australia (1992) *National Strategy for Ecological Sustainable Development.* AGPS Canberra.

COAG (1994) Water resource policy. In: *Compendium of National Competition Policy agreements.* 2nd edition. National Competition Council, Council of Australian Governments. pp. 99–109.

DLWC (1997) *NSW Government – The NSW State Groundwater Policy Framework Document.* NSW Water Reforms. NSW Department of Land and Water Conservation.

DLWC (1998a) *NSW Government – Aquifer Risk Assessment Report.* NSW Department of Land and Water Conservation.

DLWC (1998b) *NSW Government – The NSW State Groundwater Quality Protection Policy.* NSW Water Reforms. NSW Department of Land and Water Conservation.

DLWC (2000) *NSW Government – Water Management Act.*

DLWC (in draft) *NSW Government – The NSW Groundwater Quantity Policy.* NSW Water Reforms. NSW Department of Land and Water Conservation.

DLWC (2002) *NSW Government – The NSW Groundwater Dependent Ecosystems Policy.* NSW. Department of Land and Water Conservation.

DLWC (2002) *NSW Government – Draft Water Sharing Plan for the Alstonville Plateau Basalt Groundwater Source.* NSW Department of Land & Water Conservation.

DLWC (2003) *NSW Government – Water Sharing Plan for the Alstonville Groundwater Sources 2003.* NSW Department of Land and Water Conservation.

DUAP (1979) *NSW Government – Environmental Planning & Assessment Act.* NSW Department of Urban Affairs and Planning.

Drury, L.W. (1982) *Hydrogeology and Quaternary Stratigraphy of the Richmond River valley.* Ph.D. Thesis, Department of Geology, University of New South Wales, 2vol. 643pp.

EPA (1997) *NSW Government – Protection of the Environment Operations Act.* NSW Environmental Protection Agency.

Graham, M. (2001) *Water Habitats of the Tweed, Brunswick and Richmond Catchments.* Report 1, NSW North Coast Water Habitats Study, NSW National Parks and Wildlife Service.

Green, R.T., Williams, R.M. & Rumpf, C.S. (2001) *Sustainable Groundwater Management of High Risk Aquifers for the North Coast of New South Wales, Australia. New Approaches Characterising Groundwater Flow.* In: Seiler & Wohnlich. pp. 507–511.

Green, R.T. (2006) *Alstonville Groundwater Investigations.* Status report. NSW Department of Natural Resources, unpublished.

McElroy, C.T. (1962) *Geology of the Clarence-Moreton Basin. Memoirs of the Geological Survey*, NSW 9.

NSW (1912) *NSW Government – Water Act.*

Rohweder, D. (1992) *Management of Platypus in the Richmond River Catchment, Northern New South Wales.* Honours thesis, University of New England, Lismore.

Ross, J.B., Williams, R.M. & Rolfe, C. (1989) Groundwater recharge and management for the Alstonville Plateau, New South Wales. In: Sharma, M.L. (ed.) *Groundwater Recharge. Proceedings of the Symposium of Groundwater Recharge. Mandurah, 6–9 July 1987.*

SLWRMC (1996a) Allocation and use of groundwater – A national framework for improved groundwater management in Australia. Sustainable Land and Water Resource Management Committee. *Occasional Paper No. 2.*

SLWRMC (1996b) *National principals for the provision of water for ecosystems.* Sustainable Land & Water Resource Management Committee. *Occasional Paper SWR No. 3.*

Williams, J. (1998) *Interim Groundwater Management Plan for the Alstonville Plateau.* NSW Department of Land and Water Conservation, North Coast Region, Grafton.

Chapter 16

A shift in the ecohydrological state of groundwater dependent vegetation due to climate change and groundwater drawdown on the Swan Coastal Plain of Western Australia

Ray Froend[1]*, Muriel Davies*[1] *& Michael Martin*[2]

[1]*Centre for Ecosystem Management, Edith Cowan University, Australia*

[2]*Water Corporation, Western Australia*

ABSTRACT

The deep sand, unconfined aquifers of the Swan Coastal Plain in Western Australia support extensive open woodlands dominated by *Banksia* and other phreatophytes. Since the mid-1970s these ecosystems have been subject to declining annual rainfall and water-tables. In the summer of 1990/91, a phreatophytic *Banksia* woodland on the Swan Coastal Plain was subjected to increased rates of drawdown resulting in over 80% mortality of the phreatophytic overstorey species. The impacted *Banksia* woodland recovered, however, facultative phreatophyte species now dominate the overstorey, suggesting that the ecohydrological state of the site has shifted to one in which the dependence on groundwater access is reduced. A field experiment was performed over three consecutive summers, in which the recovered vegetation was subjected to further drawdown and its physiological water stress and water source partitioning compared to vegetation at reference sites.

16.1 INTRODUCTION

The Gnangara groundwater mound forms the superficial aquifer of the northern Swan Coastal Plain (SCP) in Western Australia. In order to protect the quality of this groundwater resource, several development restrictions exist, including maintenance of large areas of undeveloped land supporting native phreatophytic vegetation. The deep sand aquifers are integral to the persistence of these *Banksia* dominated woodlands, particularly during the Mediterranean-type hot, dry summer of the region. Phreatophyte use of the groundwater increases as moisture in the vadose zone is rapidly depleted (Zencich *et al.*, 2002) as the dry season progresses. However, since the mid-1970s these ecosystems have been subject to declining annual rainfall and water tables. Groundwater abstraction for horticulture and public water supply has also increased the rate of water table decline in parts of the SCP. As a consequence, impacts have been noted in a diverse array of groundwater dependent ecosystems including the phreatophytic *Banksia* woodland (Kite & Webster, 1989; Groom *et al.*, 2000, 2001).

The groundwater use of *Banksia* woodland plants has been found to vary temporally and with topography/depth to water table; variability which is significant to water resource management since it potentially determines how hydrological changes will impact upon native vegetation. Zencich *et al.* (2002) suggest that dependency on groundwater as a summer water source decreased with increasing depth to water table. *Banksia* that had developed in a habitat typified by a shallow (<3 m) water table displayed a far greater use of groundwater than individuals of the same species that developed in habitats with moderate (3–6 m) and deep (6–10 m) depths to the water table. One could extend this argument to vulnerability to groundwater drawdown, i.e. stands of *Banksia* that are more dependent on groundwater as a summer water source will be more vulnerable to drawdown. *Banksia* were shown to meet a larger proportion of their water requirements from the vadose zone, particularly at depth, and developed deeper roots to exploit soil moisture beyond the influence of summer evaporation and competition from shallow rooted species.

One of the first and most significant impacts of groundwater abstraction on native *Banksia* vegetation on the SCP was observed during the summer of 1985 within the vicinity of the Wanneroo wellfield, where up to 80% of all *Banksia* trees died (Mattiske and Associates, 1988). This prompted the then Water Authority of Western Australia to recognise the need to research groundwater abstraction impacts on phreatophytic woodlands (Kite and Webster, 1989). As part of a monitoring programme, vegetation assessment was conducted within close proximity to an abstraction bore (Pinjar bore P50) (Fig. 16.1) one year prior to becoming operational. Three years later (in February 1991), extensive overstorey death of the native vegetation occurred at P50 (Groom *et al.*, 2000; Froend & Sommer, 2010) as a consequence of both reduced rainfall and significant drawdown due to groundwater abstraction. Analysis of documented vegetation responses is critical in appraising actual and potential impacts of groundwater use.

Traditional 'Clementsian' succession (Clements, 1936), representing unidirectional change in vegetation towards a single stable climax community, may not adequately describe the process of ecosystem dynamics for groundwater dependent ecosystems on the Swan Coastal Plain (SCP). Altered environmental conditions such as climate, soils, interspecific interactions or groundwater abstraction may lead to different stable states that are inconsistent with traditional succession models. More recent models suggest that vegetation dynamics may involve transition phases between a series of different, more or less stable states (Westoby *et al.*, 1989; George *et al.*, 1992). Such models recognise transient conditions in which the vegetation does not persist but develops to another persistent state, with transitions occurring once a threshold has been exceeded, and progressing rapidly or more gradually depending on the management regimes (Westoby *et al.*, 1989; Pettit & Froend, 2001). The changes in the phreatophytic plant populations after changes to the water regime may best be described in terms of a state and transition model. The plant populations that develop under a particular water regime can be considered as separate more or less stable states. One may then consider plant populations impacted by rapidly changing water regime to represent a transitional stage between two ecohydrological states.

Prior to groundwater drawdown, the ecohydrological state at P50 was typical of shallow depths to groundwater (<3 m); dominated by obligate phreatophytes dependent on access to the water table throughout the dry summer months (Fig. 16.2). Over the last 16 years, the impacted *Banksia* woodland has 'recovered', however it has a reduced floristic similarity and plant abundance relative to the pre-impact state (Sommer & Froend, 2011). This floristic recovery is despite water tables remaining

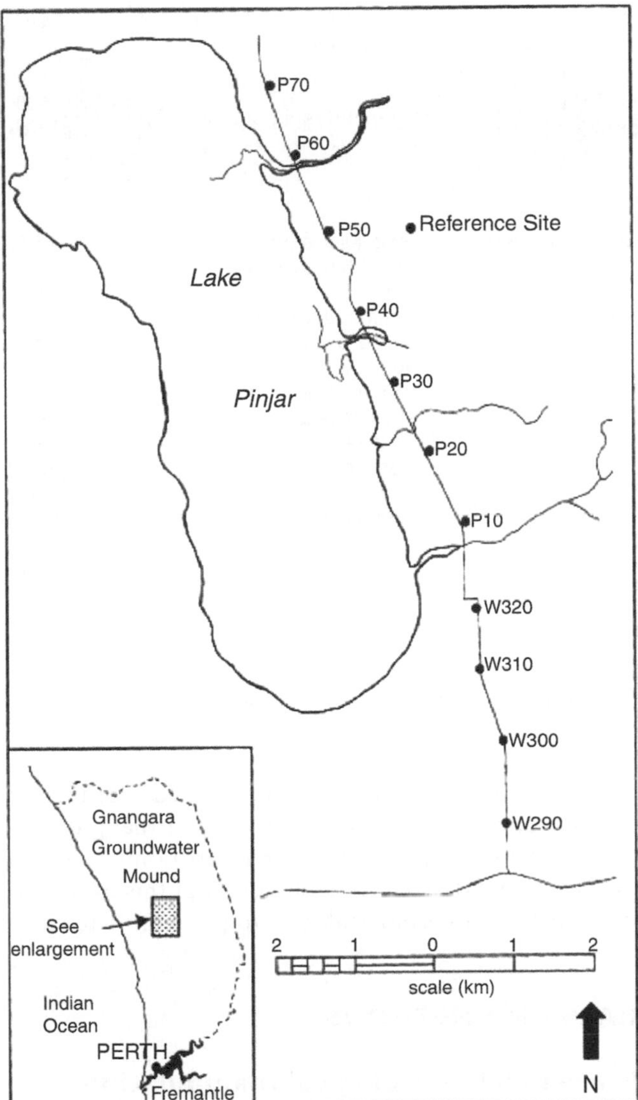

Figure 16.1 Location of the study site (bore P50) on the Pinjar Borefield near Perth, Western Australia.

up to 6 m below pre-impact levels. Relative densities of obligate phreatophyte tree species have reduced whilst the occurrence of vadophytes (plants with shallow roots restricted to the vadose zone) has significantly increased in the understorey. Facultative phreatophyte species (as opposed to obligate phreatophytes) now dominate the overstorey, suggesting that the ecohydrological habitat of the site has shifted to one in which the dependence on groundwater access is reduced relative to the pre-impact state. Therefore, one may expect that further groundwater drawdown will have less or no impact on the water status and survival of the new populations of *Banksia* at P50.

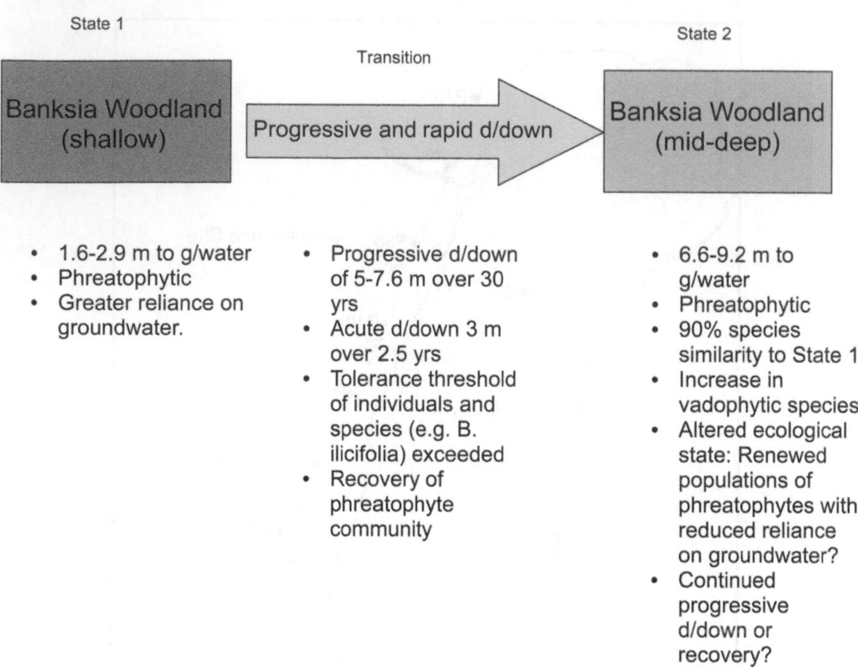

Figure 16.2 Conceptual representation of the observed ecohydrological states at P50.

To test this theory, a field experiment was performed over three consecutive summers, where the recovered vegetation was subjected to further groundwater drawdown and its physiological water status and water source partitioning compared to vegetation at reference sites. This paper describes the outcomes of this experiment and discusses the relevance of a shift in ecohydrological state to phreatophyte vulnerability.

16.2 MATERIALS AND METHODS

16.2.1 Study area and hydrological manipulation

The groundwater abstraction bore known as P50 (31°37′21″S, 115°49′11″E) is one of several bores comprising the Pinjar wellfield (Fig. 16.1), 38 km north of Perth, Western Australia. The vegetation surrounding the bore has an overstorey dominated by *Banksia attenuata*, *B. menziesii* and *B. ilicifolia* with an understorey typical of a sclerophyllous sandplain heath; consisting mainly of small shrubby species from the families Myrtaceae, Fabaceae, and Epacridaceae. Like Perth, the Pinjar borefield experiences a Mediterranean-type climate, with hot dry summers (December–March) and cool wet winters (June–August) with 870 mm average annual rainfall. The soil profile was predominately medium to coarse sand to depths well below the water table.

Vegetation directly adjacent to the production bore P50 represented the impacted site and was contrasted with a reference site over a kilometre away. The reference site was of a similar ecohydrological state (defined by phreatophytic species, depth to aquifer and soil stratigraphy) to the P50 site but was not subject to groundwater abstraction. Recovered *Banksia* woodland vegetation at the impacted site was subjected to controlled drawdown of the water table using the P50 production well, with the desired outcome being exposure to a rapid drawdown of over two metres in five months, maintained for three consecutive summers. During this period, plant water status and water source use was monitored. Initial artificial drawdown commenced at the start of the first drying phase (early summer 2005) and maintained thereafter for three consecutive drying seasons. This was believed to be sufficient for repeated measures of plant water deficit response if the plants were highly dependent on groundwater as a summer water source. Superficial groundwater levels were logged continuously via monitoring bores established at each site.

16.2.2 Plant response

Detailed assessment of physiological response was conducted on two overstorey species, *Banksia attenuata* and *B. ilicifolia*, which represent facultative and obligate phreatophytes respectively. All plant parameters were measured on three individuals of each species per site and on a seasonal basis. To examine the changes in water use patterns in response to altered water availability over the course of the trial, plant shoot water potential, transpiration and stomatal conductivity were monitored. Predawn and midday shoot water potential (Ψ MPa) was determined from each individual using a Scholander-type pressure chamber (Soil Moisture Equipment Company, Model MK3005, Santa Barbara, California, USA). Total conductance of water vapour (g_t, mmol m^{-2} s^{-1}) and transpiration rate (E, mmol m^{-2} s^{-1}) were measured from each individual under natural conditions at intervals over the daylight period (approximately 08:00–17:00) from sun-lit leaves with a steady-state porometer (LI-1600, Li-Cor, Lincoln, Nebraska, USA). Measurements were undertaken on the same day as leaf water potential determination. Total stomatal conductance to water vapour was corrected for the effect of boundary layer conductance (g_b, mmol m^{-2} s^{-1}) to yield stomatal conductance to water vapour (g_s, mmol m^{-2} s^{-1}).

The relative abundance of the stable isotope of hydrogen in the study species was assessed to clarify the sources of water utilised at intervals throughout the trial. By identifying the sources of water used, and the transitions between these over the course of the experiment, the use of groundwater by the study species can be demonstrated rather than inferred. Samples of plant twig tissue, soil moisture from the vadose zone, rainfall and groundwater were sampled for analysis. Water was extracted from soil and non-photosynthetic stems by cryogenic vacuum distillation (Dawson, 1993). Isotopic ratios of hydrogen ($^2H/^1H$) were then measured from extracted water (soil water, stem xylem water and groundwater) with a continuous flow mass spectrometer (PDZ Europa Model 20-20 Cheshire, England). The data produced were normalized and expressed relative to the V-SMOW standard:

$$\delta^2H = (R_{sample} - R_{standard} - 1) \cdot 1000(‰)$$

where R_{sample} and $R_{standard}$ are the hydrogen isotopic ratios of the water sample and the V-SMOW sample respectively.

A three component water source mixing model was applied to the δ^2H data to predict the likely contribution of a water source to that present in plant xylem (Zencich *et al.*, 2002). The three potential water sources were considered to be upper soil (surface to 4 m depth), lower soil (4–5 m) and deep soil (5–8 m). Deep soil included capillary fringe groundwater when present.

16.3 RESULTS AND DISCUSSION

16.3.1 A changed ecohydrological state

Prior to the 1990/91 abstraction impacts, a gradual inter-annual decline in water table was observed at both P50 and the reference site indicating groundwater response to declining rainfall. Rapid drawdowns are noted at P50 (Fig. 16.3) once abstraction commenced in the mid-1980s and particularly preceding and during the summer of 1990/91. Aquifer recovery was noted at P50 after the production bore was turned off in response to declining vegetation. Abstraction recommenced in 1996, inducing high rates of drawdown, increased seasonal amplitude and lower water tables than the reference site. Recovery of the vegetation has occurred during this period of highly altered hydrology relative to the pre-impact period. Both periods can be considered as alternative ecohydrological states, characterised by the water source partitioning of phreatophytes, i.e. the relative importance of groundwater as a summer water source to *Banksia* (Fig. 16.2).

16.3.2 Banksia response to the drawdown trial

Neither the obligate or facultative phreatophyte (*Banksia ilicifolia* and *B. attenuata* respectively) demonstrated a water deficit response stress during the drawdown trial.

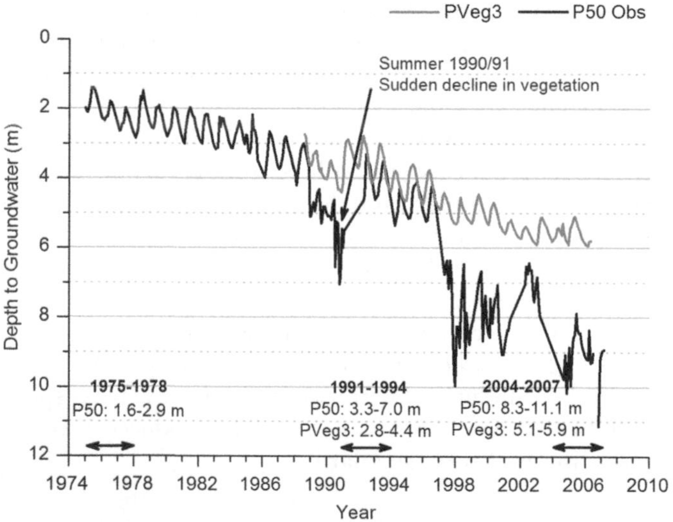

Figure 16.3 Hydrograph of the observation bore at P50 (P50 Obs) and the reference site (Pveg 3). Ranges in groundwater levels are noted for periods before drawdown impacts, immediately after and during the recent drawdown trial.

Plants at both the impacted and reference sites had predawn xylem pressure potentials well below the thresholds derived from vulnerability to xylem embolism response curves for these species (Froend & Drake, 2006) (Fig. 16.4). No significant differences were observed between P50 and the reference site. Lower Ψ_{md} was noted in *B. attenuata* compared to *B. ilicifolia* but this is typical of this species which can maintain lower leaf potentials during the day, a trait that is characteristic of facultative phreatophytes (Froend and Drake, 2006).

Banksia species under drought stress demonstrate water conservation traits such as stomatal regulation during periods of high transpirational demand, that is, closure of stomata and reduced stomatal conductance during the middle of the day (Dodd and Bell, 1993). Diurnal measurement of transpiration and stomatal conductance of both study species at P50 and the reference site (Fig. 16.5) indicates no midday stomatal closure in either species at both sites. As depicted by the plant water status measurements taken at both study sites, there is no evidence to suggest that either the obligate or facultative phreatophyte study species are under drought stress, even during summer.

Water source partitioning modelling yielded results that supported the conclusions drawn from the plant water status measurements. No differences were observed between species when comparing dates or sites (Fig. 16.6). During the dry summer (January) both species accessed water sources deeper in the soil profile and from groundwater. In contrast, plants at the reference site changed their water source use from deeper to upper soil horizon during spring when soil water content is highest. This did not occur however at the impacted site where the majority of plant water continued to be sourced from lower and deep soil. This may reflect the deeper root system of re-established populations of these species that have developed subsequent

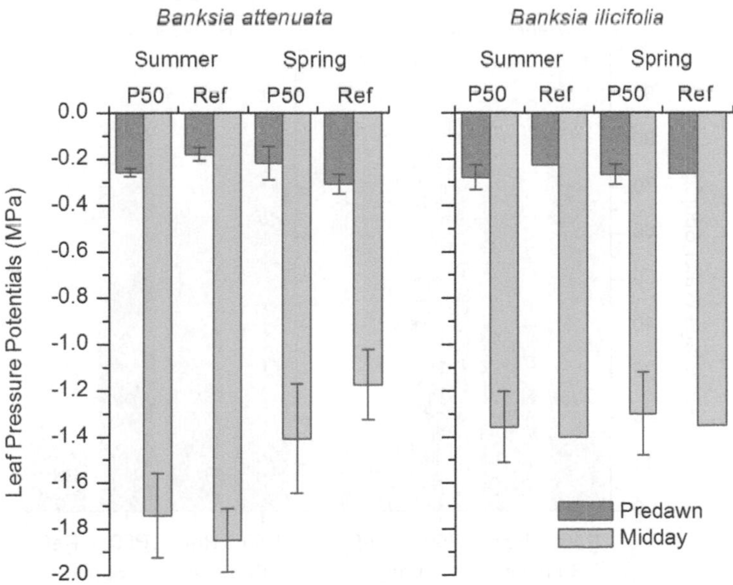

Figure 16.4 Representative predawn and midday leaf pressure potentials (Ψ) (MPa) for *Banksia attenuata* and *B. ilicifolia* during summer (January 2005) and spring (November 2005) at P50 and the reference site (n = 3, except for reference *B. ilicifolia* where n = 2).

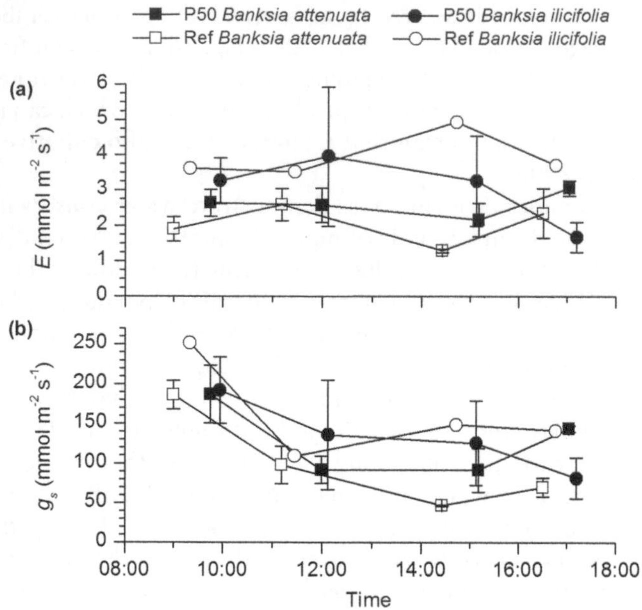

Figure 16.5 Mean (SE) diurnal (a) transpiration (mmol m^{-2} s^{-1}) and (b) stomatal conductance (mmol m^{-2} s^{-1}) during February 2007 for *Banksia attenuata* and *B. ilicifolia* at P50 and the reference site (n = 3, except for reference *B. ilicifolia* where n = 2).

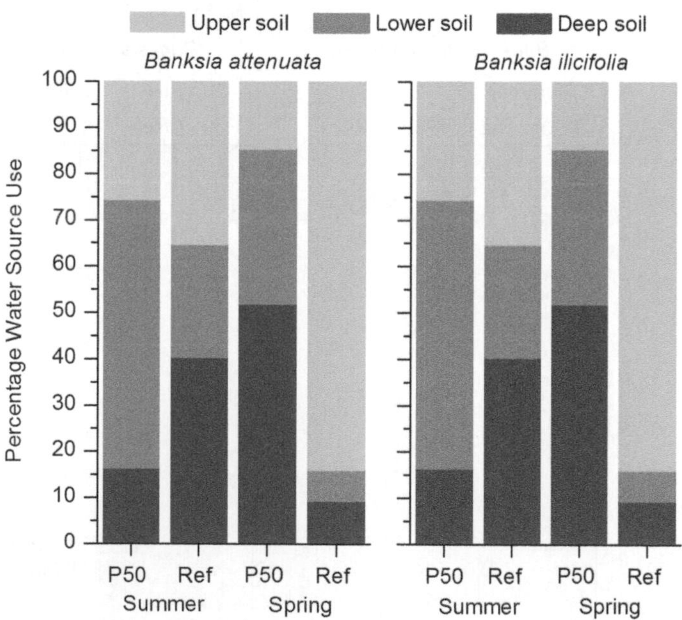

Figure 16.6 Water source partitioning for *Banksia attenuata* and *B. ilicifolia*, during January (summer) and November (spring) 2005 at P50 and the reference site. Upper soil is surface to 4 m depth, lower soil is 4–5 m and deep soil 5–8 m and includes the capillary fringe.

to the drawdown impacts and in response to lower groundwater levels. However, it indicates that plants at the impacted site are continuing to access groundwater, or deep soil water of similar isotopic signature to groundwater, despite rapid drawdown as part of this trial. Deep soil water use was noted to be lower (proportionally) in summer at P50 compared to the reference site, which would indicate a greater degree of separation from the aquifer.

16.4 CONCLUSION

Both the water source partitioning and plant water status results indicate that phreatophytes at the impacted site are not under a water deficit as a consequence of rapid groundwater drawdown. Even the more vulnerable *Banksia ilicifolia* did not display any evidence of drought stress during the trial. Recovery of the vegetation at the impacted site appears to have resulted in a different ecohydrological state that is defined by current populations of phreatophytes that are less vulnerable to groundwater drawdown (Fig. 16.7). Populations of phreatophytes that have established under this new ecohydrological state do not exhibit a water deficit response under similar water availability conditions to the threshold transition event that resulted in almost complete mortality of their parental populations. Current populations appear to meet their water requirements from deeper portions of the vadose zone suggesting a deeper, more expansive root distribution compared to the parental populations that established over a water table only 1.6 to 2.9 m deep.

The concept of a shift in ecohydrological state is supported not only by the obvious change in groundwater availability at P50 but also in the reduced vulnerability of recovered populations to further groundwater drawdown. The current, alternate ecohydrological state is likely to persist unless a significant, further change in water availability occurs, or plant response is influenced by other environmental pressures. The application of ecohydrological states in water resource management to identify likely responses of groundwater-dependent ecosystems to drawdown. This case study of phreatophytic vegetation response, recovery and tolerance to further hydrological modification of their environment, demonstrates the long-term persistence of groundwater-dependent vegetation subjected to reduced rainfall and increased water table drawdown.

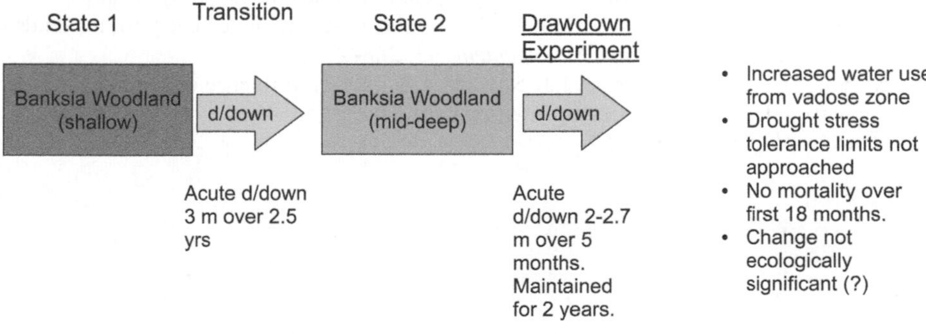

Figure 16.7 Conceptual representation of the observed ecohydrological states at P50 and the outcomes of an induced, repeated rapid drawdown event.

ACKNOWLEDGEMENTS

The authors gratefully acknowledge the financial and technical support of the Water Corporation of Western Australia.

REFERENCES

Clements, F.E. (1936) Nature and structure of the climax. *Journal of Ecology*, 24, 252–284.

Dawson, T. (1993) Hydraulic lift and water use by plants: implications for water balance, performance and plant–plant interactions. *Oecologia*, 95, 565–574.

Dodd, J. & Bell, D.T. (1993). Water relations of the canopy species in a Banksia woodland, Swan Coastal Plain, Western Australia. *Australian Journal of Ecology*, 18, 281–293.

Froend, R.H. & Drake, P.L. (2006) Defining phreatophyte response to reduced water availability: Preliminary investigations on Banksia woodland species in Western Australia. *Australian Journal of Botany*, 54, 173–179.

Froend, R. & Sommer, B. (2010) Phreatophytic vegetation response to climatic and abstraction-induced groundwater drawdown: Examples of long-term spatial and temporal variability in community response. *Ecological Engineering*, 36, 1191–1200.

George, W.R., Brown, J.R. & Clawson, W.J. (1992) Application of non-equilibrium ecology to management of Mediterranean grassland. *Journal of Range Management*, 45, 436–440.

Groom, P.K., Froend, R.H. & Mattiske, E.M. (2000) Impact of groundwater abstraction on a Banksia woodland, Swan Coastal Plain, Western Australia. *Ecological Management and Restoration*, 1, 117–124.

Groom, P.K., Froend, R.H., Mattiske, E.M. & Gurner, R.P. (2001) Long-term changes in vigour and distribution of Banksia and Melaleuca overstorey species on the northern Swan Coastal Plain. *Journal of the Royal Society of Western Australia*, 84, 63–70.

Kite J. & Webster, K. (1989) Management of groundwater resources for protection of native vegetation. *Journal of the Royal Society of Western Australia*, 81, 101–102.

Mattiske and Associates (1988) *Monitoring the Effects of Groundwater Extraction on Native Vegetation on the Northern Swan Coastal Plain.* Report to the Water Authority of Western Australia, Perth.

Pettit, N.E. & Froend, R.H. (2001) Long-term changes in the vegetation after the cessation of livestock grazing in Eucalyptus marginata (jarrah) woodland remnants. *Australian. Ecology*, 26, 22–31.

Sommer, B. & Froend, R. (2011) Resilience of phreatophytic vegetation to groundwater drawdown: Is recovery possible under a drying climate? *Ecohydrology*, 4, 67–82.

Westoby, M., Walker, B. & Noy-Meir, I. (1989) Opportunistic management for rangelands not a equilibrium. *Journal of Range Management*, 43, 266–274.

Zencich, S.J., Froend, R.H., Turner, J.T. & Gailitis, V. (2002). Influence of groundwater depth on the seasonal sources of water accessed by banksia tree species on a shallow, sandy coastal aquifer. *Oecologia*, 131, 8–19.

Chapter 17

Response of wetland vegetation to climate change and groundwater decline on the Swan Coastal Plain, Western Australia: Implications for management

Robyn Loomes[1], Ray Froend[2] & Bea Sommer[2]

[1]*Department of Water, Government of Western Australia, Perth, Western Australia*

[2]*Centre for Ecosystem Management, Edith Cowan University, Joondalup, Western Australia*

ABSTRACT

A shallow, unconfined aquifer underlies much of coastal south-western Australia. Groundwater abstraction from this aquifer has increased significantly since the 1970s in response to declining rainfall and increasing domestic and commercial demand. With the subsequent decline in groundwater levels has come evidence of ecological change and degradation. A wetland vegetation monitoring program was established in the mid-1990s to address these issues. Vegetation Response patterns appear to be the cumulative result of groundwater abstraction, reduced rainfall and fire. This paper describes the outcomes of 14 years of vegetation monitoring, examining the degree and rate of change of vegetation in wetlands of varying geomorphology, hydrology and disturbance regimes, and identifies ecological thresholds that could be used to trigger a management response.

17.1 INTRODUCTION

The Gnangara Groundwater Mound, a shallow superficial aquifer, underlies 2140 km^2 of the northern Swan Coastal Plain (SCP) of south-western Australia (Fig. 17.1). The aquifer consists mainly of deep sands, sandstone and limestone which comprise aeolian, alluvial, swamp, estuarine and shoreline sediments deposited on a gently seaward-sloping unconformity on top of Mesozoic sedimentary rocks (McArthur and Bettenay, 1960). The aquifer is maintained through diffuse recharge by local rainfall due to the porous nature of the sediments, with a small component being derived from local run-off from the Darling Scarp to the east. It stores >20 000 GL of fresh water, with the crest of the groundwater mound being 70 m above sea level (Government of Western Australia, 2003) and represents an important water source for numerous groundwater-dependent ecosystems (GDE) such as *Banksia* woodlands, wetlands, mound springs and caves.

The capital city, Perth, has a population of two million and is amongst the fastest growing regions in the country. The aquifer currently meets up to 60% of the northern SCP's domestic, horticultural, industrial and commercial needs. The region has

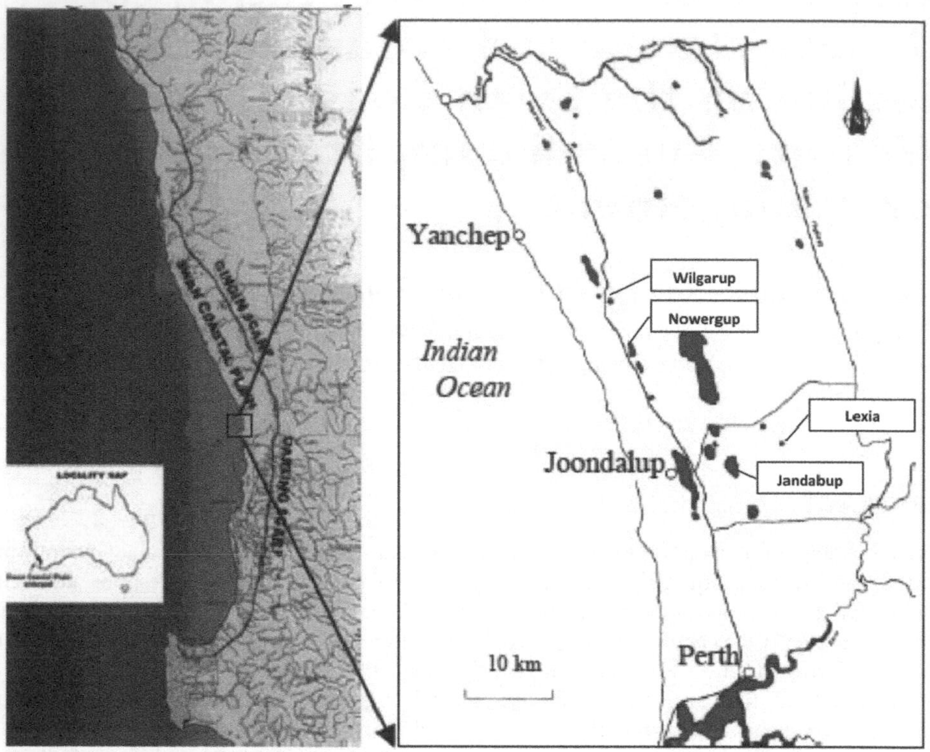

Figure 17.1 Locations of the four case study wetlands on the Gnangara Groundwater Mound, Swan Coastal Plain, Western Australia (Adapted from Balla, 1994 & Froend *et al.*, 1999).

a Mediterranean-type climate, with cool, wet winters and hot, dry summers (Beard, 1984). Hence, the aquifer is recharged during the autumn/winter (April to October) when the majority of the annual rainfall (~90%) is received. Groundwater tables fall during the summer months which are virtually rainless, although rainfall can occur following cyclonic activity.

Groundwater levels across the Gnangara Mound have been declining since the 1970s. In some areas a decline of up to 4 m has been recorded, with decreases of 1–2 m not uncommon (Fig. 17.2). Altered groundwater levels have been attributed to three factors: decline in rainfall, land use changes and abstraction of groundwater (Yesertener, 2002). Annual rainfall has been decreasing since the early 1970s (Fig. 17.3).

Land use impacts on groundwater levels arise from clearing, urban development, bushfires (controlled and wild), pine plantations and horticulture. In the northern area of the Gnangara Mound, pine plantations were established in the mid-1900s to meet the State's demands for quality sawlogs (Butcher, 1977). The replacement of native vegetation with pine plantations is thought to have contributed to the decline of groundwater levels by increasing interception of rainfall and evapotranspiration rates (Butcher, 1977; Farrington & Bartle, 1991).

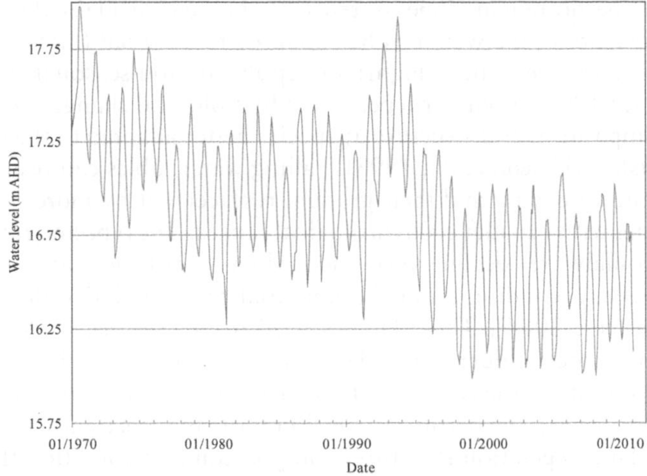

Figure 17.2 Hydrograph of a representative wetland on Gnangara Mound (Lake Joondalup) showing long-term groundwater decline (Data custodian: Western Australian Department of Water).

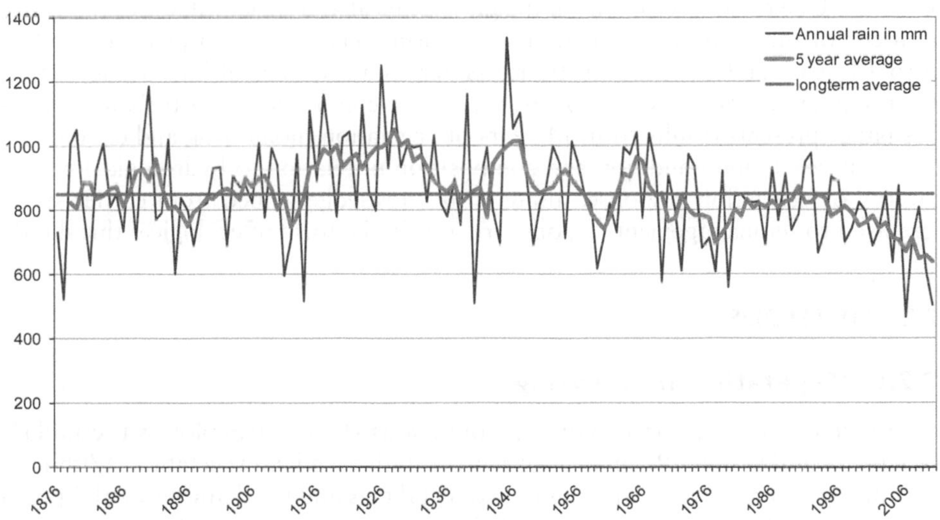

Figure 17.3 Long-term annual rainfall and five-year moving averages for Perth, Western Australia, 1876–2010 (Data sourced from the Australian Government Bureau of Meteorology [www.bom.gov.au]).

Besides numerous commercial and private bores, abstraction for public water supply represents the greatest use of groundwater in the Perth region. Abstraction can affect groundwater levels up to 0.5 km from an unconfined aquifer production bore, and cumulative impacts from confined and unconfined aquifers can extend up to 5 km from the abstraction area (Yesertener, 2002).

The Gnangara Mound supports some 400 wetlands which usually occur in depressions between the dominant dune systems of the coastal plain (Davidson, 1995). Vegetation surrounding the wetlands has adapted to seasonal fluctuations in surface and/or groundwater levels and is highly susceptible to non-seasonal changes in water regimes (Wheeler, 1999). Long-term persistent hydrological change can cause a shift in community composition and structure as species better adapted to the new conditions become established (Froend *et al.*, 1993). Falling water tables can result in the loss of species intolerant of drying and their gradual replacement by more drought-tolerant terrestrial species with broader ecohydrological ranges. The type and rate of vegetation response to groundwater decline varies according to historic water regime changes, the existing wetland condition and the potential for cumulative disturbance events (Froend & Loomes, 2004). These factors need to be considered by water resource managers because they influence the degree of resilience of wetland vegetation and determine the capacity of the ecosystem to recover in the event that water tables should return to historic levels. The pertinent question for managers is, at which point should changes in wetland vegetation structure, composition and condition that result from groundwater decline, trigger a management response?

In the mid-1990s, in recognition of the potential for ecological impacts from reduced rainfall and increased groundwater abstraction, the Western Australian water resource management agency implemented a long-term biological monitoring program for GDEs. Due to their sensitivity and high ecological and social values, wetlands were the focus of the program. Vegetation monitoring has now taken place annually for up to 14 years at 15 permanent transects across a selection of high value wetlands of varying geomorphology, hydrology, vegetation condition and disturbance regimes. This paper presents results from 14 years of vegetation monitoring and examines the change in vegetation condition and composition in relation to hydrological changes. Patterns of vegetation response, along with knowledge gained from earlier studies, were used to set management response triggers in the form of ecological thresholds.

17.2 METHODS

17.2.1 Vegetation monitoring

A 10 m wide transect composed of four contiguous 10 m × 10 m plots was established at each wetland in 1996/97 in areas of representative wetland vegetation. Within each plot, all trees were marked with numbered metal tags and the diameter of all branches was measured 1.5 m above the ground. Each tree was remeasured annually to provide an indication of growth rates. Crown health was determined by assessing foliage density, proportion of dead branches and degree of epicormic growth. All overstorey and understorey species were identified and their foliage cover estimated using the Domin-Krajina scale of cover and abundance (Kent & Coker, 1992). The abundance of seedlings, saplings and/or resprouters was also recorded for overstorey species.

17.2.2 Data analysis

Changes observed in wetland vegetation condition and composition over time were described and compared to wetland type, changes in hydrology and external influences (see below) to determine which factors have potentially had the greatest impact. Each transect was monitored annually for up to 14 years. Historical groundwater and lake level data were analysed for rate and magnitude of water level change. The following factors were examined:

- wetland type/hydrology;
- magnitude and rate of water level change;
- proximity of sites to private and public abstraction bores;
- proximity to pine plantations;
- recent fire history.

Vegetation response is described as % change in canopy health, % change in the abundance of exotic species and a similarity matrix representing the % similarity (Sørensen index) in species composition between the earliest sampling year and 2010.

17.2.3 Case studies

Four wetlands were chosen to represent differing magnitudes of water level decline, water regimes, wetland type and degree of vegetation response. Hydrological thresholds, which once breached should trigger a management response, were determined by considering the level of impact to vegetation associated with differing magnitude and rates of groundwater decline. Findings from the case studies, as well as those from the wider study and previous research (Froend *et al.*, 2004) were taken into consideration for this task.

17.3 RESULTS AND DISCUSSION

17.3.1 Wetland type and water regime

Approximately half of the monitored wetlands are permanently inundated (i.e. 'lakes') (Table 17.1). All of the monitored wetlands experienced declines in surface and groundwater levels since commencement of monitoring (Table 17.1). The greatest declines occurred at Lakes Wilgarup and Mariginiup (>1 m) and at the Lexia wetlands (0.9 m). Rates of decline ranged from 0.01 to 0.10 m/a. All of the wetlands classified as lakes continued to retain surface water throughout the year. However, two of these (Lakes Nowergup and Jandabup) are artificially supplemented with groundwater in order to maintain water levels and preserve their ecological and social values.

Sumplands retain surface water seasonally following winter rainfall and higher groundwater levels (Semenuik, 1988). Despite the low number of sumplands in the monitoring programmme (3), it is to this category of wetland that many of the previously permanently inundated systems of the SCP now belong. For example, Lake Mariginiup, having experienced a groundwater level decline of 0.93 m, now rarely retains any surface water throughout the year. The same applies to Lexia 86 which experienced a decline in groundwater level of 0.86 m (Table 17.1). Most of the wetlands on the Gnangara Mound are damplands (Semenuik, 1988); seasonally waterlogged wetlands that do not retain surface water. Four damplands are monitored under the

Table 17.1 Wetland type, water regime change, distance from abstraction and pines, fire history and vegetation change at 15 monitored Gnangara Mound transects.

			Vegetation change between earliest and latest surveys			*Proximity to abstraction bores and pines (km)*			*Water level decline*		
Wetland name	*Wetland type*	*Monitoring period*	*Canopy Vigour (% change)*	*Exotic cover (% change)*	*Sørensen Similarity*	*Public bore*	*Private bore*	*Pines*	*Magnitude (m)*	*Rate (m/a)*	*Year/s fire occurred*
Lake Wilgarup	Dampland	1997–2010	−100	800	0.00	11.3	0.4	4.8	1.34	0.10	1996, 2004–2005
Lexia 186	Dampland	1997–2010	64	500	0.50	5.9	4.2	2.5	1.06	0.08	–
Lake Mariginiup	Sumpland	1996–2009	−66	53	0.32	3.8	<0.4	2.7	0.93	0.07	1996, 1997, 2003
Lexia 94	Dampland	1997–2010	7	0	0.45	5.0	3.4	0.4	0.89	0.07	–
Lexia 132	Dampland	1997–2010	27	0	0.36	4.4	3.8	1.3	0.89	0.07	2000
Lexia 86	Sumpland	1997–2010	−16	0	0.52	5.9	3.6	2.1	0.86	0.07	2004
Lake Nowergup North	Lake	1996–2010	−9	25	0.42	9.0	0.8	3.2	0.79	0.06	–
Lake Nowergup South	Lake	1996–2010	9	13	0.49	9.0	0.8	3.2	0.79	0.06	–
EPP 173	Sumpland	1997–2010	−10	−100	0.54	6.7	1.7	2.9	0.61	0.05	–
Lake Jandabup	Lake	1996–2010	−19	−68	0.57	1.7	<0.4	0.8	0.53	0.04	–
Loch McNess	Lake	2004–2010	3	197	0.42	4.6	2.5	5.0	0.39	0.07	2004, 2009
Lake Joondalup North	Lake	1996–2009	−19	92	0.28	8.0	<0.4	6.5	0.39	0.02	2007
Lake Joondalup South	Lake	1996–2009	−34	12	0.38	8.0	<0.4	6.5	0.39	0.03	1996, 2003
Lake Goollelal	Lake	1998–2008	7	23	0.41	5.9	<0.4	6.5	0.31	0.02	–
Lake Yonderup	Lake	1997–2010	−22	1	0.49	5.9	1.7	4.8	0.09	0.01	2004, 2005

annual program (Table 17.1). Rates of water level decline were highest at wetlands that experienced the greatest magnitude of water levels decline ($r=0.91$).

17.3.2 Proximity to abstraction bores and pines

Five wetlands occur within 5 km of a public abstraction bore. In general, wetlands closest to public bores, Lake Jandabup (1.7 km), Mariginiup (3.8 km), Lexia 132 (4.4 km), and Lexia 94 (5.0 km) experienced larger magnitudes of water level decline over the study period (Table 17.1). However, despite the high magnitude of groundwater level decline experienced at Lake Wilgarup, it is furthest away from a public bore (11.3 km). Hence there was no correlation between distance to public abstraction bore and magnitude of water level decline. Most wetlands were closer to private, than to public bores (all <5 km). There was also no clear relationship between distance to private abstraction bore and groundwater level decline, however it is possible that nearby intensive horticulture has had a significant impact on Lake Wilgarup water levels. The lack of correlation between distance to abstraction bores and water level decline is probably due to differences in actual volumes abstracted and aquifer response.

Most of the monitored wetlands are within a <5 km radius from a pine plantation (Table 17.1). Unlike proximity to abstraction bores, proximity to a pine plantation was significantly correlated with the magnitude of groundwater decline ($r=-0.50$; $p<0.05$). With some exceptions such as Lake Wilgarup, pine plantations are believed to have contributed to the decline of groundwater levels across the northern Gnangara Mound (Farrington & Bartle, 1991).

17.3.3 Disturbance

Bushfires, both wild and controlled, represent the most common physical disturbance experienced at the study wetlands. Within the past 14 years seven wetlands, Lakes Mariginiup, Wilgarup, Joondalup, Yonderup, Loch McNess, Lexia Wetlands 132 and 94 have burnt (Table 17.1), with Lake Mariginiup having experienced three fires during that time. The 2004/2005 fires at Loch McNess, Lake Wilgarup and Lake Yonderup were particularly destructive, with much of the wetland vegetation at Lake Wilgarup destroyed.

17.3.4 Wetland vegetation

All of the wetlands experienced large changes in species composition (between 43% and 100%) over the monitoring period (Table 17.1). This was associated with a decrease in canopy vigour and an increase in the cover of exotic (introduced, weed) species in most cases. The cover of exotic species, however, can vary considerably from year to year, depending on rainfall and the occurrence of fire.

Lake Wilgarup experienced the greatest deterioration in canopy condition, having lost the majority of overstorey species after the 2004/05 fire. The Sørensen similarity of 0.0% (no species in common) between 1997 and 2010 represents a significant change in the vegetation (Table 17.1). The wetland also experienced the largest increase in foliage cover of exotic species of all the sites (800%). After Lake Wilgarup, Lake Joondalup showed the greatest change in species composition, with 38% and 28%

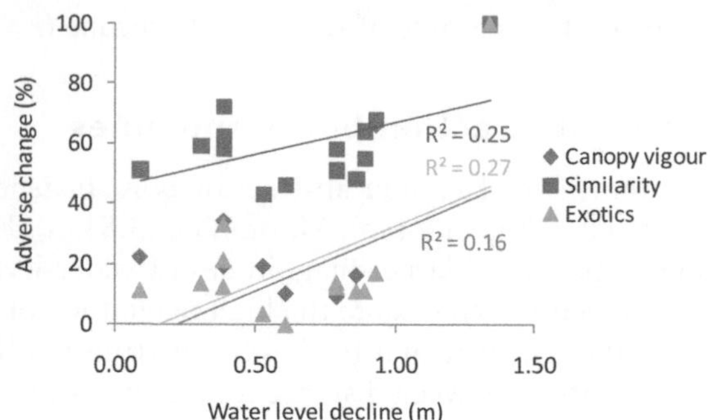

Figure 17.4 Relationship between water level decline and adverse changes relating to three vegetation indices. Adverse change = decrease in canopy vigour; decrease in similarity; increase in number of exotic species. Wetlands as per Table 17.1, excluding Lexia 186.

(for south and north respectively) similarity between 1996 and 2010. Lake Mariginiup also experienced a large change in species composition (similarity of 68%) and canopy condition (−66%), while exotic species foliage cover increased by 53%. Lexia 186 retained a comparatively large similarity in species composition (50%) and a general improvement of canopy condition (+64%) in spite of having experienced a very large decline in groundwater (Table 17.1). The improvement of canopy condition at this wetland was due to the establishment of trees less tolerant of flooding. The cover of exotic weed species, however, increased by 500% over 13 years. Lake Yonderup, where water levels declined the least (0.09 m), had unchanged exotic species foliage cover (+1%), was 49% similar in species composition and recorded 22% decline in canopy condition. Apart from Lexia 186, there was a tendency for large declines in tree health and changes in species composition to correspond with water level declines, although correlations were not significant (Fig. 17.4).

17.3.5 Case studies

The four wetlands chosen for the case studies represented differing magnitudes of water level decline, water regimes, wetland type and degree of vegetation response.

Lake Jandabup is a relatively large (~330 ha), shallow (<1.5 m), wetland surrounded by rural properties. Declining groundwater levels between 1997 and 1999 (Fig. 17.5a) and the subsequent exposure of underlying sulfidic sediments resulted in changes in the quality of surface water (including a significant lowering of pH), adversely affecting the aquatic invertebrate fauna (Sommer & Horwitz, 2001). In response, surface water levels are now being artificially supplemented. Despite pressures from nearby public and private abstraction and pine plantations, surface water levels initially showed an increasing trend since artificial maintenance commenced (Fig. 17.5a). Water quality and macroinvertebrate composition recovered, and the fringing vegetation of the wetland remained unchanged. However, in 2010 the wetland experienced well below-average winter rainfall and a reduction in the amount of water pumped into the wetland. Consequently, the maximum water level in spring 2010

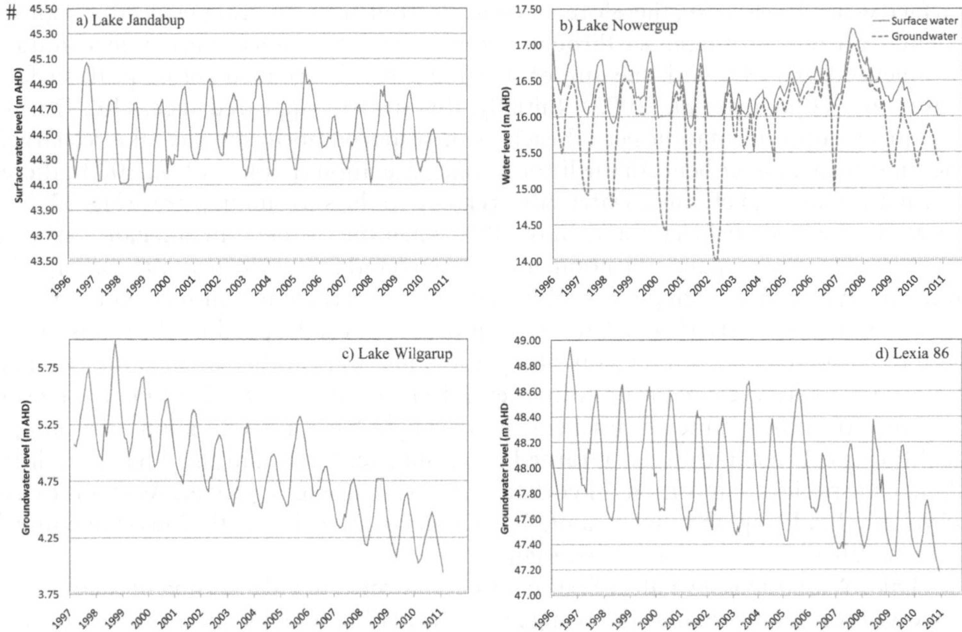

Figure 17.5 Surface and groundwater levels recorded at four Gnangara Mound wetlands from January 1996 to January 2010 a) Lake Jandabup, b) Lake Nowergup, c) Lake Wilgarup, d) Lexia Wetland 86 (Data custodian: Western Australian Department of Water, 2007).

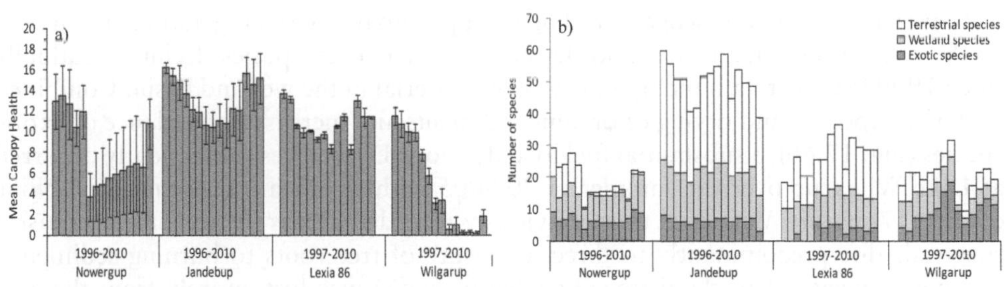

Figure 17.6 Vegetation changes at four study wetlands from 1996/97 to 2010. a) Mean canopy health of all wetland tree species (error bars are standard errors); b) species richness and composition (exotics, wetland and terrestrial species).

was the lowest on record. Although there has been some evidence of down-gradient migration of emergent macrophytes and decline in tree canopy condition (Fig. 17.6a), species composition and the abundance of exotics (Fig. 17.6b) has changed little since monitoring commenced.

As one of the deepest permanent wetlands on the SCP, Lake Nowergup represents important habitat for aquatic invertebrates and fish and is as a drought refuge for water birds (Water Authority, 1995). Climate and abstraction-induced lowering of surface water levels were, therefore, of concern from an ecological perspective. Consequently, artificial maintenance through pumping from the underlying confined aquifer commenced in 1989, with the aim of meeting a pre-determined spring peak surface water level (Loomes *et al.*, 2003). Initially, this strategy appeared successful. During the spring supplementation period each year, water pumped into the lake recharged the superficial aquifer beneath the littoral zone vegetation (Loomes *et al.*, 2003), therefore maintaining an elevated water table relative to the surrounding area (Fig. 17.5b). However, between February and May 2001 when the artificial maintenance scheme was shut down for repairs, the populations of wetland trees (*Melaleuca rhaphiophylla* and *Eucalyptus rudis*) fringing the lake suffered a decrease in canopy condition and high rates of mortality (Fig. 17.6a). Groundwater levels dropped by 2.78 m over the period. Although the requirements of aquatic fauna were still being met, the rate and magnitude of surface and groundwater level decline was too great for fringing trees to tolerate. After the maintenance scheme was brought back into operation, tree canopy health (Figs. 17.6a and b) has returned to pre-impact levels and there has been little change in species composition other than an increase in exotic species. As was the case with Lake Jandabup, the winter rainfall deficit in 2010 resulted in the lowest maximum water levels recorded at Lake Nowergup.

Lake Wilgarup is a small (16 ha) dampland formed in a depression of a limestone ridge as a result of discharge and rising groundwater levels (Semeniuk, 1988). Groundwater levels at Lake Wilgarup have been declining significantly since 1997 in response to regional drawdown and local abstraction for horticulture (Fig. 17.5c). The extensive peat deposits in the bed of the lake, suggest saturation of the sediments for a long period of time. However, no surface water has been recorded in the lake since 1998 and the peat sediments became dry and vulnerable to combustion.

When vegetation monitoring commenced, the overstorey was dominated by dense wetland trees over native sedges, grading upslope into terrestrial vegetation. The health of the overstorey (Fig. 17.6a) and density of understorey species declined gradually since 1999 leading to a build up of dry plant material in the wetland basin. Combined with the exposure and drying of organic sediments and increased abundance of exotic species (Fig. 17.6b), a substantial fuel load developed. Two fires subsequently occurred at Lake Wilgarup over the summer of 2004/05 with significant consequences. None of the 117 mature *Melaleuca* trees previously recorded across the monitored transect survived, due predominantly to direct exposure of tree roots to burning sediments. A second species of wetland tree (*Eucalyptus rudis*) was lost entirely from the site, along with three species of emergent macrophyte. Some *Melaleuca* saplings have since re-established in the lowest lying (wettest) areas of the wetland, however two terrestrial eucalypts now dominate the 'wetland' vegetation.

Lexia 86 is a small (0.5 ha), undisturbed wetland belonging to a chain of groundwater fed, seasonal sumplands and damplands located within a large area of remnant native woodland. There is a marked zonation of wetland vegetation at Lexia 86, with mono-specific concentric bands of emergent macrophytes and wetland shrubs. Above the littoral zone is a mixed community of wetland and terrestrial shrubs and trees.

Groundwater levels have declined since vegetation monitoring commenced, which is reflected in shorter periods of surface water retention (Fig. 17.5d). As a result, there has been a trend of shifting wetland plant distribution towards the centre (wetter parts) of the wetland. In 2002, the mono-specific stand of native sedge dominating the centre of the wetland basin suffered a 70% mortality of plants, leaving the soil surface bare and open to recruitment and encroachment of terrestrial species. Despite two seasons of higher than average water levels since that time, the trend of thinning sedges and down-gradient migration of shrubs and, more recently, fringing tree species has continued. Although tree canopy health improved following higher than average rainfall in 2005, there has been a longer-term decline in condition (Fig. 17.6a). There has also been an increase in the richness of both exotic and terrestrial species across the transect over time (Fig. 17.6b).

To-date many wetlands have shown a progressive decline in vegetation condition and composition that is proportional to the progressive decline in groundwater levels. Changes are typified by gradual alterations in species composition and structure through weed species invasion, encroachment of terrestrial species and reductions in cover and abundance of species most susceptible to water level decline, sometimes resulting in local extinction of these species. Although a small number of wetlands have shown little change, others have undergone threshold changes that may be irreversible. The combination of hydrological change at Lake Wilgarup and the burning of exposed (dry) organic sediments, has resulted in a threshold response in which the wetland ecosystem is being rapidly replaced with a terrestrial system. The extreme rate and magnitude of drawdown experienced at Lake Nowergup in 2001/02 also triggered a threshold response in the fringing vegetation. Conversely, the lower magnitude of groundwater level decline at Lexia 86 has to date only manifested as a gradual, progressive change in vegetation character. A cursory examination of the remaining 11 monitored wetlands suggests that declining groundwater tables have had the greatest impact on sumplands and damplands. Sumplands support ecosystems adapted to seasonal drying but are susceptible to longer term water regime change. Damplands support vegetation reliant on a high winter water table but are generally unable to tolerate inundation. On the Gnangara Mound, many sumplands and damplands are undergoing terrestrialisation as xeric species encroach into drying basins.

17.3.6 Implications for management

The results from this monitoring programme suggest that the type and rate of vegetation response to water table decline will vary according to historic water regime changes, the existing ecohydrological state of the wetland and the occurrence of fire. This study and other recent work suggest that the magnitude (m) and rate (m/a) of water level change are relevant to potential vegetation impacts (Froend *et al.*, 2004; Froend *et al.*, 2006). Therefore, trends observed in wetland vegetation response to water level changes can be used to define hydrological criteria that result in an adverse ecological response. This can be represented by a scale of the 'risk of impact' (ROI) to wetland vegetation from drawdown at differing rates and magnitudes that is informed by the observations whilst adopting a precautionary approach.

The relatively minor responses of wetlands experiencing a magnitude of water level decline less than 0.25 m during 1996/97–2010 suggests this degree of decline may represent a low level of impact. Declines between 0.25 m and 0.5 m generally

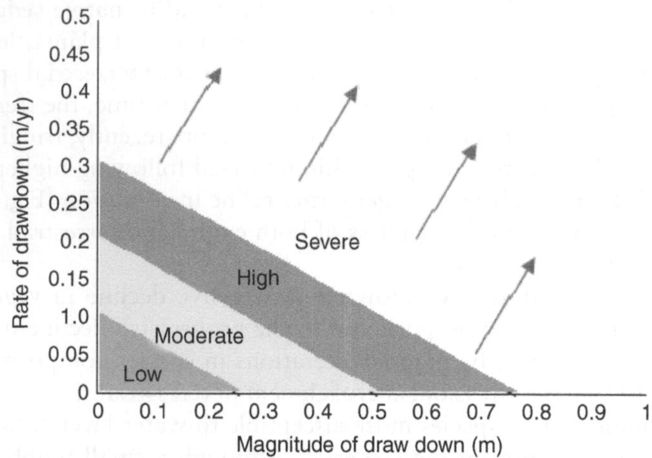

Figure 17.7 Risk of impact categories for wetland vegetation based on rate and magnitude of groundwater level change and observed vegetation responses at 15 Gnangara Mound wetlands 1996/97–2010.

corresponded to moderate changes in vegetation condition, and magnitudes of water level change between 0.5 m and 0.8 m related to significant or high levels of impact. Beyond a 0.80 m decline the impact to vegetation was seen as severe, as evidenced by the nature of the vegetation response at Lakes Wilgarup, Mariginiup and Lexia wetlands 94, 132, and 86. Although the rates of decline across all wetlands only ranged from 0.01 to 0.10 m/a, previous studies (Froend *et al.*, 2004) have suggested rates below 0.1 m/a are likely to cause a low impact (if the magnitude of decline is also low), rates between 0.1 and 0.2 m/a a moderate impact, 0.2–0.3 m/a high and above 0.3 m/a a severe impact (even if the magnitude of decline is comparatively low) (Fig. 17.7). It should be noted that there is also much more research required to fully understand the resistance and resilience of the wetland vegetation to hydrologic change and that a numeric ROI framework represents an approximation to assist current management information requirements.

Nevertheless, the ROI approach represents a rapid, easily applied method of alerting water resource management agencies of elevated risks to groundwater dependent wetland ecosystems. The appropriate management response will vary according to the perceived cause of water level decline and may involve temporary or permanent decommissioning of public bores, stricter licensing conditions for private bore users, thinning of pine plantations within the catchment, or artificial maintenance of wetland water regimes.

ACKNOWLEDGEMENTS

The authors gratefully acknowledge the support of the Western Australian Department of Water.

REFERENCES

Balla, S. (1994) *Wetlands of the Swan Coastal Plain Volume, 1, Their nature and Management.* Water Authority of Western Australia/Department of Environment, Perth.

Beard, J.S. (1984) Biogeography of the Kwongan. In: Pate, J.S. & Beard, J.S. (eds.) *Kwongan, Plant Life on the Sandplain.* Perth, University of Western Australia Press. pp. 1–26.

Butcher, T.B. (1977) Impact of moisture relationships on the management of *Pinus pinaster* Ait. plantations in Western Australia. *Forest Ecology and Management*, 1, 97–107.

Davidson, W.A. (1995) Hydrogeology and groundwater resources of the Perth region, Western Australia. *Geological Survey of Western Australia, Bulletin*, 147, 3–15.

Farrington, P. & Bartle, G.A. (1991) Recharge beneath a Banksia woodland and a *Pinus pinaster* plantation on coastal deep sands in south Western Australia. *Forest Ecology and Management*, 40, 101–118.

Froend, R.H., Farrell, R.C.C., Wilkins, C.F., Wilson, C.C. & McComb, A.J. (1993) *Wetlands of the Swan Coastal Plain. Vol. 4. The Effect of Altered Water Regimes on Wetland Plants.* Water Authority of Western Australia and the Environmental Protection Authority, Perth.

Froend, R., Groom, P., Nickoski, S. & Gurner, R. (1999) *Phreatophytic Vegetation and Groundwater Interactions on the Gnangara Groundwater Mound.* A report to the Water and Rivers Commission and the Land and Water Resources Research and Development Corporation. Edith Cowan University, Joondalup.

Froend, R. & Loomes, R. (2004) *Approach to Determination of Ecological Water Requirements of Groundwater Dependent Ecosystems in Western Australia.* Edith Cowan University, Joondalup.

Froend, R., Loomes, R. & Bertuch, M. (2006) *Determination of Ecological Water Requirements for Wetland and Terrestrial Vegetation – Southern Blackwood and Eastern Scott Coastal Plain. Baseline Monitoring Results and Monitoring Protocol.* A report to the Department of Water. Edith Cowan University, Joondalup.

Froend, R.H., Rogan, R., Loomes, R., Horwitz, P., Bamford, M. & Storey, A. (2004) *Study of Ecological Water Requirements on the Gnangara and Jandakot Mounds under Section 46 of the Environmental Protection Act; Parameter Identification and Monitoring Program Review.* A report to the Water and Rivers Commission. Edith Cowan University, Joondalup.

Government of Western Australia (2003) *A State Water Strategy for Western Australia. Securing Our Water Future.* Government of Western Australia, Perth.

Kent, M. & Coker, P. (1992) *Vegetation Description and Analysis: A Practical Approach.* London, Belhaven Press.

Loomes, R., Lam, A. & Froend, R. (2003) *Assessment of the Management of Lake Nowergup.* A report to the Water and Rivers Commission. Edith Cowan University, Joondalup.

McArthur W.M. & Bettenay, E. (1960) *The Development and Distribution of the Soils of the Swan Coastal Plain, Western Australia.* Melbourne, CSIRO.

Semenuik, C.A. (1988) Consanguineous wetlands and their distribution in the Darling System, Southwestern Australia. *Journal of the Royal Society of Western Australia*, 70, 69–87.

Sommer, B. & Horwitz, P. (2001) Water quality and macroinvertebrate response to acidification following intensified summer droughts in a Western Australian wetland. *Marine and Freshwater Research*, 52, 1015–21.

Water Authority (1995) *Review of Proposed Changes to Environmental Conditions: Gnangara Mound Groundwater Resources (Section 46).* Water Authority of Western Australia, Perth.

Wheeler, B.D. (1999) Water and plants in freshwater wetlands. In: Baird, A.J. & Wilby, R.L. (eds.) *Eco-Hydrology: Plants and Water in Terrestrial and Aquatic Environments.* London, Routledge Press. pp. 127–180.

Yesertener, C. (2002) *Declining Water Levels in the Gnangara and Jandakot Groundwater Mounds (Stage 1).* Hydrogeology report HR199. Water and Rivers Commission Hydrogeology and Water Resources Branch, Perth.

REFERENCES

Chapter 18

Hydrogeochemical processes in the Pateira de Fermentelos lagoon (Portugal) and their impact on water quality

Clara Sena[1] *& M. Teresa Condesso de Melo*[2]

[1] *CESAM – Centre for Environmental and Marine Studies, Geosciences Department, University of Aveiro, Aveiro, Portugal*

[2] *Geo-Systems Centre/CVRM, Instituto Superior Técnico, Lisbon, Portugal*

ABSTRACT

Hydrogeochemical processes were investigated in the Pateira de Fermentelos lagoon, a shallow freshwater body with important ecosystems. Besides rainwater, at least three main types of natural waters contribute to the water balance and water quality in the lagoon: (1) the Cértima River with a mineralised Ca-HCO_3 water and circum-neutral pH; (2) the Cretaceous aquifers with a Na-Cl water and acidic pH; (3) the Triassic aquifers with a mixed water type, and slightly alkaline pH; (4) and, possibly, the Águeda River with a diluted Na-Cl water. The hyporheic and riparian sediments associated to the lagoon are rich in clays, iron oxyhydroxides and organic matter. Water-rock interactions play a major role in the hydrochemistry of water bodies, leading to a wide range of the mineralisation of natural waters and to the occurrence of trace elements such as arsenic, lead and manganese. Additionally, anthropogenic activities are deteriorating the water quality of surface and groundwater bodies.

18.1 INTRODUCTION

The hydrochemistry of shallow freshwater lagoons strongly depends on the climatic setting, hydrological regime, groundwater-surface water interactions, biogeochemical processes and water-rock interactions. In a shallow freshwater lagoon two interface zones play a major role in the partitioning of chemical elements: (1) the hyporheic zone, composed by the water-saturated sediments beneath the surface water bodies where biogeochemical processes exert a great control on the partitioning of organic and inorganic elements between the solid matrix and porewater (Gandy *et al.*, 2007; Hancock *et al.*, 2005); and, (2) the riparian zone, composed by unsaturated sub-aerial sediments deposited along the margins of surface water bodies where exuberant vegetation may develop, and nutrients are transferred between water, solid matrix, biomass and the atmosphere (Hoffmann *et al.*, 2006; Rivett *et al.*, 2008).

Pateira de Fermentelos is a shallow freshwater lagoon which hosts important aquatic and terrestrial ecosystems and has well-developed hyporheic and riparian zones. It is the largest natural freshwater lagoon of the Iberian Peninsula. Due to the seasonal occupation by several migrating wild birds, Pateira de Fermentelos is included in a Special Protection Area (SPA) under the European Union (EU) Directive on the Conservation of Wild Birds (Law-Decree 140/1999). However, intense agricultural activity and inefficient urban and cattle wastewater treatment are gradually leading to its eutrophication and, therefore, threatening the sustainability of the dependent ecosystems (Teles *et al.*, 2007). For this reason and taking into account the implementation of Directive 91/271/CEE (European Commission, 1991) the lagoon and corresponding catchment are included in the list of the Portuguese Sensitive Areas with respect to nutrient loads.

In the present work, the hydrochemical, geochemical and biogeochemical processes that control water quality in the Pateira de Fermentelos lagoon, its affluent streams and surrounding aquifers are identified and described based on the monthly monitoring of water levels and hydrogeochemical data.

18.2 STUDY AREA

Pateira de Fermentelos is a natural shallow lagoon (with an average depth between 0.8 and 2 m), located in the NE of Portugal (40°29′N, 8°36′W), at a distance of 20 km from the Atlantic coast. The lagoon is the most downstream part of the Cértima River before its confluence with the Águeda River. The Cértima River has a catchment area of about 545 km^2 and is 40 km long, the last 5 km of which corresponding to the Pateira de Fermentelos lagoon.

The Pateira de Fermentelos lagoon lies on a topographical lowland area of the Cértima River catchment. However, in terms of regional hydrogeology, the area surrounding the lagoon is part of the recharge area of the Aveiro Cretaceous Groundwater Body (ACGWB).

From the wet season (December to March) to the dry season (June to September), the area of the lagoon may change from 9 km^2 to 3 km^2, respectively. The large difference between wet and dry season area is explained by (i) the flat topography of the riparian zones adjacent to the lagoon; (ii) the tendency for the water in the lagoon to flush out into the Águeda River; and, (iii) the relatively high dependency of the lagoon and adjacent streams (mainly the Cértima River and Pano stream) on rainfall, surface and subsurface runoff.

Under heavy rain events, the Águeda River has shorter response time lags than the Cértima River. For this reason, the Águeda River may temporarily flush into the Pateira de Fermentelos lagoon. Taking into account this natural process, and since the lagoon tends to dry out during summer, the local Water Authorities build a dike every year, from July to October, immediately downstream of the confluence between the lagoon outlet and the Águeda River. Therefore, Pateira de Fermentelos turns into an endorheic catchment when (i) the dike is built in summertime and (ii) the heavy rain events trigger the natural inflow of the Águeda River to the lagoon.

On the west side of Pateira de Fermentelos, Cretaceous siliciclastic sandstones crop out, while on the east side the surrounding elevations are made up of iron-oxyhydroxide-rich Triassic sandstones (Barbosa *et al.*, 1981, Teixeira & Zbyszewski, 1976). Upstream of the lagoon, and within the Cértima River catchment, Jurassic carbonate rocks with trace amounts of gypsum and pyrite crop out. These carbonate rocks were also intersected in depth by boreholes located southwest of the lagoon (Barbosa *et al.*, 1981; Marques da Silva, 1990) (Fig. 18.1).

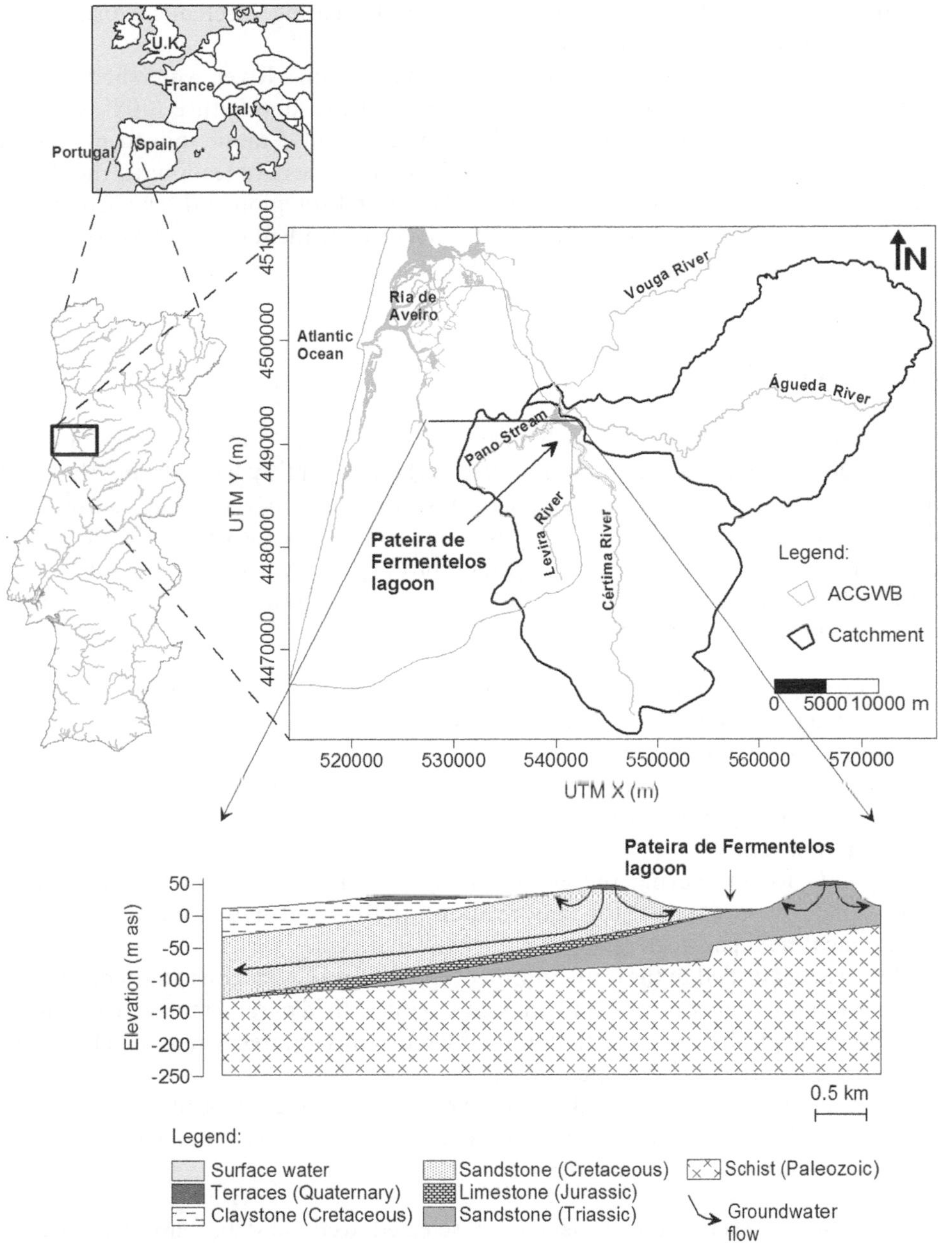

Figure 18.1 Location of Pateira de Fermentelos and geological cross section from Pateira de Fermentelos to the Atlantic Ocean.

The lagoon and adjacent wetlands are characterised by well developed aquatic and riparian vegetation. Water-hyacinth (*Eicchornia crassipes*), water milfoil (*Myriophyllum sp.*), common reed (*Phragmites australis*) and poplar (*Populus sp.*) are some of the most abundant plants. Due to the invasive nature of water-hyacinth, the lagoon's surface may be seasonally invaded by huge agglomerations of water-hyacinths, but since 2007 the Águeda municipality is implementing mechanical processes to control the water-hyacinth population.

The main human activities developed in the Cértima River watershed are wine production (13% of the soil is occupied by vineyards), agriculture (30% soil occupation, other than wine production), electroplating and ceramic industries (2% soil occupation) (Painho and Caetano, 2006). These activities are leading to the deterioration of the water quality in the Pateira de Fermentelos region and represent a major threat to (i) the water quality of the lagoon; (ii) the dependent ecosystems; and (iii) the infiltrating water that recharges the ACGWB.

18.3 METHODS

In order to assess the hydrogeochemical processes that prevail in the Pateira de Fermentelos region, 112 points were monitored on a monthly basis during the hydrologic year of 2005/06 (Sena, 2007). Groundwater and surface water samples were taken for major anion (Cl^-, SO_4^{2-}, NO_3^-) analysis, and measurement of physico-chemical parameters in the field (temperature; pH; electrical conductivity (EC); redox potential (Eh); dissolved oxygen (DO) concentration; and alkalinity). The 112 monitored points correspond to (Fig. 18.2):

- 12 springs where the groundwater discharge was measured, 4 of which were also sampled for major anion (Cl^-, SO_4^{2-}, NO_3^-) analysis, and measurement of physico-chemical field parameters;
- 4 points where the elevation of the lagoon's water surface was measured. Physico-chemical field parameters were also measured, and samples were taken for major anion analysis;
- 4 points for measuring the elevation of the water surface of the stream, two of which were sampled for major anion analysis and measurement of physico-chemical field parameters;
- 2 boreholes drilled in the semi-confined Triassic and Cretaceous aquifers where the major anion concentration and physico-chemical field parameters were monitored;
- 1 borehole drilled in the semi-confined Cretaceous aquifer where the elevation of the piezometric level was measured;
- 89 wells in the phreatic aquifers for measuring the elevation of the water table, one of which was also sampled for major anion analysis and measurement of physico-chemical field parameters.

Piezometric levels were measured in selected wells and boreholes in order to understand the hydrodynamic relations between surface and groundwater bodies. The piezometric surface elevation and surface water levels were measured with an electrical sounding device, while spring discharges were quantified measuring the time required to fill a predefined volume.

In April 2006 a broader field campaign was carried out to study the spatial hydrochemical variability of the study area. Water samples from surface water and groundwater bodies were collected from 28 sampling points (including the ten monthly

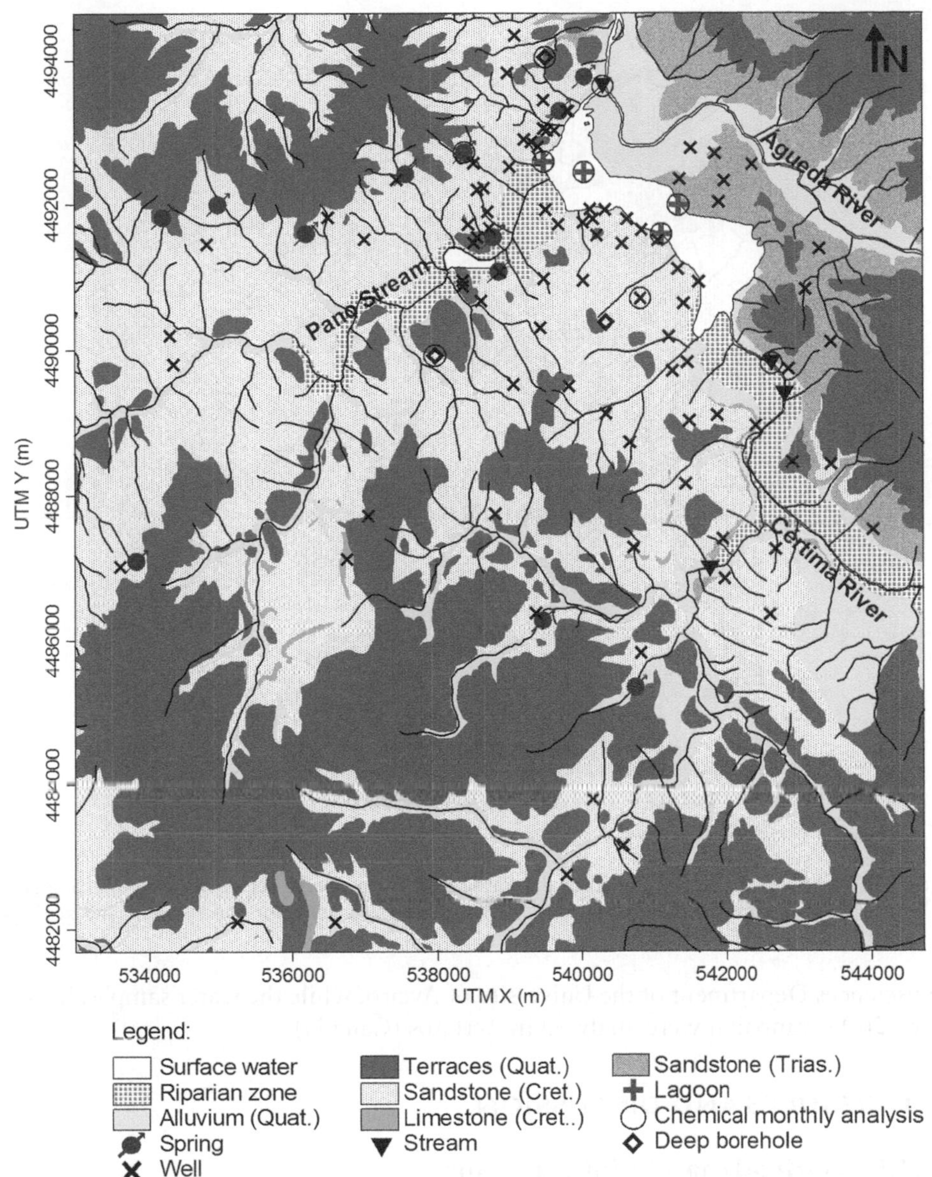

Figure 18.2 Simplified geological map of the Pateira de Fermentelos region showing the different types of points monitored on a monthly basis. The points that were used to obtain water samples for chemical analysis are indicated with a circle around the symbol. Geology digitised from Barbosa *et al.* (1981) and Teixeira and Zbyszewski (1976).

monitored points) for major, minor and trace element analysis (Fig. 18.3). Water samples were collected directly from the surface water bodies or collected with a flux cell (for groundwater) to PVC bottles. Samples for cation analysis were collected in separate bottles acidified with nitric acid. Monthly water samples were analysed in the

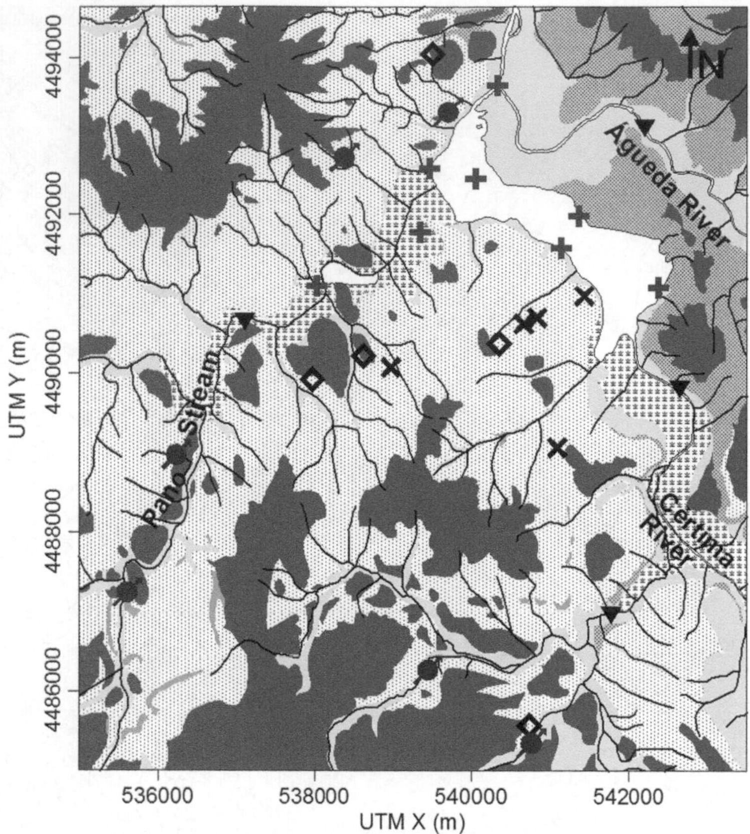

Figure 18.3 Simplified geological map of the Pateira de Fermentelos region showing the points that were sampled during the April 2006 campaign. Legend for the symbols and geology is shown in Figure 18.2.

Geosciences Department of the University of Aveiro, while the water samples from the April 2006 campaign were analysed in ActLabs (Canada).

18.4 RESULTS AND DISCUSSION

18.4.1 Hydrodynamic interactions

Monthly piezometric maps were built after gathering the field data for the elevation of the water table and surface water levels. These maps show that the Pateira de Fermentelos lagoon and its feeding streams drain the phreatic aquifers throughout the year, generating water table depressions in the vicinity of the main surface water bodies (Fig. 18.4). Inversion of the flux transfer may occur only in the lowland riparian zones adjacent to the lagoon during heavy rain events. Nevertheless, the loss of surface water to the surrounding riparian zones is most likely transferred back to the lagoon, within a couple of days, due to the high hydraulic gradient imposed by the elevated areas of the phreatic aquifers (Sena, 2007).

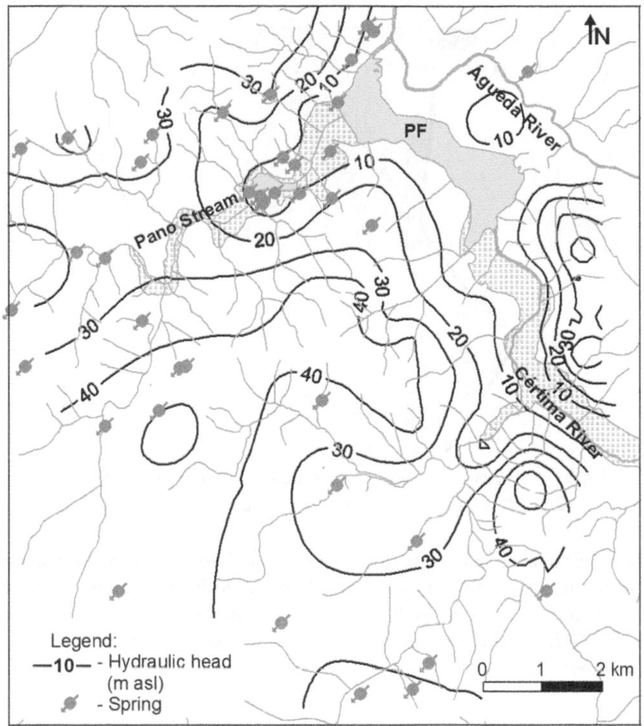

Figure 18.4 Contour map of hydraulic heads around the Pateira de Fermentelos lagoon in August 2006.

Since the surface water bodies drain the phreatic aquifers, the hydrochemistry of the Pateira de Fermentelos lagoon and adjacent streams is expected to be strongly influenced by the geochemistry of these aquifers.

A high density of springs was found in the Cretaceous outcrops to the southwest of the Pateira de Fermentelos lagoon. 51 springs were identified within an area of 90 km^2, of which only two were natural springs (Fig. 18.4). The remaining springs are 'fountains' with constructed drains that take advantage of the relatively shallow water table combined with permeable aquifer layers.

18.4.2 Hydrochemical patterns

In the Pateira de Fermentelos region, both natural and anthropogenic processes lead to the increase of solutes in groundwater and surface water. Results were obtained from the monthly campaigns and during the broader field campaign of April 2006.

18.4.2.1 Major elements

Rainwater in the studied region is a highly diluted Na-Cl water (ionic strength 5×10^{-4} mol/l) with a slightly acidic pH (6 pH units). In this region, rainwater originates mainly from the Atlantic Ocean, and it is the main contributor for the recharge of the aquifers surrounding the Pateira de Fermentelos lagoon (Condesso de Melo *et al.*,

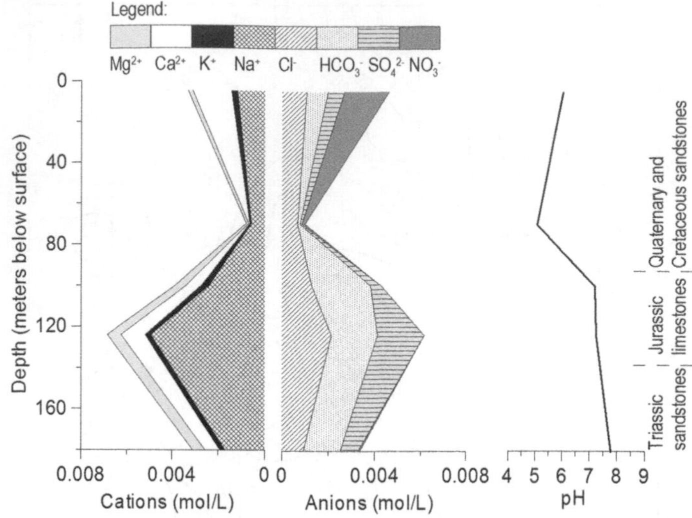

Figure 18.5 Synthetic vertical profile of groundwater hydrochemistry showing the distribution of major ions concentration with depth at the Pateira de Fermentelos region based on chemical data collected in shallow wells and deep boreholes. Concentrations of ions are plotted cumulatively from the centre to the outer borders of the plot. pH evolution with depth is also shown.

1999). Nevertheless, agricultural runoff and sewage seepage also recharge the phreatic aquifers. This is confirmed by the high nitrate content in many of the samples taken from the phreatic aquifers. This nitrate-contaminated shallow groundwater is present at the top of the synthetic vertical profile built for the studied region (Fig. 18.5).

When rainwater infiltrates, the decomposition of organic matter leads to an increase in the partial pressure of CO_2 which in turn acidifies the porewater and triggers the dissolution of pH sensitive minerals like carbonates and silicates (Appelo and Postma, 2005). The abundant silicate minerals of the Cretaceous sandstones are relatively insoluble and, therefore, pristine deep groundwater is often of Na-Cl type with acidic pH (around 5 pH units in Fig. 18.5).

Underlying the Cretaceous sandstones, Jurassic limestones, with more soluble carbonate and sulphate minerals, provide bicarbonate, sulphate, calcium and magnesium to groundwater, making it more mineralised and increasing its pH to circum-neutral values. Calcium and magnesium may then be exchanged by sodium which is present in clay minerals of the aquifer matrix. In the deepest parts of the synthetic profile, where the iron-oxyhydroxide rich Triassic sandstones lie, groundwater has a slightly alkaline pH with a mixed water type (Fig. 18.5).

The monthly monitoring of the main water bodies in the Pateira de Fermentelos region reveals that the widest range of major anions concentration occurs in surface water bodies and in the shallow Cretaceous aquifer, whilst the narrowest range of major anions concentration occurs in the deep Cretaceous aquifer (Fig. 18.6). Surface water bodies and the shallow Cretaceous aquifer are more vulnerable to external driving

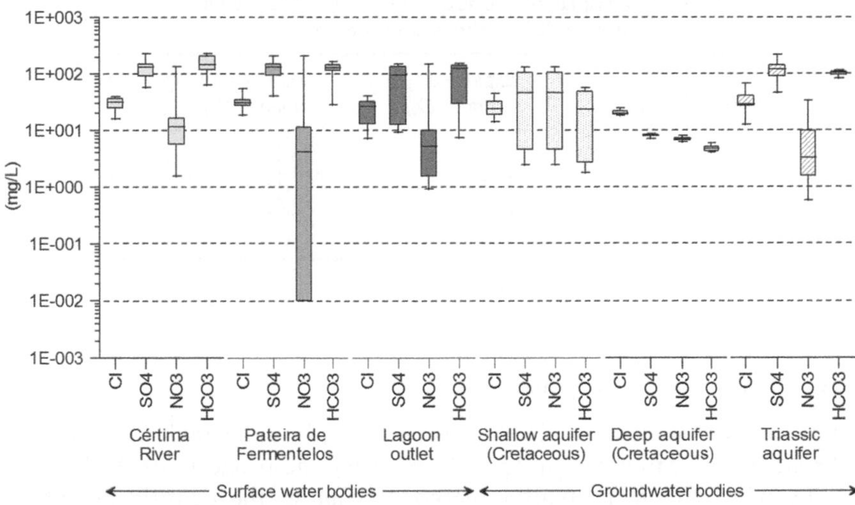

Figure 18.6 Box plots for major anions concentrations obtained from the monthly monitoring of the main water bodies in the Pateira de Fermentelos region, during the year 2005/06. Horizontal line inside each box stands for the median value; lower limit of each box stands for the first quartile; upper limit of each box stands for the third quartile; bottom extension of each box stands for the minimum value; and, top extension of each box stands for the maximum value.

forces such as climatic conditions, biological processes and anthropogenic activities than the deep Cretaceous aquifer which is naturally more protected and less prone to hydrochemical variations.

In all water bodies chloride concentrations show the smallest range of variation which could be explained by the fact that all the monitored water bodies share the same origin for chloride – rainwater and dry deposition. An exception is made by the deep Triassic aquifer formations where bicarbonate has the narrowest variation range reflecting the equilibrium with carbonate minerals present in this aquifer.

Nitrate has a considerable range of concentrations in all water bodies, except in the deep Cretaceous aquifer which indicates that somehow most of the water bodies are influenced by anthropogenic activities such as agriculture, cattle or domestic wastewater sewage. Sulphate has a narrower range of concentrations and its occurrence could be linked to different processes; mainly dissolution of sulphate minerals and trace amounts of pyrite, but also agricultural activities and treated electroplating wastewaters (where sulphate is added to hinder the aqueous mobility of metals, as observed in one stream of the study area).

The Cértima River has a narrower range of major anion concentrations then the lagoon outlet, and here the range of concentrations of chloride, sulphate and bicarbonate is wider than in the lagoon, which is related to the periods when the (less mineralised) Águeda River discharges into the (more mineralised) lagoon. The Águeda River has a diluted water due to the predominance of schist and granite rocks, typically with low reactivity minerals, in the corresponding catchment.

Table 18.1 Variance of the four axes of the PCA which explain at least 70% of the cluster inertia.

Axis of the PCA	*Explained variance (%)*	*Accumulated explained variance (%)*
1	36.12	36.12
2	15.65	51.77
3	10.30	62.07
4	9.25	71.32

18.4.2.2 Minor and trace elements

Principal Component Analysis (PCA) was performed on the analytical data obtained in the broader field campaign in April 2006, using the programme ANDAD (Sousa, 2002). The PCA is a factorial analysis method that allows the identification of relations between variables; i.e. the physico-chemical parameters and the similarities between individuals; i.e. the sampling points. The selection of the dimension of the subdomain incorporated the axes that explain at least 70% of the cluster inertia (Table 18.1).

The projection of the variables and individuals on the first three planes allowed the identification of the main trends that influence the chemical composition of surface and groundwaters in the Pateira de Fermentelos region (Fig. 18.7). Sulphate, bicarbonate, calcium, magnesium, sodium, chloride, pH and EC are projected on the negative side of the first axis (with values between −1 and −0.5). In addition, the second, third and forth axes separate these variables (between the corresponding negative and positive sides) in different ways. Exception is made for sodium and chloride which always plot very close to each other, indicating a common origin for both ions. Their origin is most likely rainwater (and dry deposition) and also, in the deeper aquifer layers old seawater that may be trapped in the sediments (Condesso de Melo, 2002). Sulphate, bicarbonate, calcium and EC appear very close to each other on the third plane, indicating that the dissolution of sulphate and carbonate minerals leads to the mineralisation of water. In addition, calcium, bicarbonate, EC, pH, and several surface water samples plot close to each other on the second plane which indicates that the dissolution of calcite contributes for the mineralisation of surface waters and to the increase of pH.

Eh and the partial pressure of CO_2 are plotted on the positive side of Axis 1. Eh has a relatively narrow range and it is mainly oxidising in all of the 28 samples, varying from 175 to 368 mV. Surface water samples in the Pateira de Fermentelos region are naturally oxidising as all of them are relatively shallow. On the other hand, the redox sensitive minerals that are present in the shallow aquifers are mainly iron oxyhydroxides which impose an oxidising porewater. Deep Jurassic aquifer layers which are intersected by one of the sampled boreholes of this campaign, may have trace amounts of pyrite that could lead to more reducing conditions, nevertheless, this borehole is multi-screened and, therefore, abstracts groundwater from shallow (oxidising) and deep (more reducing) aquifer layers.

The partial pressure of CO_2 projected in the PCA was calculated with the geochemical code PHREEQC (Parkhurst & Appelo, 1999). The projection of the partial

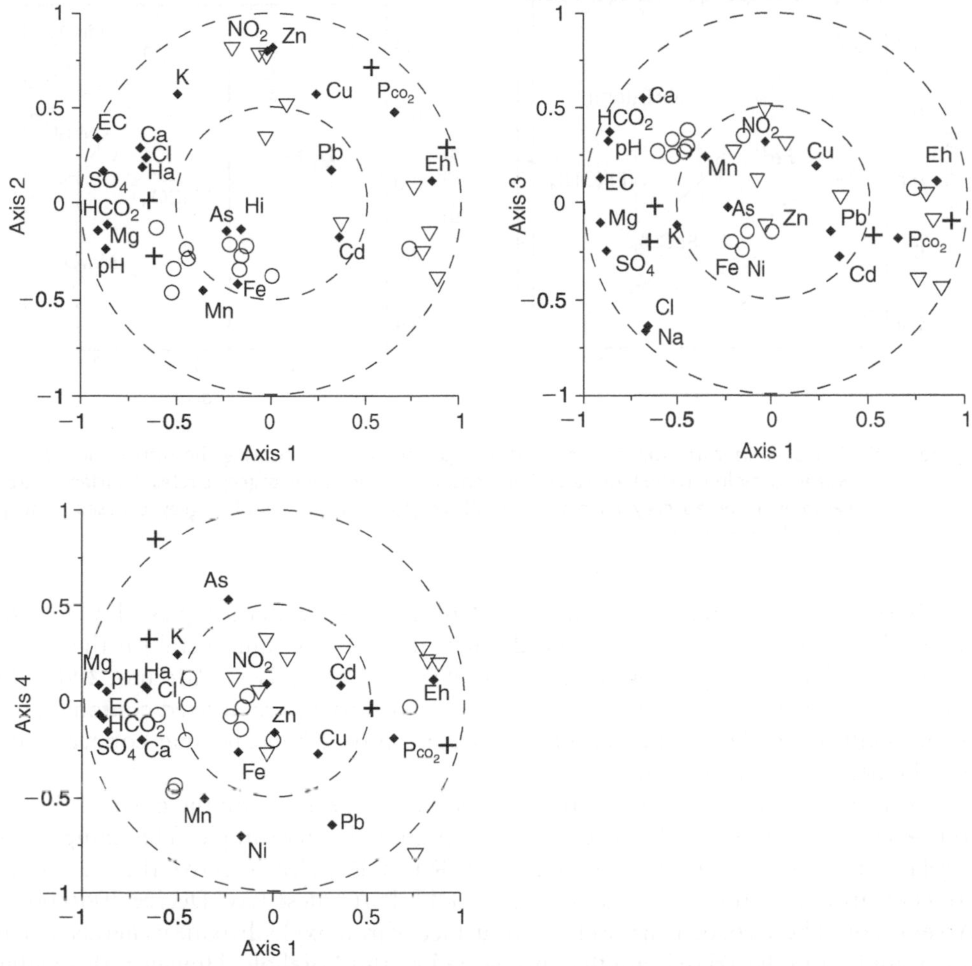

Figure 18.7 Projection of the variables (physico-chemical parameters) and individuals (water samples) selected from the broader field campaign of April 2006 in the first three planes of the PCA. Legend: black dots – position of the variables; grey circles – position of surface water samples; inverted grey triangles – position of the shallow groundwater samples; grey crosses – position of the deep groundwater samples.

pressure of CO_2 is clearly opposite to pH in the three planes studied which reflects the role of CO_2 in the acidification of natural waters. There is an important amount of organic matter in the lagoon sediments that favours a relatively high partial pressure of CO_2 in surface waters (varying from $10^{-2.80}$ to $10^{-1.66}$ atm which is clearly above that of the atmosphere; $10^{-3.45}$ atm). However, the highest partial pressure of CO_2 is calculated for the deep groundwater samples where CO_2 produced from organic matter degradation has limited possibilities to escape to the atmosphere.

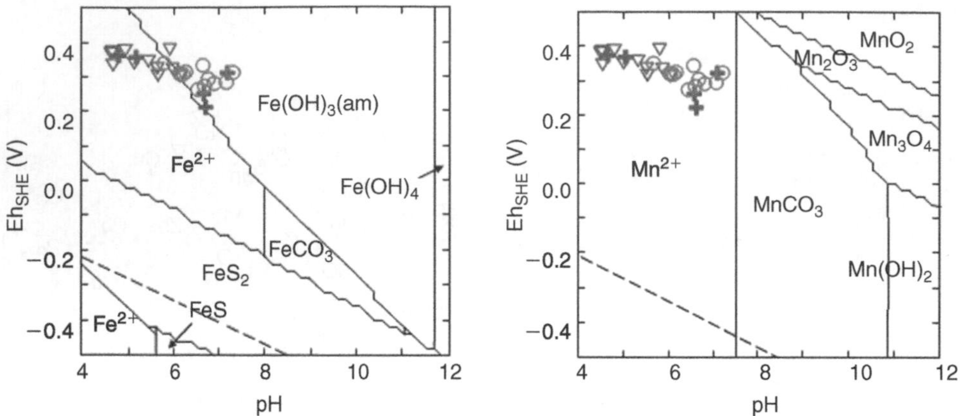

Figure 18.8 Pourbaix diagrams for the iron and manganese systems, showing the distribution of the waters sampled in the Pateira de Fermentelos region. Legend: grey circles – surface water samples; inverted grey triangles – shallow groundwater samples; grey crosses – deep groundwater samples.

Nitrate and zinc, together with three shallow groundwater samples plot close to each other on the positive side of Axis 2. Since it is an important metal for the development of plants, zinc is often a constituent of synthetic fertilisers (Adriano, 2001). In this context, Axis 2 is explaining the occurrence of agriculture derived contaminants in the shallow aquifers. Although not so evident, Axis 2 indicates that copper could also be derived from agriculture.

Arsenic is projected isolated from the rest of the variables, on the positive side of axis 4 where it lies relatively close to one deep groundwater sample. This sample has a relatively high concentration of arsenic (1.08 μmol/l), clearly above the maximum concentration for drinking standards (0.7 μmol/l, Portuguese Law-Decree 306/2007). Arsenic could be adsorbed on the charged surface of iron oxyhydroxide minerals which are abundant in the Triassic aquifer, intersected by this borehole. However, this water sample has an iron concentration which is below the detection limit (0.18 μmol/l). This water sample is in the field of stability of amorphous ferrihydrite which limits the concentration of soluble Fe(II) to low values (Fig. 18.8). The presence of arsenic in solution is likely to be related to the competition with other adsorbable species, like aqueous silica and bicarbonate that are relatively abundant in this groundwater sample. This competition has been reported as a plausible process in other works (Piqué *et al.*, 2010; Postma *et al.*, 2007).

Although the surface water samples lie within the stability field of ferryhydrite (Fig. 18.8), most of them show an iron concentration above the detection limit, varying from 0.54 to 23.3 μmol/l. This is probably due to the high content of aqueous carbonates in these waters which favours the mobility of iron. Iron and cadmium are not explained by any axis of the PCA since they always plot below the coordinates 0.5 of any axis.

Manganese, nickel and lead seem to be explained by axis 4 in Figure 18.7. These elements may form aqueous complexes that are adsorbed on the surface of iron

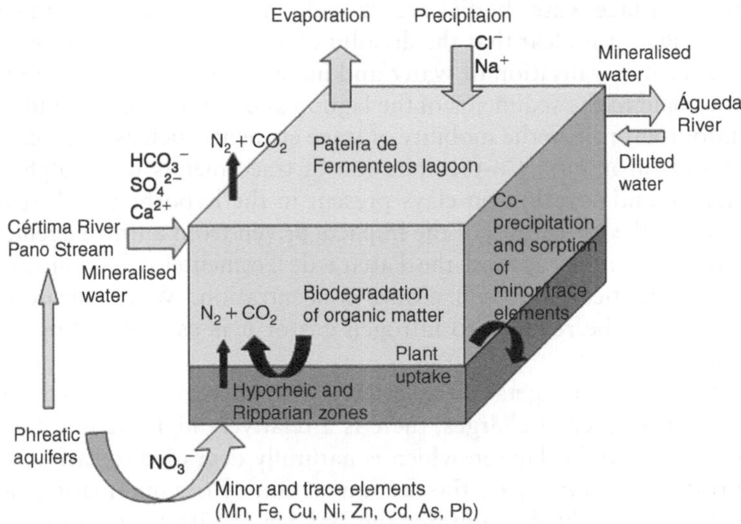

Figure 18.9 Conceptual model of the Pateira de Fermentelos lagoon summarising the main hydrogeochemical processes identified. Legend: grey arrows indicate fluxes of water; black arrows indicate biogeochemical reactions involving inorganic and organic substances.

oxyhydroxides (Dzombak and Morel, 1990) and desorption from these minerals could be a possible common origin. Although all the water samples are relatively oxidising, their circum-neutral to acidic pH is enough to promote the occurrence of aqueous Mn(II) (Fig. 18.8), and this element is present at relatively high concentrations in the studied region (ranging from 3.6×10^{-3} to 4.8 μmol/l).

Chromium is above the detection limit (0.01 μmol/l) solely in one sampling point (0.044 μmol/l) and, therefore, chromium was excluded from PCA. Knowing that in the study area there is a relatively important electroplating industrial activity, which frequently uses chromium by-products, it is quite surprising that chromium appears below the detection limit in most of the water samples. Most probably, sorption reactions on the surfaces of iron oxyhydroxides and clays, and also plant uptake are playing a major role in the attenuation of chromium concentrations in the aqueous phase.

18.5 CONCLUSIONS

The Pateira de Fermentelos lagoon is a shallow freshwater lagoon with complex hydrogeochemical patterns. The main water inflow to the lagoon is the Cértima River which has a mineralised water due to the presence of Jurassic limestones. Pateira de Fermentelos discharges to the Águeda River which dilutes the water due to the low reactivity of the minerals present in the schist and granite rocks that crop out in this catchment. Under certain conditions the Águeda River discharges to the Pateira de Fermentelos lagoon leading to the dilution of the concentration of ions in the downstream part of the lagoon.

Since the lagoon and adjacent streams drain the phreatic aquifers, the hydrochemical patterns of surface water bodies are strongly linked to the geochemistry of these aquifers. From PCA it is clear that the dissolution of sulphate and carbonate minerals contribute to the mineralisation of water and increase of pH. The iron-oxyhydroxide minerals, abundant in the sediments of the lagoon and in the Triassic sandstones, seem to play an important role in the mobility of trace elements such as arsenic, manganese, nickel, lead and chromium. Co-precipitation of trace metals with sulphate and carbonate minerals, and sorption on clays present in the hyporheic and riparian zones also contribute to the mitigation of the impacts driven from anthropogenic activities.

The phreatic aquifers around the Pateira de Fermentelos lagoon influenced by anthropogenic activities have high nitrate concentration. When nitrate reaches the lagoon it is likely to be reduced to nitrogen gas or it is assimilated by riparian and aquatic plants (Fig. 18.9).

Due to the exuberant vegetation associated with the riparian zones and the domestic and cattle wastewater discharges, there is a relatively high organic matter load in the Pateira de Fermentelos lagoon which is naturally degraded to carbon dioxide. By reducing nitrate to nitrogen gas, the decomposition of organic matter, catalysed by denitrifying bacteria, is likely to buffer the leakage of nitrate-rich groundwater and subsurface runoff in the lagoon.

ACKNOWLEDGMENTS

The authors are grateful to the Portuguese Ministry of Science, Technology and Education for a PhD Grant to Clara Sena (POCI 2010, BD/16647/2004), and for funding the Project Ecowet (POCI/CTE/GEX/58951/2004), which supported this research. We thank the reviewers for the valuable comments that helped to improve this manuscript.

REFERENCES

Adriano, D. (2001) *Trace Elements in Terrestrial Environments – Biogeochemistry, Bioavailability, and Risks of Metals.* 2nd edition. New York, Springer-Verlag.

Appelo, C.A.J. & Postma, D. (2005) *Geochemistry, Groundwater and Pollution.* Rotterdam, A.A. Balkema Publishers.

Barbosa, B., Bordalo da Rocha, R. & Ferreira Soares, A. (1981) Geological map of Portugal – Sheet 16-C (Vagos) and corresponding note. Geological Survey of Portugal.

Condesso de Melo, M.T. (2002) *Flow and Hydrogeochemical Mass Transport Model of the Aveiro Cretaceous Multilayer Aquifer (Portugal).* Ph.D. dissertation thesis. University of Aveiro, Portugal. http://hdl.handle.net/10773/2753

Condesso de Melo, M.T., Marques da Silva, M.A. & Edmunds, W.M. (1999) Hydrogeochemistry and Flow Modelling of the Aveiro Multilayer Cretaceous Aquifer. *Physics and Chemistry of the Earth*, 24 (4), 331–336.

Dzombak, D.A. & Morel, F.M.M. (1990) *Surface Complexation Modelling.* New York, Wiley Interscience. 431 pp.

European Commission (1991) Collection, treatment and discharge of urban waste water and the treatment and discharge of waste water from certain industrial sectors (Council Directive 91/271/EEC). *European Parliament and Commission, Official Journal*, 21 May.

European Commission (2008) Environmental quality standards in the field of water policy (Directive 2008/105/EC). *European Parliament and Commission, Official Journal of the European Union*, 16 December.

Gandy, C.J., Smith, J.W.N. & Jarvis, A.P. (2007) Attenuation of mining-derived pollutants in the hyporheic zone: A review. *Science of the Total Environment*, 373, 435–446.

Hancock, P.J., Boulton, A.J. & Humphreys, W.F. (2005) Aquifers and hyporheic zones: Towards an ecological understanding of groundwater. *Hydrogeology Journal*, 13, 98–111.

Hoffmann, C.C., Berg, P., Dahl, M., Larsen, S.E., Andersen, H.E. & Andersen, B. (2006) Groundwater flow and transport of nutrients through a riparian meadow – Field data and modelling. *Journal of Hydrology*, 331, 315–335.

Law-Decree 140/99 on 24 April (1999) Appendix, A.2. *Special Protection Area of Ria de Aveiro*. Portuguese National Press.

Law-Decree 306/2007 on 27 of August (2007) *Drinking Water Quality Standards*. Portuguese National Press.

Marques da Silva, M.A. (1990) *Hydrogeology of the Lower Vouga Multilayer Cretaceous aquifer – Aveiro (Portugal)*. Ph.D. dissertation thesis [in Spanish]. University of Barcelona, Spain.

Painho, M. & Caetano, M. (2006) *Cartography of Soil Occupation in the Mainland of Portugal 1985-2000 – CORINE Land Cover 2000* [in Portuguese]. Portuguese Environmental Agency.

Parkhurst, D.L. & Appelo, C.A.J. (1999) *User's Guide to PHREEQC (version 2) – A Computer Program for Speciation, Batch-Reaction, One-Dimensional Transport and Inverse Geochemical Calculations*. U.S. Geological Survey Water Resources investigations report 99-4259. 312 pp.

Piqué, A., Grandia, F. & Canals, A. (2010) Processes releasing arsenic to groundwater in the Caldes de Malavella geothermal area, NE Spain. *Water Research*. doi:10.1016/j.watres.2010.07.012.

Postma, D., Larsen, F., Hue, N.T.M., Duc, M.T., Viet, P.H., Nhan, P.Q. & Jessen, S. (2007) Arsenic in groundwater of the Red River floodplain, Vietnam: Controlling geochemical processes and reactive transport modelling. *Geochimica et Cosmochimica Acta*, 71, 5054–5071.

Rivett, M.O., Buss, S.R., Morgan, P., Smith, J.W.N. & Bemment, C.D. (2008) Nitrate attenuation in groundwater: A review of biogeochemical controlling processes. *Water Research*, 42, 4215–4232.

Sena, C. (2007) *Groundwater/Surface Water Interactions in Pateira de Fermentelos Region (Portugal)*. Master thesis presented at the University of Aveiro (in Portuguese). http://hdl.handle.net/10773/2701

Sousa, J. (2002) *Program ANDAD Version 7.10 – User's Manual. CVRM – Geosystems Centre*. Technical University of Lisbon [in Portuguese].

Teixeira, C. & Zbyszewski, G. (1976) *Geologic Map of Portugal – Sheet 16-A (Aveiro) and Corresponding Note*. Geological Survey of Portugal.

Teles, M., Pacheco, M. & Santos, M.A. (2007) Endocrine and metabolic responses of *Anguilla anguilla* L. caged in a freshwater–wetland (Pateira de Fermentelos – Portugal). *Science of the Total Environment*, 372, 562–570.

Chapter 19

Relationship between dry and wet beach ecosystems and *E. coli* levels in groundwater below beaches of the Great Lakes, Canada

Allan S. Crowe & Jacqui Milne

National Water Research Institute, Environment Canada, Burlington, Ontario, Canada

ABSTRACT

There are two types of beaches along the shores of the Great Lakes in Canada: "Dry Beaches" (dry sand at surface, deep water table, presence of beach grass) and "Wet Beaches" (damp sand at surface, shallow water table, presence of phreatophytes and turf grass). *E. coli* is never detected in groundwater below dry beaches but *E. coli* is consistently detected in groundwater below wet beach. Phreatophytes and turf grass at wet beaches provide an environment and source of food that attracts geese. Goose feces are a source of *E. coli.* The shallow water table and high moisture content of the sand at a wet beach enables *E. coli* from fecal material to infiltrate to the water table. At dry beaches, only beach grass is present, and within the dunes, and this does not attract geese. The deep water table and low moisture content in surficial sand inhibits infiltration of *E. coli* to the water table.

19.1 INTRODUCTION

The bacterial contamination of water adjacent to the shoreline is a common and pervasive problem at swimming beaches of the Great Lakes in North America (Fogarty *et al.*, 2003; Haack *et al.*, 2003; Whitman & Nevers, 2003; Whitman *et al.*, 2003, 2004; Edge & Hill, 2005; Sampson *et al.*, 2005; Alm *et al.*, 2006; Scopel *et al.*, 2006). *Escherichia coli* (*E. coli*) are often detected at levels considerably higher than the recreational water quality guidelines. Exceeding this guideline results in a posting (warning) of high levels of bacteria in the waters and restricts the use of recreational beaches.

Beaches of the Great Lakes exhibit very high levels of *E. coli* in the sand and groundwater below the swash zone within a few metres of the lake (Kinzelman *et al.*, 2003; Whitman & Nevers, 2003; Whitman *et al.*, 2003; Alm *et al.*, 2003, 2006; Edge and Hill, 2005; Beversdorf *et al.*, 2007; Ishii *et al.*, 2007). However, studies have not investigated *E. coli* levels in groundwater below beaches further from the shoreline, nor studied groundwater as a mechanism for the delivery of *E. coli* to the shoreline.

The objectives of this paper are to discuss (1) levels of *E. coli* in groundwater below beaches along the shores of the Great Lakes, (2) relationships between beach ecosystem (both physical features of the beach and the plant communities) and *E. coli* levels in the groundwater below beaches, and (3) potential sources of *E. coli* in groundwater below a beach. The beach environment, groundwater characteristics and *E. coli* levels

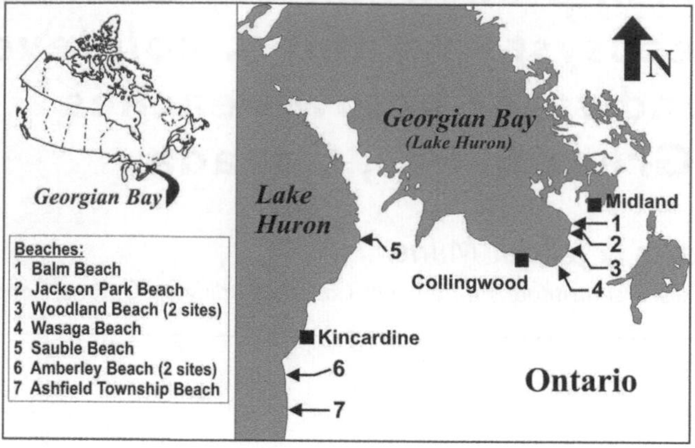

Figure 19.1 Location of the beach sites.

in groundwater were studied at nine sites at seven beaches along the shores of Lake Huron, including its major bay, Georgian Bay, Ontario, Canada (Fig. 19.1).

19.2 STUDY SITES

The Balm Beach (Fig. 19.1) site exhibits natural beach conditions with dry coarse sand, well developed sand dunes, and beach grass and other native vegetation within the dunes. The beach is over 30 m wide, and an extensive area of sand dunes is present. The dunes rise 9 m above the beach. Residences are located within the dunes, approximately 80 m from shore.

Jackson Park Beach is a public beach located in a sheltered bay, with houses along the leeward side of the beach. The beach is approximately 50 m wide. Sand dunes and beach grass are absent, and the maximum elevation of the ground surface is only 0.85 m above the lake. This beach typically exhibits wet sand and ponded water. Phreatophytes and turf grass are common on the beach.

The Woodland Beach south site has a beach approximately 40 m wide. The sand dunes and native vegetation have been removed by the beach-front residents and retaining walls erected at the edge of the beach where the dunes used to exist. The maximum elevation of the ground surface across the beach is 1.5 m above the lake. The beach exhibits wet sand at ground surface and numerous springs. Phreatophytes and turf grass have become established along the beach.

The Woodland Beach north site is located 330 m north of the Woodland beach south site. Woodland Beach north has 10–15 cm of dry sand on the surface, well developed sand dunes, and native beach grass. The beach is approximately 30 m wide followed by sand dunes that rise 5 to 6 m above the lake. Residences are located within the dunes, approximately 55 m from the shoreline.

Wasaga Beach is a public beach approximately 45 m wide. Low sand dunes rise about 2 m above the beach, but the maximum elevation of the ground surface of the beach is only 0.5 m above the lake. This beach typically exhibits wet sand and ponded water. Phreatophytes completely cover the beach to the shoreline. Beach front residences are present at this site.

Two sites were studied at Amberley beach. The sites are approximately 370 m apart and have similar characteristics. The beach is approximately 7 m wide between Lake Huron and sand dunes, and has approximately 5 to 10 cm of dry sand and gravel at surface. The sand dunes are 4 to 5 m above the lake and contain beach grasses. Beach front residences are located 50 m from the shoreline.

Ashfield Township Beach is a public beach, approximately 60 m wide and composed of dry sand and gravel at surface. The surface of the beach is approximately 1 to 1.5 m above the lake. A series of low dunes (2.5 m high) are present at the leeward side of the beach. There are no beach front residences at this site. Beach grass and native vegetation are present within the dunes.

Sauble Beach is a public beach, approximately 100 m wide. The surface of the beach is composed of moist fine sand, and rises 1.7 m above the lake. Low dunes (2.5 m high) are present at the leeward side of the beach. There is a road between the residences and the dunes, with the residences 25 m from the beach. Phreatophytes are present on the beach and native beach grass present within the dunes.

19.3 METHODS

Groundwater conditions below the beaches were investigated through a series of boreholes located in a grid at the beach sites. The spacing between these boreholes was 5 m perpendicular to the lake and either 5 m or 10 m parallel to the shoreline. The number of boreholes at each site varied between 25 and 42. The number of boreholes and the borehole spacing at four sites are represented in Figure 19.2. The locations of all holes were referenced to a benchmark at each site to ensure that locations of water table measurements could be exactly compared at different times. The depth to the water table was determined by measuring from the top of a plank securely placed across the hole. Ground surface elevations were obtained by leveling the top of the plank at where the water table depth was measured, using an electronic survey level, and referencing these elevations to a local temporary bench mark at each of the beaches. The elevation of the water table was determined by subtracting the depth to the water from the ground surface elevation. The elevation of the surface of the lake was leveled at each sampling time. Only a single line of boreholes perpendicular to the shoreline was completed at Wasaga Beach, Sauble Beach, Amberley Beach south, and Ashfield Township Beach.

The boreholes were used to collect groundwater samples for analyses of *E. coli*. Care was taken to avoid cross-hole contamination. The boreholes were dug by first removing the upper few centimeters of sand at ground surface (sand exposed to surface sources of *E. coli*) before the holes were dug. The equipment used to dig the bore holes was cleaned in a bleach solution to remove any bacteria, and then completely rinsed with sterile distilled water to remove all bleach from the equipment. Each borehole was dug to 10–15 cm below the water table to allow groundwater to seep into the borehole. The groundwater samples for *E. coli* analyses were obtained by inserting by hand a sterile 250 ml HDPE bottle below the water level at the base of a borehole,

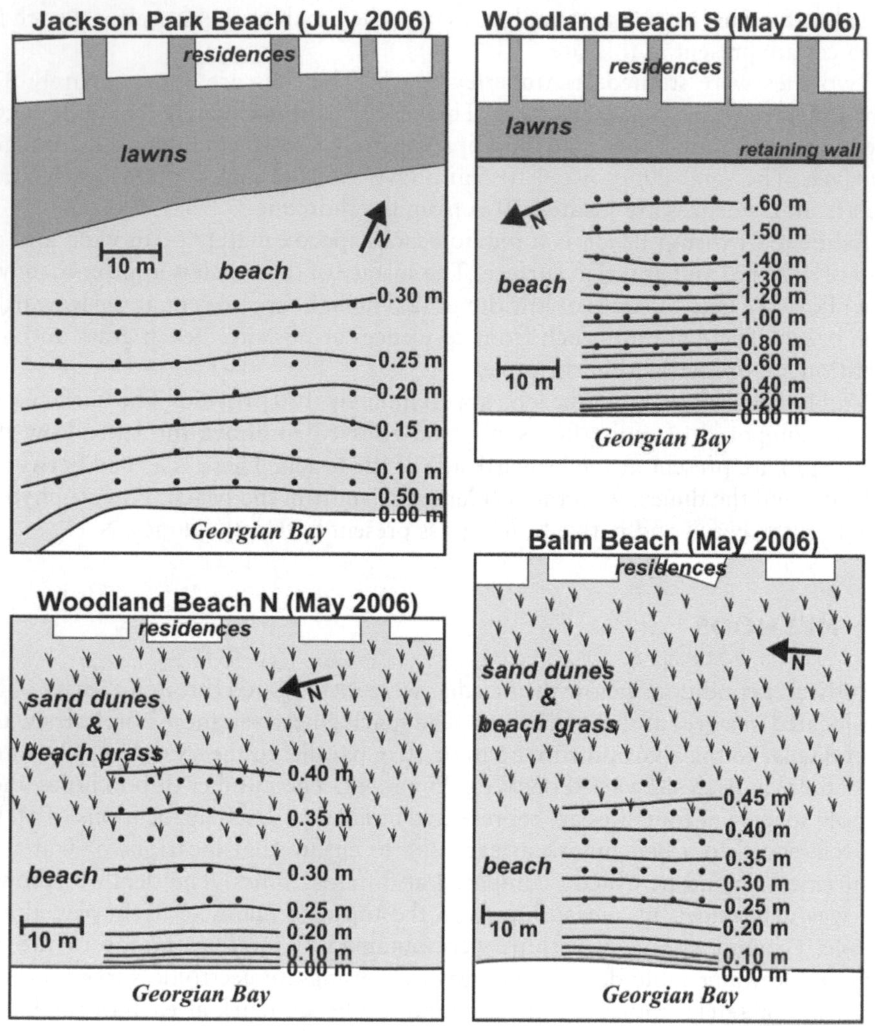

Figure 19.2 The elevation of the water table relative to the elevation of the surface of the lake at four of the beach sites. The black dots indicate the location of the boreholes.

and allowing groundwater to flow into the bottle. Clean disposable latex gloves were worn during the sampling. The sample bottles were stored in a cooler and kept chilled with ice until delivery to the laboratory.

E. coli analyses were undertaken using the ColiPlate-400™, manufactured by Environmental Bio-Detection Products Inc. (Brampton, Ontario). These plates consist of 96 wells containing a nutrient-rich media to stimulate *E. coli* growth. Water from a sample is poured over the plate, filling each well. The ColiPlates™ are incubated at 35°C for 24 hours. Following incubation, an *E. coli* count is determined by counting the number of wells that exhibit fluorescence under a long wavelength UV light, and converting this count to a Most Probable Number (MPN) per 100 ml using a conversion table from Environmental Bio-Detection Products. This method is limited to a minimum count of 3 MPN *E. coli*/100 ml (no fluorescence = <3 MPN *E. coli*/100 ml) and a maximum count without dilution of 2424 MPN *E. coli*/100 ml (fluorescence in all 96 wells = >2424 MPN *E. coli*/100 ml).

19.4 GROUNDWATER FLOW BELOW BEACHES OF THE GREAT LAKES

There is little information that specifically describes groundwater characteristics and groundwater quality below beaches of the Great Lakes. Existing studies have focused on assessing or quantifying the seepage of groundwater into the Great Lakes on a local scale (Harvey *et al.*, 1997a, 1997b), and regional scale (Cherkauer & Hensel, 1986; Cherkauer & McKreeghan, 1991; Sellinger, 1995). Crowe & Meek (2009) examined groundwater characteristics below beaches of Lake Huron/Georgian Bay, including some of the beaches discussed in this paper.

Figure 19.2 shows the elevations of the water table across four of the beach sites and illustrates similar groundwater characteristics seen among all beaches. The water table at all beaches slopes towards the lake, and hence groundwater flow is perpendicular and towards the lake throughout the year. Cross sections through the centre line of boreholes perpendicular to the lake at all beach sites (Fig. 19.3) show that with the exception of Woodland Beach south, the slope of the water table and the hydraulic gradient are uniform across a beach. However, hydraulic gradients vary from beach to beach (Table 19.1). There is a rapid decline in the water table adjacent to the shore corresponding to the zone of groundwater discharge at the seepage face. The water table rises and falls approximately 20 to 50 cm throughout the year in response to seasonal fluctuations in the elevation of the Lake Huron. Lake Huron fluctuates 20 to 50 cm annually with lowest lake levels during February and highest lake levels during July. However, the water table maintains a consistent slope throughout the year at each beach site, and hence the hydraulic gradient at each beach is consistent throughout the year.

Differences in the water table conditions among beaches is shown in Figure 19.3. At some beaches (Woodland north, Balm, Ashfield, Amberley north and south) the water table is relatively deep because of a higher elevation of the surface beach due to the presence of the sand dunes at the leeward side of the beaches relatively close to the shore. At other beaches (Woodland south, Wasaga, Jackson, Sauble) the water table is relatively shallow because of a lower ground surface of the beach and sand dunes are absent or far from the shore. Table 19.1 summarises both the change in the elevation of the ground surface and the depth to the water table moving inland from the shoreline. The elevation of the ground surface at all beaches varies considerably among the beaches at both 40 m and 25 m from the shoreline. However the depth to the water table increases inland considerable more at Woodland north, Balm, Amberley, and Ashfield Township than at Jackson Park, Sauble, Wasaga, and Woodland south, where the water table at these latter beaches is less than 0.6 m deep across the entire beach.

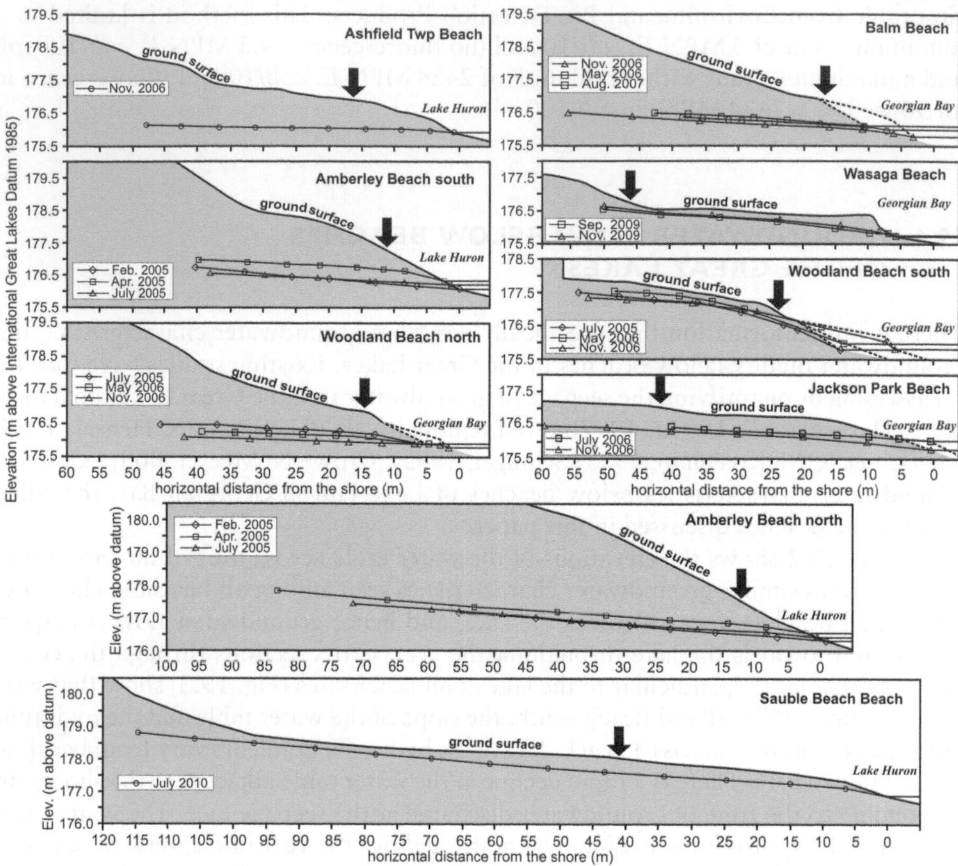

Figure 19.3 Cross sections perpendicular to the lake illustrating water table depths and seasonal fluctuations in the water table and surface of the lake. The arrows indicate where the beach is 1 m above the lake. Dashed lines represent the change in the position of the shoreline as the lake level rises and falls.

19.5 CHARACTERISTICS OF DRY AND WET BEACHES

Based on the depth to the water table, the beaches can be grouped into two types; those with the water table greater than 0.8 m across the beach and those with the water table consistently less than 0.8 m. The first type of beach (Woodland north, Balm, Amberley north and south, and Ashfield Township) is characterised relative deep water tables (>~0.8 m), and a surficial layer of dry sand about 10 to 15 cm thick on the beach surface. Vegetation is absent on this type of beach, but native plants adapted to the dry environment, such as American Beach Grass (*Ammophila breviligulata*) and Great Lakes Wheat Grass (*Elymus lanceolatus*), and other vegetation such as Wormwood (*Artemisia campestris*), Sand Cherry (*Prunus pumila*) are present in the adjacent dunes. Because of the dry conditions, deep water table, and vegetation suited for dry sand

Table 19.1 Physical and hydrogeological characteristics of the beach sites.

Site	Beach Type	Surface elev. above lake at 10 m[a]	25 m[a]	40 m[a]	Depth to water table at 10 m[a]	25 m[a]	40 m[a]	Sand Texture	K cm/s	Hydraul. Gradient
Wood N	dry	0.8 m	1.5 m	3.3 m	0.7 m	1.3 m	2.3 m	medium	4.0×10^{-2}	6.7×10^{-3}
Balm	dry	1.0 m	1.8 m	2.9 m	0.9 m	1.5 m	2.4 m	coarse	8.7×10^{-2}	1.1×10^{-2}
Amb N	dry	0.9 m	2.5 m	3.7 m	0.8 m	2.2 m	3.2 m	gravel	3.7×10^{-2}	1.4×10^{-2}
Amb S	dry	1.0 m	2.1 m	3.6 m	0.8 m	1.7 m	3.1 m	gravel	3.7×10^{-2}	1.3×10^{-2}
Ashfld	dry	0.6 m	0.9 m	1.7 m	0.5 m	0.8 m	1.6 m	gravel	2.9×10^{-2}	2.8×10^{-3}
Sauble	wet	0.5 m	0.9 m	1.1 m	0.5 m	0.8 m	1.6 m	v fine	9.6×10^{-3}	1.5×10^{-2}
Wood S	wet	0.4 m	1.4 m	1.8 m	0.3 m	0.3 m	0.4 m	fine	4.5×10^{-2}	1.4×10^{-2}
Jackson	wet	0.7 m	0.8 m	1.1 m	0.4 m	0.5 m	0.6 m	fine	3.2×10^{-2}	9.3×10^{-3}
Wasaga	wet	0.6 m	0.8 m	1.0 m	0.3 m	0.2 m	0.3 m	fine	2.5×10^{-2}	1.6×10^{-2}

a: average surface elevation and water table depth at three distances inland from the shoreline.

Figure 19.4 Photograph of a typical dry beach (Balm Beach) illustrating the presence of dry sand, sand dunes with native beach grass, and the absence of vegetation on the beach.

dunes, this type of beach is referred to as a "Dry Beach"; an example of a dry beach is shown in Figure 19.4.

The second type of beach (Jackson Park, Woodland south, Wasaga, Sauble) is characterised by relatively shallow water tables (<~0.8 m), and damp/wet sand at the ground surface across the beach, or if dry sand is present, it has a thickness of only 1 to 3 cm. Springs and ponded water are often present. The surface of the beach is quite flat and sand dunes may be absent. Native dune vegetation is generally absent. Native phreatophytes; sedge (*Carex* spp.), bulrush (*Scirpus validus*), common cattails (*Typha latifolia*) are prevalent across the beach. Recently, the invasive common reed (*Phragmites australis*) has become established on these beaches. Because of the wet conditions and phreatophyte vegetation, this type of beach referred to as a "Wet Beach"; an example of a wet beach is shown in Figure 19.5.

Figure 19.5 Photograph of a typical wet beach (Jackson Park Beach) illustrating the absence of sand dunes and native beach grass, and presence of wet sand, ponded water, and phreatophytes on the beach.

Many wet beaches along the shores of Lake Huron were originally dry beaches that were altered by beach-front residents. The natural sand dunes and native beach grass have been removed and surface of the beach lowered. In some cases retaining walls were built where the dunes formerly existed, and each year the residents remove sand that has blown against the wall. The removal of sand dunes and beach sand has lowered the ground surface and resulted in a shallow water table. Also, without beach grass to hold the sand, beach sand has blown away resulting in a lower ground surface. Many residents have replaced the native vegetation (e.g., beach grass) with lawns (turf grass), which migrates to the wet sand on the beach, thus, altered the natural beach ecosystem. Woodland Beach south is an example of a degraded beach, where the natural sand dunes were lowered and replaced by retaining walls. The ground surface had a profile similar to Woodland Beach north; at approximately 40 to 45 m from the lake the surface of the beach was formerly 2 to 2.5 m higher, and about 0.5 m higher 25 m from the shoreline. Removal of the sand has lowered the ground surface close to the water table creating wet-sand conditions that enabled phreatophytes and turf grass to become established.

19.6 *E. COLI* IN GROUNDWATER BELOW BEACHES

E. coli were always detected in groundwater below the swash zone (within 10 m of the shoreline) at all dry beach and all wet beach sites. Levels of *E. coli* exhibited considerable variability at each site, ranging from 100 s to 10000 s of *E. coli*/100 ml over a few meters. The range of *E. coli* levels and its variability was consistent among all sites.

At distances greater than ~10 m from shore, *E. coli* were not detected (<3 MPN *E. coli*/100 ml) in groundwater below the beach or adjacent dunes at dry beach sites.

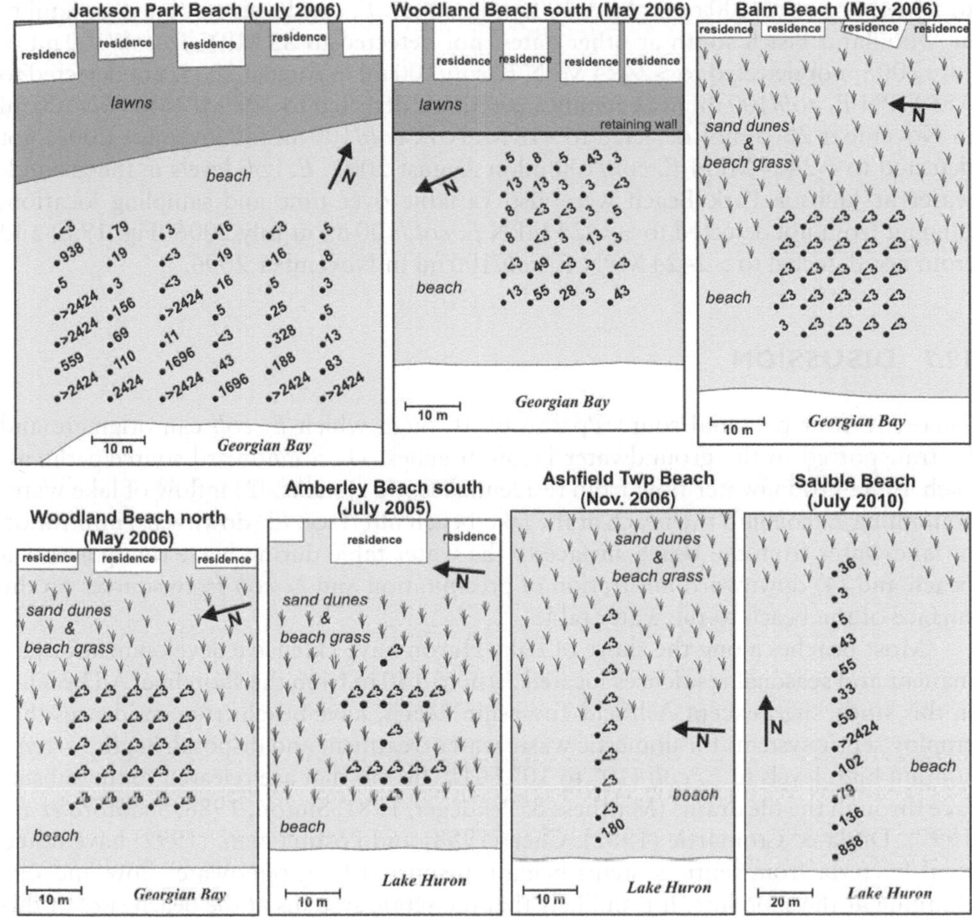

Figure 19.6 *E. coli* levels (MPN *E. coli*/100 ml) in groundwater below three wet beaches (Sauble, Jackson Park, Woodland south) and four dry beaches (Woodland north, Balm, Amberley south, Ashfield Township). Sample locations (boreholes) are indicated by the black dot.

Figure 17.6 shows the absence of *E. coli* below Woodland Beach north, Balm Beach, Amberley Beach north and south, and Ashfield Township Beach on one date each. Similar patterns of no *E. coli* in the groundwater away from the shore at these beaches also occurred at other sampling dates (not shown); Woodland north in July 2005, August 2007; Balm in August 2007; Amberley north and south in October 2004, February 2005, April 2006.

E. coli were consistently detected in groundwater below the wet beach sites away from the shore at Sauble Beach, Jackson Park Beach and Woodland Beach south. *E. coli* analyses were not undertaken at Wasaga Beach. Levels of *E. coli* in the groundwater at these wet beach sites were variable throughout the beach and over time. *E. coli* levels ranged from 3 MPN *E. coli*/100 ml to >2424 MPN *E. coli*/100 ml) at Sauble Beach in

July 2010 (Fig. 19.6). At Woodland Beach south, *E. coli* levels ranged from not detected to 55 MPN *E. coli*/100 ml in May 2006 (Fig. 19.6). *E. coli* levels exhibited variability at Woodland Beach south at other dates: not detected to 83 MPN *E. coli*/100 ml in July 2005; not detected to >2424 MPN *E. coli*/100 ml in August 2005; not detected to 555 MPN *E. coli*/100 ml in September 2005; not detected to 298 MPN *E. coli*/100 ml in November 2005; not detected to 418 MPN *E. coli*/100 ml in November 2006; not detected to >2424 MPN *E. coli*/100 ml in August 2007. *E. coli* levels in the groundwater at Jackson Park Beach were also variable over time and sampling location, ranging from not detected to >2424 MPN *E. coli*/100 ml in July 2006 (Fig. 19.6) and from not detected to >2424 MPN *E. coli*/100 ml in November 2006.

19.7 DISUSSION

There are four potential sources/pathways through which *E. coli* can originate and be transported to the groundwater below beaches: (1) a landward source/pathway, such as via groundwater flow from residential septic systems, (2) inflow of lake water containing *E. coli* into the beach at the lake-beach interface, (3) downward infiltration of lake water from the beach surface to the water table during wave runup across a beach and (4) downward infiltration of precipitation and *E. coli* from sources on the surface of the beach to the water table.

Most beaches along the shore of Lake Huron have extensive development of permanent and seasonal residences located 30 m to 150 m from the shoreline. All beaches in this study sites, except Ashfield Township Beach, have beach-front residences that employ septic systems for domestic waste water treatment and disposal. Septic systems contain high levels of *E. coli* (10^4 to 10^9 FCU/100 ml) that are released to the subsurface through the tile drains (Matthess & Pekdeger, 1981; Sinton, 1986; Shadford *et al.*, 1997). Duda & Cromartie (1982), Chen (1988) and Postma *et al.* (1992) have noted fecal bacteria from septic systems being transported by groundwater flow and discharging at the shoreline. It is unlikely that the septic systems at the beach sites of this study are the source of *E. coli* in the groundwater below these beaches. Septic systems are used by all beach-front residences, and hence if septic systems are contributing *E. coli* then *E. coli* should be seen, and at essentially the same levels, at all beach sites, whether the beach is wet or dry. Also, *E. coli* levels would decrease from the septic systems towards the lake. But, as shown by Figure 19.7A, the frequency of *E. coli* detection in the groundwater at both wet and dry beach site are not the same, nor do *E. coli* levels consistently decrease away from the septic systems. For example, *E. coli* levels at Woodland Beach south and Jackson Park Beach have the same range (not detected to >2424 MPN *E. coli*/100 ml) and frequency of detection at both 5–10 m and 40–50 m from septic systems. Only at Woodland Beach south, where the groundwater samples were collected within 5 m of the septic system tile beds, was there an indication that the *E. coli* in the groundwater was from a septic system (Fig. 19.6).

Groundwater flows towards the lake and discharges at the shoreline throughout the year (Crowe and Meek, 2009; Horn, 2002). Thus, the outward flow would prevent migration of lake-water and associated *E. coli* into a beach. Figure 19.3 shows that the water table slopes towards the lake throughout the year and at all beaches. Thus, lake-water is not flowing into the beaches. The distributions of *E. coli* in the groundwater

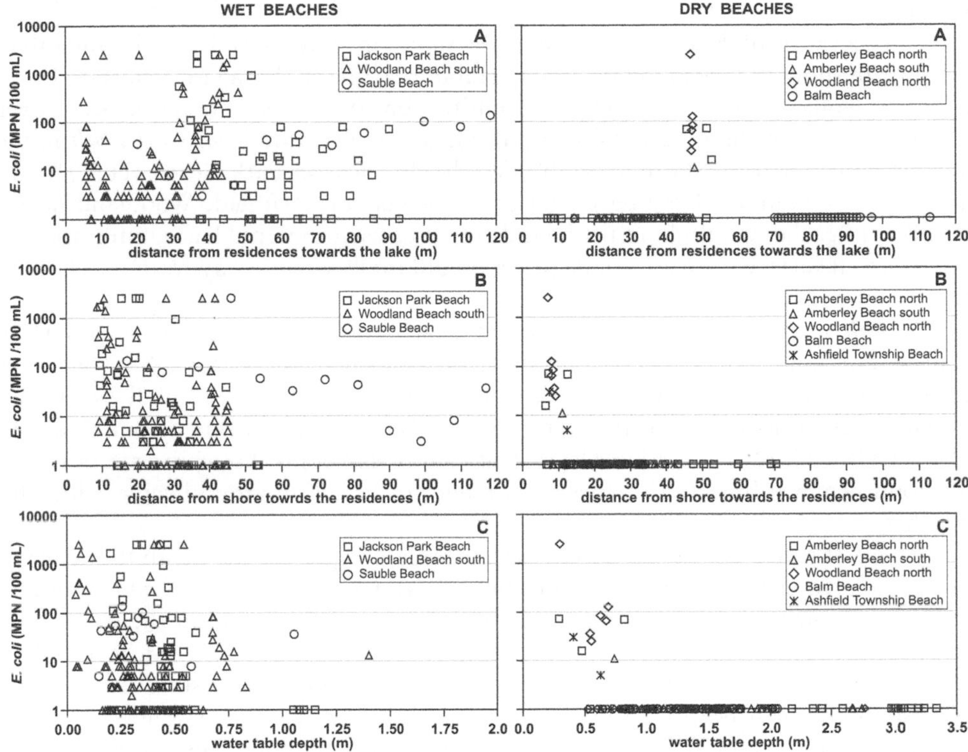

Figure 19.7 Distribution of *E. coli* in the groundwater below wet and dry beaches (A) away from residences (septic systems), (B) inland from the shore, and (C) versus depth to the water table. Data for each beach are a composite of several sampling times. Data within 10 m of the shoreline (swash zone) are not plotted. Data plotted at 1 *E. coli* MPN/100 ml axis represent non-detections.

below the beaches confirm that lake water is not flowing into the beaches; *E. coli* is not similarly distributed inland from the shore at both wet and dry beaches (Fig. 19.7B).

The high levels of *E. coli* commonly seen at the swash zone (Fogarty *et al.*, 2003; Haack *et al.*, 2003; Whitman *et al.*, 2003, 2004; Edge & Hill, 2005; Sampson *et al.*, 2005; Scopel *et al.*, 2006) are due to the infiltration of lake water during wave runup. Either the infiltrating lake water contains *E. coli*, or infiltrating lake-water transports *E. coli* from sources along the shoreline such as gull and geese feces. Wave runup across the shore during storm events will infiltrate lake water into the beach face, raising the water table at the shore causing inland movement of lake water into a beach (Li *et al.*, 1997, 2002; Baird *et al.*, 1998; Gibbs *et al.*, 2007). However, the landward movement of infiltrated lake water will be restricted to a few metres because groundwater below the beach will continue to flow towards the shoreline and the rapid exfiltration of most of the infiltrated lake water (Li *et al.*, 1997, 2002; Baird *et al.*, 1998; Gibbs *et al.*, 2007). The inland distribution and levels of *E. coli* in the groundwater depends on several factors including, the distance from the shoreline affected by wave runup (elevation of the beach surface, wave intensity) and the frequency and duration of wave runup.

The elevation of the ground surface is less than 1 m above the lake level greater than 40 m from shore at Jackson Park Beach, Sauble Beach, and Wasaga Beach (Fig. 19.3). A storm would cause wave runup and infiltration to the water table to occur over a large portion of these beaches, could result in the elevation levels of *E. coli* in the groundwater far from shore at Sauble Beach and Jackson Park Beach (Fig. 19.7B). The higher elevation of the beach surface at Amberley, Ashfield, Woodland north, and Balm Beaches would limit the inland extent of wave runup. Hence, infiltration of *E. coli* from surface sources or lake water would rarely occur away from the shore at these latter beaches (Fig. 19.7B).

E. coli in the groundwater below the beach could also originate as a surface source (gull and geese feces) and infiltrate downward to the water table with infiltrating precipitation. Studies have shown that infiltration from rainfall events can transport microorganisms to the water table through the unsaturated zone (Balkwill *et al.*, 1998). There are four interdependent factors that control the presence of *E. coli* in groundwater below wet beaches and absence of *E. coli* in groundwater below dry beaches.

The first factor is the depth to the water table; the greater the depth to the water table, the greater the distance that precipitation and associated *E. coli* must infiltrate from the ground surface to the water table. Beaches with *E. coli* in the groundwater have a shallow water table (<~0.8 m) and beaches without *E. coli* have a deep water table (>~0.8 m) (Fig. 19.7C). In general, a higher ground surface leads to a corresponding increased depth to the water table (as is the case at Amberley, Balm, Woodland north, and Ashfield Beaches). However, it may not always be the case; Woodland Beach south consistently exhibits *E. coli* in the groundwater even though the elevation of the surface of the beach above the lake is comparable to the beaches without *E. coli* (Table 19.1). But, the water table is shallow across this beach. Thus, *E. coli* is consistently detected in the groundwater below beaches where the depth to the water table is shallow.

The second factor is the size of the sand grains and associated soil moisture content of the sand between the water table and ground surface. Under unsaturated conditions, water is held in the pores by capillary forces (Stephens, 1996). As the texture of the sand becomes increasingly coarser, there is a corresponding decrease in the volumetric moisture content of the sand, and a corresponding decrease in hydraulic conductivity (op.cit). Thus, coarser sand or sand with a low moisture content require greater of infiltration to raise the water content of the entire soil profile to a level where infiltration, and associated *E. coli*, will move from surface to the water table. Beaches that do not have *E. coli* in the groundwater have a surficial 10 to 15 cm layer of dry sand, and the sand below this dry layer has a low moisture content; these are dry beaches. Beaches that have *E. coli* in the groundwater are beaches in which surficial sand is wet or damp, or only dry for 2 to 3 cm, and the sand below has a high moisture content; these are wet beaches.

The third factor is the type of vegetation that grows on wet and dry beaches. Dry beaches, with a low moisture content will support only native plants that can survive in this dry environment, such as beach grass. The shallow water table and the high moisture content of the sand at a wet beach can support the growth of phreatophyte vegetation. Phreatophytes will not survive in the dune and dry sand environment found at dry beaches. In areas where the sand dunes have been destroyed by beach-front residents, beach grass is often replaced by a lawn. The turf grass will migrate onto the beach and become established within the moist sand conditions found at wet beaches.

The fourth factor is the presence of a surface source of *E. coli*, such as geese feces. Geese are a major source of *E. coli* found at recreational beaches within the Great Lakes (Whitman & Nevers, 2003; Edge & Hill, 2005; Ishii *et al.*, 2007) and are proficient at producing large amounts of fecal material. The phreatophyte vegetation and turf grass that is present at wet beaches provide an environment and source of food that attracts geese. The lack of vegetation on dry beaches and the presence of beach grass within the dunes at dry beaches do not attract geese; geese do not consume beach grass. The shallow depth to the water table and higher water content in the sand at wet beaches provide conditions that are conducive for the infiltration of precipitation and *E. coli* to the water table. Geese tend to avoid beaches with beach grass. Even if fecal material is present on the surface at dry beaches, the dry sand at surface and deep water table severely limits the infiltration. Thus, it is likely that *E. coli* from fecal material on the beach surface could infiltrate to the water table at wet beach sites. But at a dry beach sites there is little likelihood of infiltration of *E. coli* to the water table.

Unrestricted residential development at beaches along the shores of the Great Lakes may have a detrimental impact on *E. coli* levels in groundwater below these beaches. Our investigations show that septic systems at beach front residence are not sources of *E. coli* in groundwater below beaches. However, there is a link between the destruction of the natural dune-beach grass ecosystem and increased levels of *E. coli* in the groundwater below beaches. In particular, the replacement of natural dunes and native beach grass with lawns on the beach provides an attractive environment for geese and a favourable groundwater setting for infiltration of their *E. coli* to the water table. Not all beaches along the shores of the Great Lakes are composed of natural sand dunes and wide expanses of dry sand; many beaches are naturally wet, without dunes, and the native vegetation are phreatophytes. But where the natural beach environment is the dry beach, sand dune environment, beach front residents must maintain and protect the natural dune and beach grass ecosystem in order to ensure that the groundwater below the beach remains free of *E. coli.* Beach grass will thrive within the dry sand and dune environment, and its presence will both dissuade geese from beaches, and accumulate sand, further building the elevation of the dunes and beaches and increase the distance to the water table.

19.8 CONCLUSIONS

The presence or absence of *E. coli* in the groundwater relates to the physical and hydrologic characteristic of the beaches and to the types of vegetation present at the beaches. The beaches where *E. coli* were never detected in the groundwater are characterised by a high surface elevation dunes close to the shore, a surficial layer of dry sand about 10 to 20 cm thick, and a water table greater than $\sim$0.8 m deep. Vegetation is absent on this beach, but native beach grass is present within the dunes. These beaches are referred to as "dry beaches". Beaches where *E. coli* were always detected in the groundwater, referred to as "wet beaches", are characterised by a low and generally flat beach surface, damp sand at the surface, and a depth to the water table of less than 0.8 m. Beach grass is generally absent, and replaced by phreatophytes, including sedge grass, phragmites, and cattails. Many wet beaches are actually degraded beaches where the natural sand dunes and beach grass have been removed by beach-front residents and replaced with lawns of turf grass, which migrates on the wet sand on the beach. The shallow depth to the water table and moist sand at surface of wet beaches provides conditions that are conducive for the growth of hydrophyte vegetation and turf grass. This in turns attracts geese, a major source of *E. coli*, and *E. coli* from their fecal material will infiltrate to the water table because the hydrologic conditions here are conducive for infiltration of precipitation and surface contaminants to the water table. Dry beaches, are not conducive to the growth of phreatophytes or turf grass, and hence do not attract geese that produce surficial sources of *E. coli*. Dry beaches are not conducive to flooding of the beach during wave runup because of their relatively high surface elevations, nor are they conducive to infiltration of precipitation and/or lake water, from the beach surface to the water table.

REFERENCES

Alm, E.W., Burke, J. & Spain, A. (2003) Fecal indicator bacteria are abundant in wet sand at freshwater beaches. *Water Research*, 37, 3978–3982.

Alm, E.W., Burke, J. & Hagan, E. (2006) Persistence and potential growth of the fecal indicator bacteria, *Escherichia coli*, in shoreline sand at Lake Huron. *Journal of Great Lakes Research*, 32, 401–405.

Baird, A.J., Mason, T. & Horn, D. (1998) Validation of the Boussinesq model of beach groundwater behavior. *Marine Geology*, 148, 55–59.

Balkwill, D.L., Murphy, E.M., Fair, D.M., Ringelberg, D.B. & White, D.C. (1998) Microbial communities in high and low recharge environments: implications for transport in the vadose zone. *Microbial Ecology*, 35, 156–171.

Beversdorf, L.J., Bornstein-Forst, S.M. & McLellan, S.L. (2007) The potential for beach sand to serve as a reservoir for *Escherichia coli* and physical influences on cell die-off. *Journal of Applied Microbiology*, 102, 1372–1381.

Chen, M. (1988) Pollution of ground water by nutrients and fecal coliforms from lakeshore septic tank systems. *Water, Air and Soil Pollution*, 37, 407–417.

Cherkauer, D.S. & Hensel, B.R. (1986) Ground water flow into Lake Michigan from Wisconsin. *Journal of Hydrology*, 84, 261–271.

Cherkauer, D.S. & McKereghan P.F. (1991) Ground-water discharge to lake: focusing in embayments. *Ground Water*, 29, 72–80.

Crowe, A.S. & Meek, G.A. (2009) Groundwater – lake interaction beneath beaches of Lake Huron, Ontario, Canada. *Journal of Aquatic Ecosystem Health Management*, 12, 444–455.

Duda, A.M. & Cromartie, K.D. (1982) Coastal pollution from septic tank drainfields. *ASCE Journal of Environment Engineering*, 108, 1265–1279.

Edge, T.A. & Hill, S. (2005) Occurrence of antibiotic resistance in *Escherichia coli* from surface waters and fecal pollution sources near Hamilton, Ontario. *Canadian Journal of Microbiology*, 51, 501–505.

Fogarty, L.R., Haack, S.K., Wolcott, M.J. & Whitman, R.L. (2003) Abundance and characteristics of the recreational water quality indicator bacteria *Escherichia coli* and enterococci in gull faeces. *Journal of Applied Microbiology*, 94, 865–878.

Gibbs, B., Robinson, C., Li, L. & Lockington, D. (2007) Measurement of hydrodynamics and pore water chemistry in intertidal groundwater systems. *Journal of Coastal Research*, SI-, 50, 884–894.

Haack, S.K., Fogarty, L.R. & Wright, C. (2003) *Escherichia coli* and Enterococci at beaches in Grand Traverse Bay, Lake Michigan: sources, characteristics, and environmental pathways. *Environmental Science and Technology*, 37, 3275–3282.

Harvey, F.E., Lee, D.R., Rudolph, D.L. & Frape, S.K. (1997a) Locating groundwater discharge in large lakes using bottom sediment electrical conductivity mapping. *Water Resources Research*, 33, 2609–2615.

Harvey, F.E., Rudolph, D.L. & Frape, S.K. (1997b) Measurement of hydraulic properties in deep lake sediments using a tethered pore pressure probe: applications in the Hamilton Harbour, western Lake Ontario. *Water Resources Research*, 33, 1917–1928.

Horn, D.P. (2002) Beach groundwater dynamics. *Geomorphology*, 48, 121–146.

Ishii, S., Hansen, D.L., Hicks, R.E. & Sadowsky, M.J. (2007) Beach sand and sediments are temporal sinks and sources of *Escherichia coli* in Lake Superior. *Environmental Science and Technology*, 41, 2203–2209.

Kinzelman, J.L., Whitman, R.L., Byappanahalli, M., Jackson, E. & Bagley, R.C. (2003) Evaluation of beach grooming techniques on *Escherichia coli* density in foreshore sand at North Beach, Racine, WI. *Lake and Reservoir Management*, 19, 349–354.

Li, L., Barry, A., Pattiaratchi, C.B. & Masselink, G. (2002) Beach water table fluctuations due to wave run-up: Capillary effects. *Water Resources Research*, 33, 935–945.

Li, L., Barry, A., Parlange, J.Y. & Pattiaratchi, C.B. (1997) BeachWin: modelling groundwater effects on swash sediment transport and beach profile changes. *Environmental Modelling Software*, 17, 313–320.

Matthess, G. & Pekdeger, A. (1981) Concepts of a survival and transport model of pathogenic bacteria and viruses in groundwater. *Science of the Total Environment*, 21, 149–159.

Postma, F.B., Gold, A.J. & Loomis, G.T.W. (1992) Nutrient and microbial movement from seasonally-used septic systems. *Journal of Environmental Health*, 55 (2), 5–10.

Sampson, R.W., Swiatnichki, S.A., McDermott C.M. & Kleinheinz, G.T. (2005) *E. coli* at Lake Superior recreational beaches. *Journal of Great Lakes Research*, 31, 116–121.

Scopel, C.O., Harris, J. & McLellan S.L. (2006) Influence of nearshore water dynamics and pollution sources on beach monitoring outcomes at two adjacent Lake Michigan beaches. *Journal of Great Lakes Research*, 32, 543–552.

Sellinger, C.E. (1995) Groundwater flux into a portion of eastern Lake Michigan. *Journal of Great Lakes Research*, 21, 53–63.

Shadford, C.B., Joy, D.M., Lee, H., Whitely, H.R. & Zelin, S. (1997) Evaluation and use of a biotracer to study groundwater contamination by leaching bed systems. *Journal of Contaminant Hydrology*, 28, 227–246.

Sinton, L.W. (1986) Microbial contamination of alluvial gravel aquifers by septic tank effluent. *Water, Air and Soil Pollution*, 28, 407–425.

Stephens, D.B. (1996) *Vadose Zone Hydrology*. Lewis Publishers, CRC Press. 347 pages.

Whitman, R.L. & Nevers, M.B. (2003) Foreshore sand as a source of *Escherichia coli* in nearshore water of a Lake Michigan beach. *Applied Environmental Microbiology*, 69, 5555–5562.

Whitman, R.L., Nevers, M.B., Korinek, G.C. & Byappanahalli, M. (2004) Solar and temporal effects on *Escherichia coli* concentrations at a Lake Michigan swimming beach. *Applied Environmental Microbiology*, 70, 4276–4285.

Whitman, R.L., Shively, D.A., Pawlik, H., Nevers, M.B. & Byappanahalli, M. (2003) Occurrence of *Escherichia coli* and Enterococci in *Cladophora* (Chlorophyta) in neashore water and beach sand of Lake Michigan. *Applied Environmetal Microbiology*, 69, 4714–4719.

Chapter 20

Surface water, groundwater and ecological interactions along the River Murray. A pilot study of management initiatives at the Bookpurnong Floodplain, South Australia

Volmer Berens[1], *Melissa G. White*[1], *Nicholas J. Souter*[1], *Kate L. Holland*[2], *Ian D. Jolly*[2], *Michael A. Hatch*[3], *Andrew D. Fitzpatrick*[4], *Tim J. Munday*[4] & *Kerryn L. McEwan*[2]

[1] *South Australian Department for Water*
[2] *CSIRO Land and Water, South Australia*
[3] *University of Adelaide, South Australia*
[4] *CSIRO Exploration and Mining, Western Australia*

ABSTRACT

Salinisation is a major economic and environmental concern in Australia. It impacts the most significant river in Australia, the River Murray, and the sectors of agriculture, industry and recreation that it supports. It is estimated that 40 000 ha (40%) of floodplain vegetation on the lower River Murray in South Australia is now dead, dying or stressed. This severe dieback in native floodplain vegetation is largely due to floodplain salinisation, a result of saline regional groundwater discharge, decreased natural flood frequencies, permanently held weir (river) levels and a severe current drought. In an effort to manage current and future decline in tree health, the Bookpurnong Living Murray project was developed to investigate management initiatives such as engineered salt interception schemes and other technologies that artificially manipulate flow regimes. The study integrates the scientific disciplines of hydrogeology, hydrology, ecology and geophysics to examine the interplay between surface water, groundwater and vegetation. The outcomes will aid in the development of governmental policy to guide management decisions for the floodplains and their ecological status.

20.1 INTRODUCTION

The River Murray and its floodplains are an integral part of south-eastern Australia's landscape, supporting agriculture, recreation and industry. However, over the past 150 years floodplains in Australia have been progressively alienated from their parent rivers. Floodplains that were once flooded three or four years out of five are often now only wetted once in 10 or 12 years whilst some have not received any flooding for over two decades (Mussared, 1997). The health of these systems is in decline with the dieback of native woodlands attributed to rapid advancement of floodplain (soil) salinisation (Slavich *et al.*, 1999). This escalated salinisation is attributed to a number of factors including (1) the lack of natural high flow events, which would typically leach and flush salts from the system, (2) irrigation districts adjacent the river, which induce groundwater mounding and the flow of regional saline groundwater towards the river alluvium and (3) river regulation with locks, which has altered the natural temporal interaction between surface water and groundwater.

The Murray Darling Basin Commission (MDBC) in 2002 established The Living Murray Initiative (TLMI, Living Murray, 2007), which aimed to protect and improve the health of six River Murray Icon Sites. The severely stressed 17 700 ha Chowilla floodplain Icon Site, is one of the last remaining parts of the lower River Murray floodplain not bounded by irrigation and thus retains much of the natural character and attributes (White *et al.* 2006). In order to achieve the ecological floodplain objectives, TLMI has invested in a range of operational 'structural works and measures' to manipulate floodplain and wetland processes. Clark's floodplain (Fig. 20.1) at Bookpurnong (~80 km downstream of Chowilla), with exiting power and groundwater extraction and diversion infrastructure, presented as an ideal pilot investigations site from which outcomes would feed into the management of the more remote Chowilla Floodplain.

The work at Bookpurnong investigated three different floodplain management initiatives aimed at finding a strategy that is transferable to the Chowilla Floodplain (AWE, 2005). These are (1) fresh river water injection into a saline aquifer, (2) saline groundwater extraction to induce enhanced bank storage of fresh water and (3) artificial surface flooding to reduce root zone salinity. These strategies are measured for their success in improving tree community health by using a variety of vegetation health assessments and are supported by surface and groundwater investigations that included numerical modelling, soil analysis, groundwater and surface water monitoring and a range of land, in-river and airborne geophysical surveys.

20.2 INVESTIGATION SETTING

The River Murray is the natural sink for the regional groundwater within the Loxton – Bookpurnong area. The Monoman Sand Formation (~10 m thick) and the overlying Coonambidgal Clay Formation (up to 5 m thick) are the floodplain sedimentary sequences into which the channel of the River Murray is incised. The lithology and hydraulic properties of the Monoman are highly variable consisting of fine sands to coarse gravels with varying amounts of silts and clays. The floodplain vegetation is dominated by three tree species; River Red Gum (*Eucalyptus camaldulensis*), Black Box (*Eucalyptus largiflorens*), River Cooba (Acacia stenoplylla). The Bookpurnong trials target areas where there is a notable decline in the condition of these tree species, aiming to provide condition benefits for the stressed tree communities (White *et al.* 2006).

The regional hydro-stratigraphy is schematically represented in Figure 20.2 and displays the flow mechanisms that have been enhanced by the presence of highland irrigation. The 500 ha Clark's floodplain is adjacent the 1200 ha semi-arid Bookpurnong irrigation district (Fig. 20.1). Irrigation, typically using River Murray water, has increased the hydraulic and salinity pressure on the river and floodplains. Drainage from inefficient irrigation recharges the underlying aquifer resulting in localised groundwater mounding. The Bookpurnong groundwater mound has a head approximately 10 m above river pool level and is centred within 1.3 km of the river and 0.5 km of Clark's floodplain.

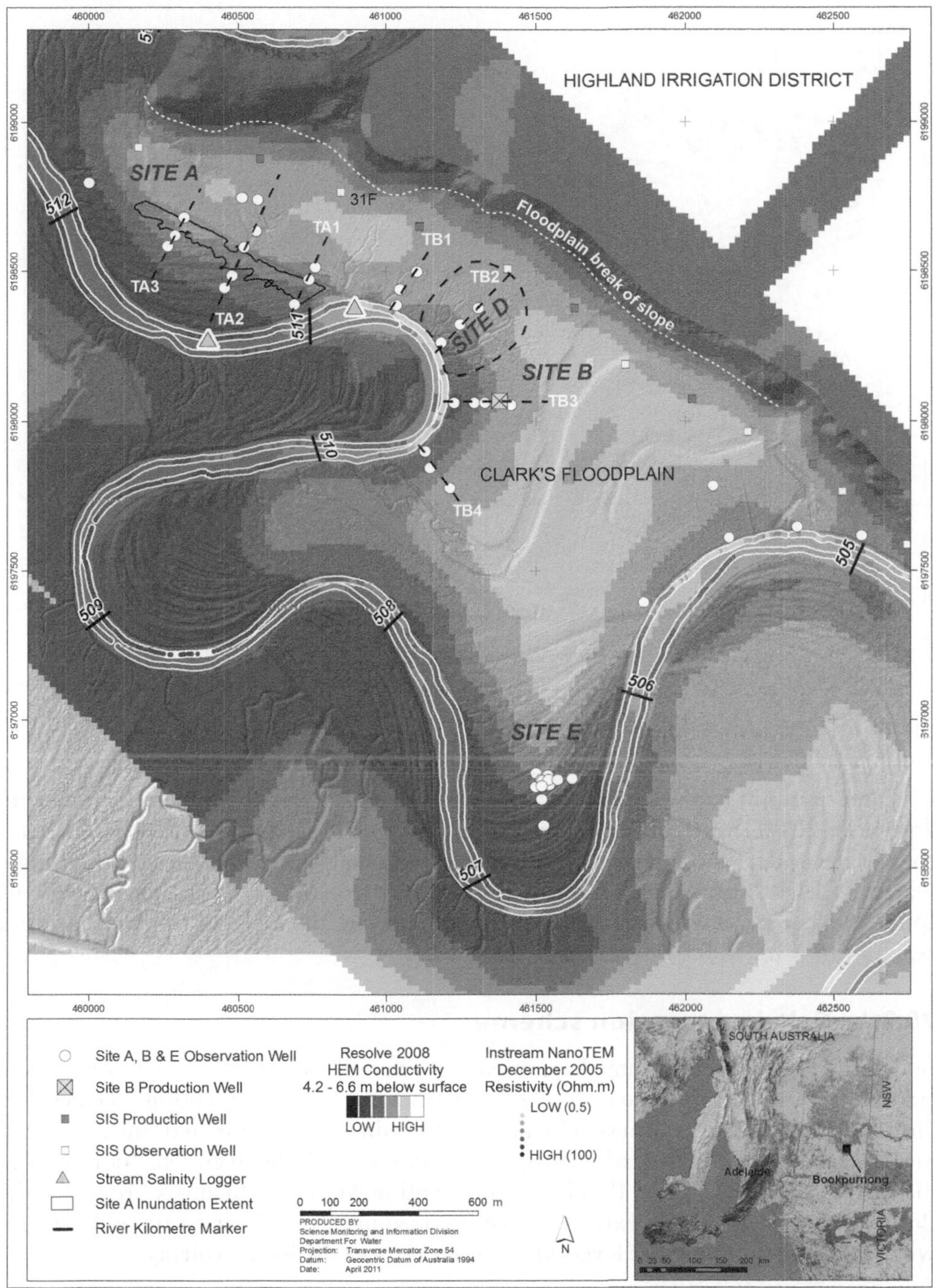

Figure 20.1 Clark's Floodplain, Bookpurnong, South Australia. The locations of trial sites A, B, D and E are identified. High resolution Light Detecting And Ranging (LiDAR) data highlight riverine morphological features. Airborne and in-river geophysical conductivities provide information of salinity distribution.

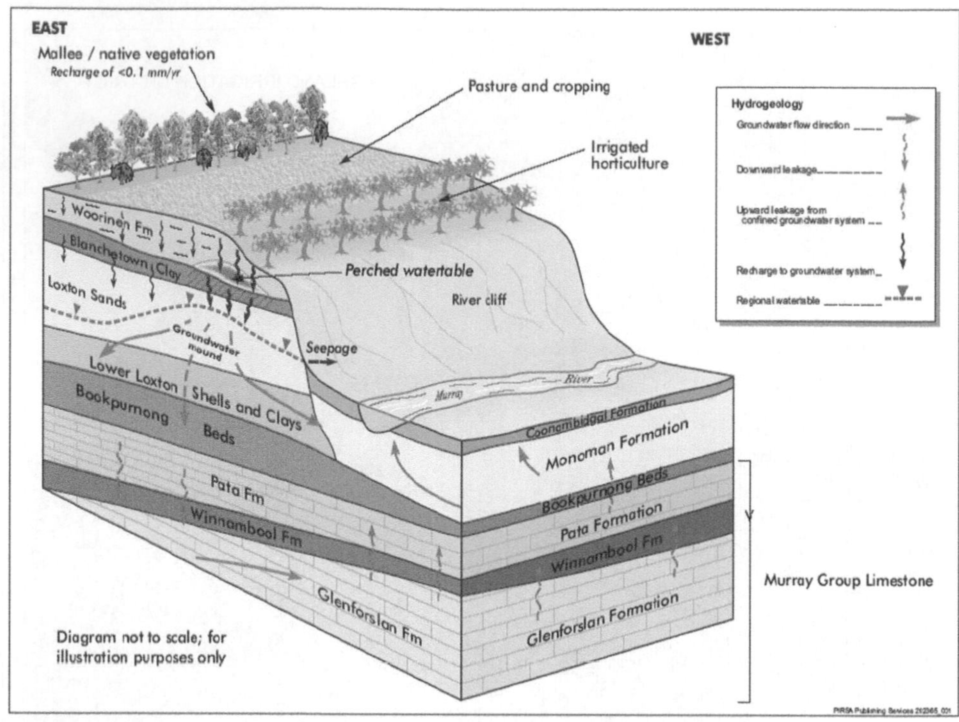

Figure 20.2 Schematic of characteristic hydrogeology for the study area.

The resultant hydraulic gradient displaces groundwater from the regional saline aquifer (>20 000 mg/l) and enhances the natural rate of saline groundwater discharge to the river and its floodplains. The results of a nationally accredited numerical groundwater model estimated that, prior to groundwater extraction intervention, around 100 tonnes of salt per day discharged into the river along the 18 km Bookpurnong river reach (Yan *et al.*, 2005).

20.2.1 Salt interception scheme

In an effort to reduce the immediate impact of river salt accession, there has been considerable investment into the design and construction of salt interception schemes (SIS) along the River Murray. These schemes aim to reduce/reverse the hydraulic gradients and intercept the movement of saline groundwater towards the river. The Bookpurnong SIS has six highland and 16 floodplain interception wells, of which, six are on Clark's floodplain. Floodplain SIS extraction commenced in July 2005 at a rate of 2–3 l/s per well, reducing groundwater levels and salt impacts to the River Murray.

20.2.2 Bookpurnong trials

The Bookpurnong trials on Clark's floodplain actioned several individual research concepts/sites (Fig. 20.1), which were all monitored for spatial and temporal variation in soil moisture and salinity, groundwater level and salinity and correlated with floodplain vegetation response.

20.2.2.1 Site A

Chosen as a location where artificial flooding could be trial. Artificial flooding of a 3.7 ha topographic depression and its River Red Gum forest (White *et al.* 2009). The aim to leach salt from the soil profile and improve the salinity condition of the root zone. Of the numerous investigations completed on Clark's floodplain, this paper focuses on the results of the Site A trial and aims to discuss key ecological outcomes.

20.2.2.2 Site B

Reported in Berens *et al.* (2009a), at Site B, a 'Living Murray' groundwater extraction well was installed to purposely induce the lateral movement of fresh river water through the adjacent floodplain aquifer to create a fresh water lens. 17 Piezometers were installed in 4 transects (TB1, TB2, TB3, TB4 on Fig. 20.1) to observe the surface and groundwater interactions across four transect extending from the river.

Transect 1 was designed to observe SIS only response, Transect 2 to incorporate Site D and the distant response to groundwater extraction, Transect 3 to measure the immediate response from the extraction well and Transect 4 as a control. Within 3 months (from August 2006) and pumping at a modest 4 l/s, a freshwater lens/wedge was induced in the aquifer extending to observation well B9. Required maintenance to the larger SIS scheme resulted in lengthy interruptions to Site B groundwater extraction, however no significant reduction in the fresh water lens was observed. Pumping recommenced in May 2007 and by the end of the trial in December 2008, lower salinity groundwater was observed at the production well.

20.2.2.3 Site D

Site D utilised Site B, Transect 2 infrastructure to monitor the artificial flooding of a dried creek system and compare vegetation response to surface flooding as opposed to Transect 3 groundwater freshening (Doody *et al.*, 2009; Holland *et al.*, 2009). Minor earthworks were required to enclose the creeks and prevent drainage back to the river. 10 Ml of river water was used to fill the creek, which had an observed effect on the nearby groundwater levels and salinity with a tree condition improvement monitored.

20.2.2.4 Site E

At Site E, fresh river water was injected into a moderately saline aquifer via a 5-point injection array to monitor the vegetation health response of a marginal tree community (Berens *et al.*, 2009b). This trial had the most uncertainty, as success was reliant on the ability to inject a sufficient volume of water. Although technically well implemented and serviced, the trial was not successful in injecting adequate volumes and only created very limited localised freshening. After 48 days of injection a number of the wells had become clogged resulting in increased aquifer pressure and breeches of the confining clays. The outcomes highlighted the physical difficulties and design shortcomings for such a shallow and hydraulically poor aquifer.

20.3 SITE A RESULTS

The morphology of Site A is that of a flood runner wetland (depression) that would normally fill during natural high river flow events and had previously occurred in 2000 (pers. comm. Clark 2005). To hold the diverted river water for the flooding trial, an 80 m earthen levy bank was constructed along the south-eastern boundary of the depression, creating a 3.7 ha and 10.7 Ml capacity inundation zone (Fig. 20.1).

Two watering periods occurred approximately one year apart; the first occurring over winter to early spring of 2005 applying 43 Ml over a three-month period. The second watering applied 52 Ml over a similar time period commencing in September of 2006. Thirteen observation wells, in three transects perpendicular to the length of the inundation zone (TA1, TA2, TA3) were constructed on Site A between 18 and 27 June 2005. The first watering commenced on 30 June allowing for only limited baseline data collection.

20.3.1 Groundwater levels

Groundwater levels from manual and logged time series data indicate that floodplain groundwater levels are responsive to river level fluctuations at all sites, with the response magnitude decreasing with increased distance from the river (White *et al.* 2009). This indicates good connectivity between river and floodplain groundwater levels. Vertical groundwater recharge via surface water inundation and groundwater level fluctuation as a result of SIS operation were observed in the data. Coincident river level rise, notably during the first watering period, complicated the distinction of groundwater response due to water level application or river level rise. Due to the numerous data collected, only some representative data is presented herein.

A control observation well (A1) was installed 200 m northwest of the inundation trial and 70 m from the river. The location is distant from the notable influence of both the Site A watering and SIS pumping. The A1 groundwater levels were consistently 0.2–0.3 m below river level, which support the understanding of a losing stream and low groundwater salinity at that location.

As a result of the initial water application prior to the commencement of SIS operation and any significant river level change, an increase in groundwater level was observed at all wells excluding the control piezometer A1. This indicates aquifer recharge due to soil water infiltration translating in a rise of groundwater levels (White *et al.* 2009). Groundwater response to the surface water application was greatest at wells nearest the inundation extent, specifically A2 where a rise of more than 0.6 m was measured in June 2005 (Fig. 20.3).

The study also aimed to examine the effect of floodplain SIS operation on groundwater levels. Independently operated, the SIS commenced operation mid-way during the first inundation period and was stopped for maintenance mid-way during the second inundation period. Data from the SIS mid-point observation well 31FO recorded groundwater levels to be above river level prior to SIS commencement, subsequently decreasing during SIS groundwater extraction and increasing post SIS shutdown (Fig. 20.3). Similar groundwater trends to 31FO were also observed at wells located between the inundation zone and the SIS extraction wells (A2, A5, A6, A9, A10, A12 & A13); these wells have a water level response dominated by SIS groundwater extraction, whereas wells riverside of the inundation zone have a greater influence due to river level variations (A3, A4, A7, A8 & A11) (White *et al.* 2009).

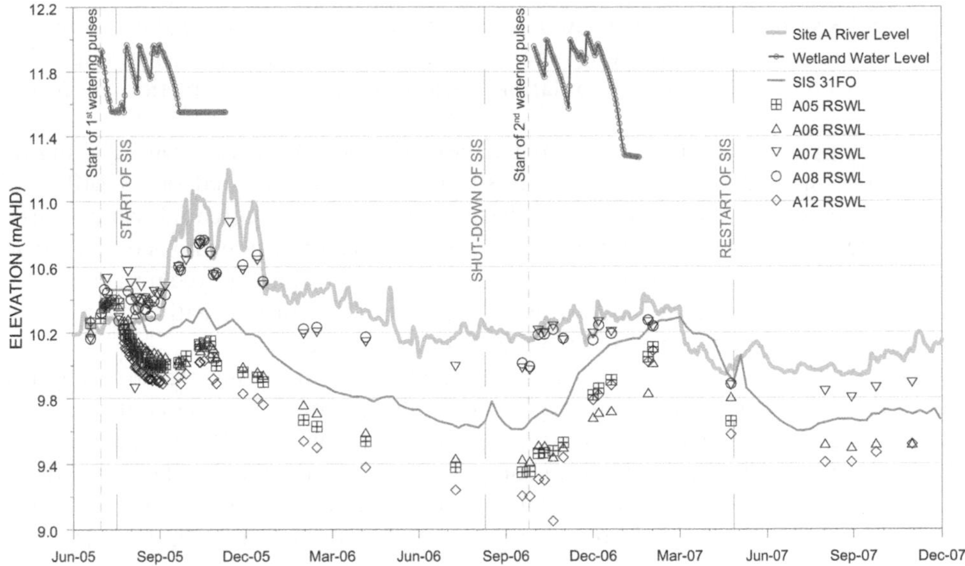

Figure 20.3 Site A Groundwater levels, river level and wetland level.

20.3.2 Groundwater salinity

Site A groundwater salinity was not adequately monitored during the trial due to resource limitations. The screened intervals were not ideal in all observation wells and shortfalls in data collection and equipment calibration made analysis troublesome. Adequate monitoring was not in place prior to the commencement of the first inundation period, with salinity data not collected until the second pulse of water.

From the available data, an increasing salinity gradient away from the river is apparent in all transects and increases in salinity were observed at A2, A9, A10 & A11 as a result of the artificial inundations (White *et al.* 2009). This is consistent with salt mobilisation processes; A2 and A10 are on the very edge of the inundation zone and hence drainage through the saline soil profile and vertical recharge increased the groundwater salinity (and groundwater level) at these locations.

Observation wells A9 and A11, although beyond the inundation extent, displayed an increase in salinity. These wells are down gradient hydraulically from the rest of the observation wells and represent the approximate region where groundwater would discharge to the river.

The river salinity impact of the trial was measured with two instream salinity loggers, the first installed immediately downstream of the watering trial and the second installed 500 m upstream (Fig. 20.1). Due to the high river flows that occurred during the first inundation period, no clear observation could be made. However, the salinity time series was successful in observing a minor localised river salinity increase (20 μS/cm, 8%) over the course of the second inundation period, which reinforces the interpretation of salt mobilisation from the floodplain to the river (White *et al.* 2009).

20.3.3 Soil salinity

Soil samples were collected at 0.5 m intervals of the unsaturated zone at each observation well location. Site A sampling was conducted on a biannual frequency, coinciding with pre and post wetland inundation periods until trial completion in March 2008.

Soil samples were analysed for matric potential (ψ, MPa, soil dryness) using the filter paper method (Greacen *et al.*, 1989) and osmotic potential (ψ_{TT}, MPa, soil salinity). The sum of these values provides the total soil water potential, which is used to indicate soil water availability for root uptake; a clear relationship is derived from the knowledge of the pre-dawn water potential of a tree, an integrated measure of soil water availability to vegetation (Eamus *et al.*, 2006). The more negative the total soil water potential, the harder it is for plants to source water.

Total chloride is measured by ion chromatography and converted to a chloride concentration in the soil solution ($mg\,l^{-1}$) using the gravimetric water content. Gravimetric water content ($g\,g^{-1}$) was measured by oven drying at 105°C for 24 hours. Osmotic potential is estimated from the chloride concentration of the soil solution and using the Van't Hoff equation, which assumes that all salts are present as sodium chloride and that the concentration used to calculate the relationship is appropriate for the range of soil salinities encountered by floodplain trees.

Notable observations of a salinity reduction in the unsaturated zone (greater osmotic potential) were limited to sites A2 & A10 (Fig. 20.4), where data shows downward drainage from the inundation zone leached salts from the unsaturated zone. The observation is supported by an increase in groundwater salinity seen at these two sites, confirming the interpretation of salt mobilisation out of the soil profile, into the groundwater and towards the river (White *et al.* 2009).

20.3.4 Geophysics

Land, airborne and in-river geophysical data was collected at all Bookpurnong trial sites. Airborne electromagnetic (AEM) surveys were conducted in July 2005, September 2006 and July 2008 (Fig. 20.1). Both the Resolve frequency domain system (2005 & 2008) and SkyTEM duel moment system (2006) where utilised (Fitzpatrick *et al.*, 2007; Sorensen & Auken, 2004; Munday *et al.*, 2007).

Airborne systems allow floodplain scale observation to be made and provide versatile datasets allowing interpretation from numerous depth intervals. For the River Murray floodplain environments it has been shown that regions of low conductivity are interpreted as areas lower floodplain salinity and conversely, high conductivities represent saline floodplain materials. Low conductivities adjacent to the main river channel infer losing river conditions and represent the extent of the flushed zone (low salinity groundwater). At Bookpurnong, temporal AEM system deployments have successfully monitored floodplain salinity changes in response to SIS and Living Murray groundwater extraction e.g. At Site B, Transect 3, where low salinity river water has been induced into the aquifer.

In-river transient electromagnetic (NanoTEM) data sets were collected in 2003, 2004 and 2005 (Fig. 20.1). Similarly these data are able to indicate whether a river reach is gaining (high conductivity) or losing (low conductivity) by the measure of

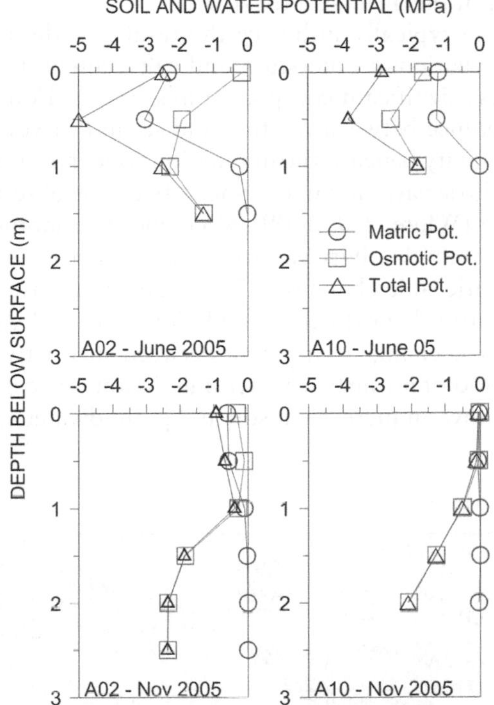

Figure 20.4 Soil profiles for observation sites closest to the artificial watering inundation zone, highlighting the change in soil condition pre and post initial intervention. Data indicates an increase in soil moisture (squares), an improvement in soil salinity (circles) and a total improvement in soil water availability for tree uptake.

bulk conductivity of the river sediments and their pore water. An in-river drilling and sediment coring program ground truthed the 2003 dataset (Berens & Hatch, 2006), concluding a strong correlation between bulk conductivity and pore water salinity. Research completed by Tan *et al.* (2007) and Tan *et al.* (2009) reported that with the Murray alluvium consisting mainly of sands and localised clay of similar porosity, the water content in the saturated environment is most likely consistent, leaving salinity as the main driver of conductivity variability.

Land surveys include conventional static NanoTEM, continuous towed NanoTEM (Sites A, B & E; Hatch *et al.*, 2007) and Geonics EM31 conductivity meter surveying (Sites A & B). Only Site A EM31 data are presented herein.

EM31 data over Site A was collected on 6 occasions; June 2005, February 2006, September 2006, February 2007, September 2007 and March 2008. The surveys were carried out before and after each inundation periods, with two post inundation (dry period) datasets also collected (Fig. 20.5). As described in McNeill (1980), the EM31 operates on the principles of electromagnetic induction to sense the grounds ability to conduct (or resist) electrical current. The EM31 has an effective depth of penetration of around 2–6 m and the resultant apparent conductivity (or resistivity) is a bulk representation of this near surface zone.

Variables that may typically influence the results of the EM31 surveys include groundwater level, variations in moisture and salt content and the amount of clay content (which is a less significant component in saline environments).

The distinct difference between the first and second surveys is a reduction of the bulk apparent conductivity beneath the inundation zone in response to the fresh water application, whereas little variation in conductivity can be observed beyond the extent of the inundation area (White *et al.* 2009). SIS induced variation in groundwater level have been attributed to conductivity variations seen northeast of the inundation zone. However, the most noticeable changes in EM conductivity are due to a reduction in soil salinity concentration beneath the inundation zone and variations in the saline groundwater level. Little change is apparent between survey two and survey five, but after an extended period of no surface water inundation, the conductivity from survey six is interpreted to show an increase in soil and groundwater salinity.

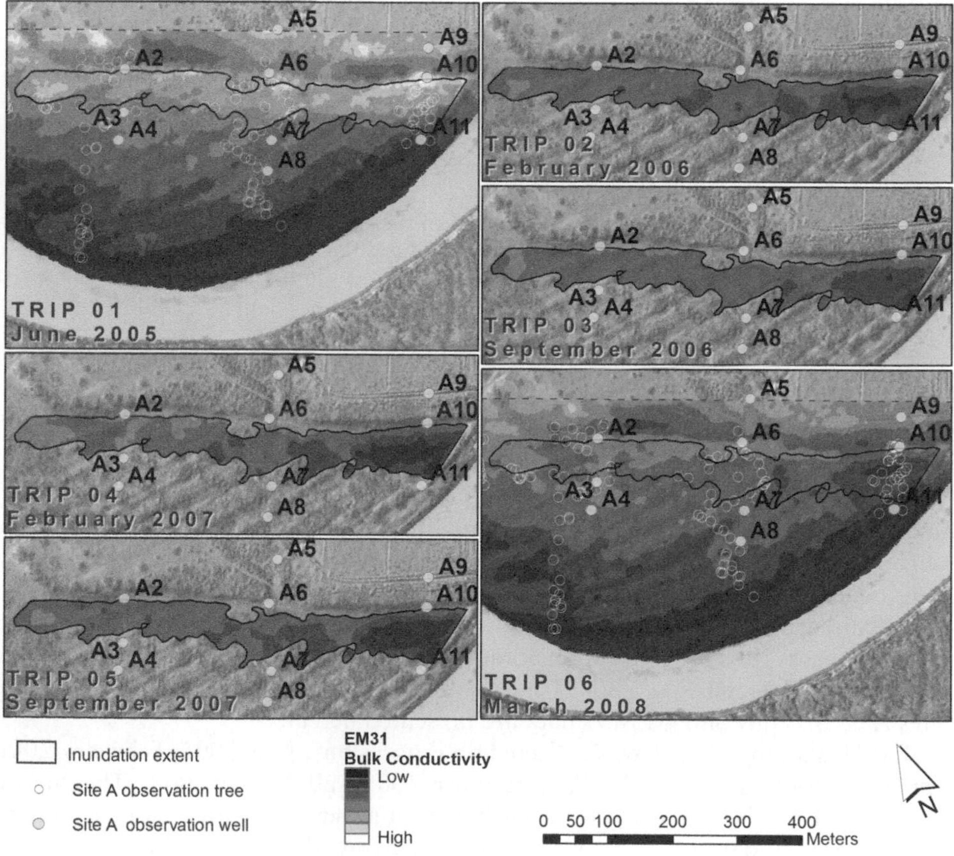

Figure 20.5 Temporal EM31 conductivity grids. Data collected prior and post both inundation events.

20.4 BOOKPURNONG VEGETATION RESPONSE

The DFW visual tree health assessment method is based on a conceptual model of declining tree health due to prevailing environmental conditions and behaviour in response to management intervention. Condition is assessed as crown extent and density on a six-category scale (White *et al.* 2009). Response is measured as behavioral reaction to environmental changes. Assessment of epicormic growth follows and bark condition was also recorded.

20.4.1 Site A

35 inundated and 85 control river red gums were compared at Site A (Figs. 20.5 and 20.6). The artificial watering of the Site A red gum forest resulted in watered trees producing significantly more epicormic growth than trees that remained dry (White *et al.* 2009). This peak in epicormic growth occurred after the first inundation period, whilst it took until after the second inundation period for an increase in crown condition to be observed in inundated trees. Whilst some control trees did respond over the course of the study, watered trees had a higher likelihood of increasing in crown condition than control trees (White *et al.* 2009). However after a final dry summer in March 2008 the frequency of watered trees in the highest crown condition category (76–100% cover) was not sustained suggesting that initial watering of trees in generally poor condition may require watering more than two consecutive occasions if the initial response is to be sustained (White *et al.* 2009).

20.4.2 Site B

Along transect 1, there was no discernible effect of SIS groundwater lowering on tree crown condition for river red gum, black box and cooba, when compared to control trees along transect 4 (Berens *et al.*, 2009a). Groundwater freshening along Transect 3 at Site B resulted in a measurable improvement in Black Box health (Doody *et al.*, 2009). River Red Gum derived less benefit as they primarily occurred close to the edge of the river and had access to less saline water. River Cooba water stress did not improve significantly regardless of its location in the floodplain. Prior to freshening, trees at some distance from the river tended to have lower visual health scores and much lower pre-dawn leaf water potentials (Holland *et al.*, 2009) than those closer to the river. Following freshening, the visual health of trees at an intermediate distance from the river increased. Black Box at an intermediate distance from the river increased in both pre-dawn leaf water potential and visual health, while River Cooba improved in visual health class only (Doody *et al.*, 2009; Berens *et al.*, 2009). Trees at a greater distance from the river remained in poor health and continued to be water stressed as the effects of groundwater freshening had not reached them at the time of measurement.

Tree water use measurements showed that net groundwater discharge was restricted to a narrow band (~50 m wide) of trees nearest the river, riparian vegetation further from the river relied on rainfall and periodic flooding (Doody *et al.*, 2009). Tree water source data showed the riparian vegetation relied on water sourced from the capillary fringe during dry periods, indicating that maintaining a freshwater lens is crucial for the long term survival of these vegetation communities (Holland *et al.*, 2009).

Figure 20.6 Example of *Eucalyptus camaldulensis* in a lower crown extent category (left) in transition to a higher crown extent category (right) over the duration of the study period.

20.4.3 Site D

At Site D the response of trees (river red gum, black box and cooba) within 15 m of the flood runner and likely to be affected by the inundation were compared to trees located >15 m away. Whilst the majority of trees across the site showed little change in tree crown condition over the survey period, a small proportion of inundated trees did improve when compared to non-inundated trees (Berens *et al.* 2009a). The response of trees at Site D was much more muted than was observed at the Site A surface flooding trial. This was likely due to the Site D trees being in a worse condition than Site A trees and having experienced an extra year of water stress (drought) prior to watering intervention.

20.4.4 Site E

The assessment of river red gum response at Site E was hampered by the limited extent of freshening around the injection bores. Of the sixty trees initially assessed only eight were likely to have been able to access water from the injection bores. There was no difference in the condition of the treatment and control trees over the course of the study (Berens *et al.*, 2009b). When analysed independently the condition of the treatment trees declined over the course of the study. Above average rainfall in the summer of 2007 was however found to be correlated with an improvement in tree condition across the entire site (Berens *et al.*, 2009b).

20.5 BOOKPURNONG TRIAL CONCLUSIONS

The Living Murray trials have provided an opportunity to demonstrate the ecological benefits of manipulated environmental flows in the short to medium term. The monitoring of groundwater has indicated a good connection between the River Murray and the aquifer across this floodplain. The artificial flooding has had an impact on the zone immediately below the inundation zone reducing root zone salinity. The extraction trial at Site B has been successful in altering the aquifer condition by creating an extensive freshwater lens, whereas, the trial of river water injection was not able to alter groundwater conditions to any significant degree and was greatly hampered by poor injection yields.

The trials provide important information/validation of floodplain rehabilitation concepts, which will help to underpin integrated policy and planning for floodplains on the lower river Murray. Overall, preliminary tree condition analyses from all sites suggests that surface water flooding such as Site A and Site D provides the best environment for a large-scale response and improvement in tree crown condition. These conditions in turn increases the capabilities of flowering, seeding and recruitment of trees on these stressed floodplains, a requirement for long term survival. The project identified preferred management and rehabilitation regimes to be implemented across salt affected floodplains.

ACKNOWLEDGEMENTS

The Bookpurnong trials were funded by the Murray Darling Basin Commission Living Murray Initiative. The Authors would also like to acknowledge the efforts of the DWLBC River Murray Hydrometric Services, DWLBC Groundwater Technical Services, Berri to Barmera Local Action Planning group, S.A. Water, the Bookpurnong/Lock4 Environmental Association, Ecophyte Technologies and the DWLBC Infrastructure and Business Group. The authors would especially like to thank Steve Clark for supporting and allowing access to the investigation site.

REFERENCES

AWE (2005) *Bookpurnong Floodplain Living Murray Pilot Project.* Unpublished report for the Department of Water, Land and Biodiversity Conservation prepared by Australian Water Environments.

Berens, V., White, M.G. & Souter, N.J. (2009a) *Bookpurnong Living Murray Pilot Project: A Trial of Three Floodplain Water Management Techniques to Improve Vegetation Condition.* DWLBC Report 2009/21, Government of South Australia, through the Department of Water, Land and Biodiversity Conservation, Adelaide.

Berens, V., White, M.G. & Souter, N.J. (2009b) Injection of fresh river water into a saline floodplain aquifer in an attempt to improve River Red Gum (*Eucalyptus camaldulensis Dehnh.)* condition. *Hydrological Processes*, 23 (24), 3464–3473. doi:10/1002/hyp.7459.

Berens, V. & Hatch, M.A. (2006) Instream geophysics (NanoTEM): A tool to help identify salt accession risk. *Australian Society of Exploration Geophysics Preview*, 120, 20–25.

Doody, T.M., Holland, K.L. & Benyon, R.G. (2009) Effect of groundwater freshening on riparian vegetation water balance. *Hydrological Processes*, 23 (24), 3485–3499. doi:10/1002/hyp7460.

Eamus, D., Hattin, T., Cook, P. & Colvin, C. (2006) *Ecohydology: Vegetation Function, Water and Resource Management*. VIC, Australia, CSIRO publishing.

Fitzpatrick, A.D., Munday, T.J., Berens, V. & Cahill, K. (2007) An examination of frequency domain and time domain HEM systems for defining spatial processes of salinisation across ecologically important floodplain areas: Lower Murray River, South Australia. *Proceedings of the 20th Annual SAGEEP Symposium, SAGEEP, Denver.*

Greacen, E.L., Walker, G.R. & Cook, P.G. (1989) Evaluation of the filter paper method for measuring soil water suction. *International Meeting on Measurement of Soil and Plant Water Status, University of Utah*. pp. 137–145.

Hatch, M.A., Berens, V., Fitzpatrick, A., Heinson, G., Munday, T. & Telfer, A. (2007) Fast sampling EM applied to the River Murray and surrounding floodplains in Australia. *Proceedings of Near Surface, the 13th meeting of Environmental and Engineering Geophysics, Istanbul.*

Holland, K.L., Doody, T.M., McEwan K.L., Jolly, I.D., White, M.G., Berens, V. & Souter, N.J. (2009) Response of the River Murray floodplain to flooding and groundwater management: Field investigations. CSIRO Water for a Healthy Country Flagship, Adelaide. 65 p.

McNeill J.D. (1980) Electromagnetic terrain conductivity measurement at low induction numbers. Geonics Limited Technical Note 6.

Munday, T.J., Fitzpatrick, A., Reid, J., Berens, V. & Sattel, D. (2007) Frequency or time domain HEM systems for defining floodplain processes linked to the salinisation along the Murray River. *Proceedings of the 19thInternational Geophysical Conference and Exhibition of the Australian Society of Exploration Geophysicists, Perth, Australia.*

Mussared, D. (1997) *Living on Floodplains*. Canberra, The Cooperative Research Centre for Freshwater Ecology and The Murray Darling Basin Commission.

Slavich, P.G., Walker, G.R., Jolly, I.D., Hatton, T.J. & Dawes, W.R. (1999) Dynamics of *Eucalyptus largiflorens* growth and water use in response to modified watertable and flooding regimes on a saline floodplain. *Agricultural Water*, 39, 245–264.

Sorenson, K.I. & Auken, E. (2004) SkyTEM – A new high resolution helicopter transient electromagnetic system. *Exploration Geophysics*, 35, 191–199.

Tan, K.P., Berens, V., Hatch, M. & Lawrie, K. (2007) *Determining the Suitability of In-Stream NanoTEM for Delineating Zones of Salt Accession to the River Murray: A Review of Survey Results from Loxton, South Australia*. Cooperative Research Centre for Landscape Environments and Mineral Exploration, Open file report 192.

Tan, K.P., Munday, T., Halas, L. & Cahill, K. (2009) Utilising airborne electromagnetic data to map groundwater salinity and salt store at Chowilla, SA. *Proceedings of the 20th Geophysical Conference and Exhibition, Australian Society of Exploration Geophysicists 2009.*

The Living Murray (2007) Murray Darling Basin Commission. http://www.thelivingmurray.mdbc.gov.au

White, M.G., Berens, V., Souter, N.J., Holland, K.L., McEwan, K.L., Jolly, I.D. (2006) Vegetation and groundwater interactions on the Bookpurnong Floodplain, South Australia. *Proceedings of the 10th Murray Darling Basin Commission Groundwater Workshop*, Canberra, Australia, 8–10 December 2006.

White, M.G., Berens, V., Souter, N.J. (2009) Bookpurnong Living Murray Pilot Project: Artificial inundation of *Eucalyptus camaldulensis* on a floodplain to improve vegetation condition, DWLBC Report 2009/19, Government of South Australia, through the Department of Water, Land and Biodiversity Conservation, Adelaide.

Yan, W., Howles, S., Howe, B. & Hill, T. (2005) *Loxton – Bookpurnong Numerical Groundwater Model 2005*. Department of Water Land and Biodiversity Conservation Report 2005/17. Report prepared for the Murray Darling Basin Commission.

Chapter 21

Hydrodynamic interaction between gravity-driven and over-pressured groundwater flow and its consequences on soil and wetland salinisation

Judit Mádl-Szőnyi[1], *Szilvia Simon*[1] & *József Tóth*[2]

[1] *Department of Physical and Applied Geology, Eötvös Loránd University, Hungary*

[2] *Department of Earth and Atmospheric Sciences, University of Alberta, Alberta, Canada*

ABSTRACT

The study's objective was to identify the source of salts and the controls and mechanism of their distribution in a sedimentary basin with a coupled gravity-driven and over-pressured flow regime. The generalised salinisation and flow model was built up based on multidisciplinary data interpretation in the Duna-Tisza Interfluve, Hungary. In the basin a gravity-driven flow regime characterised by (Ca, Mg)-$(HCO_3)_2$-type water is perched hydraulically upon an over-pressured saline flow regime. The salt for the surface salinisation originates partly from the NaCl-type groundwater of the Pre-Neogene basement and the deep-basin sediments, and in addition from the $NaHCO_3$-type water of the Neogene sequence. These saline waters mix and ascend due to tectonically controlled overpressure close to the surface. The Cl^- as a conservative element traces the appearance of this deep saline water in the near surface. The salinisation pattern on the surface is controlled by the salt-distributing effect of the fresh, shallow gravity flow-systems, which contacts the ascending saline groundwater by diffusion.

21.1 INTRODUCTION

Saline groundwater and soil salinisation cause huge problems worldwide, not only in drinking water supply but also in agriculture. The continental surface salinisation can originate from water-sediment interaction, dissolution of evaporites; concentration of shallow groundwater or rainwater by evaporation; or from deep groundwater source and anthropogenic pollution (Yechieli & Wood, 2002). One of the most important consequences of the existence of basinal scale groundwater flow systems (Tóth, 1963) is that groundwater acts as a geological agent (e.g. Tóth, 1984, 1999; Ingebritsen *et al.*, 2006). The moving groundwater dissolves, transfers and accumulates matter and energy and systematically alters its environment (Fig. 21.1). Among many other phenomena groundwater and soil salinisation are the surface manifestations of moving groundwater (Hayashi *et al.*, 1998; Paine, 2003; Avisar *et al.*, 2003; Tesmer *et al.*, 2007). It has been recognised that saline wetlands and salt accumulations are often connected to groundwater discharge. The saline wetlands appear in case of positive water balance, however, salt accumulation is caused by negative local water balance (Tóth, 1971). The intensity of surface salinisation can extend from only analitically detectable weak salinisation to salt efflorescence, playas, raw material scale salt accumulation (Yaalon, 1963; Wallick, 1981). The salinisation pattern controlled by groundwater

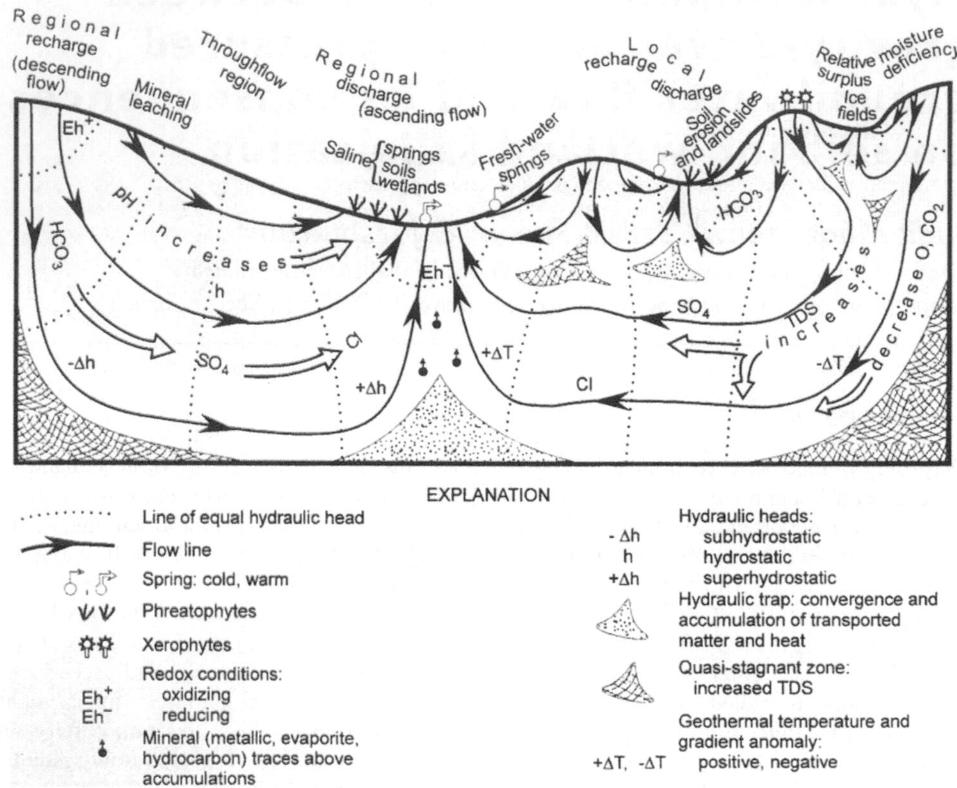

Figure 21.1 Surface and subsurface manifestations of gravitational flow systems (Tóth, 1999, modified after Tóth, 1980).

flow influences the habitats and affects the distribution of phreatophyta, xerophyta and halophyta vegetation (e.g. Engelen & Kloosterman, 1996; Klijn & Witte, 1999; Batelaan *et al.*, 2003).

The Duna-Tisza Interfluve in Hungary has severe problems of soil and wetland salinisation. The genesis and amelioration of salt-affected soils of the arable farmlands have been studied for more than two centuries (Arany, 1956; Balogh, 1840; Bíró, 2003; Erdélyi, 1976; Kiss, 1979, 1990; Kovács, 1960; Kovács-Láng *et al.*, 1999; Kuti, 1977; Molnár & Kuti, 1978; Molnár *et al.*, 1979; Sigmond, 1923; Tessedik, 1804; Tóth, 1999; Treitz, 1924; Várallyay, 1967). These studies provided important results in the delineation of salt affected surface regions, in the understanding of the role of groundwater level fluctuation and in the evaluation of the genesis of saline soils. Kovács (1960) and Várallyay (1967) recognised the effect of moving groundwater in the transport of salts. Erdélyi (1979) described an upper gravitational flow system in the Great Hungarian Plain, but neglected the role of over-pressures underneath. Saline groundwater discharges were also recognised by Kiss (1979, 1990).

In addition to the previously identified unconfined gravitationally driven flow (Erdélyi, 1979), an over-pressured confined flow-regime was recognised by Tóth and

Almási (2001) for the whole Great Hungarian Plain. Mádl-Szőnyi & Tóth (2009) recognised the connection between groundwater regimes and high salinity as well as the controlling factors of the salinisation. These results were validated and extended on a local scale to Lake Kelemenszék by Simon (2010) and Simon *et al.* (2011).

The purpose of this paper is to extend the results of the multidisciplinary studies of the Duna-Tisza Interfluve. A generalised salinisation and flow model for stratified sedimentary basins with coupled gravity-driven and over-pressured flow regimes allows transfer of information to other areas.

21.2 THE HYDROGEOLOGIC ENVIRONMENT OF THE AREA

The Duna-Tisza Interfluve region is located in Central Hungary. The topographic elevations vary between 80 to 95 m asl in the valleys up to 120–130 m asl on the divide (Fig. 21.2).

The Duna-Tisza Interfluve is situated in the central part of the Pannonian Basin. The basin is underlain by a large set of orogenic terrains (Brezsnyánszky *et al.*, 2000; Hámor *et al.*, 2001). The crustal terrains reached their present positions during closure of the Tethys Ocean. The tectogenesis of the sub-basins of the Pannonian system was controlled by the collision effects of the African (Adriatic) and European plates. The Duna-Tisza Interfluve part of the basin was developed above the collision sutures (Royden & Horváth, 1988, Rumpler & Horváth, 1988; Haas *et al.*, 1999, Hámor *et al.*, 2001, Nemcok *et al.*, 2006). The Pre-Neogene basement is made up of "flysch" sediments, carbonates, and metamorphics.

The Neogene tectonic evolution of the Study Area can be characterised by four stages: i) Early Miocene, right lateral wrench-type rejuvenation of the earlier 'sutures'; ii) Middle Miocene formation of a network of synrift transtensional troughs; iii) Late Miocene-Pliocene post-rift (thermal) subsidence of the broader interior sag basin of the Great Hungarian Plain; iv) latest plate convergence generated by late structural inversion and Pliocene-Quaternary left lateral strike slip-movements (Pogácsás *et al.*, 1989). The Neogene, semi- to un-consolidated marine, lake-deltaic, lacustrine and fluviatile clastic sediments thicken from ≈600 m in the Duna-Valley on the west, to over ≈4000 m at the Tisza Valley in the east (Fig. 21.3) (Juhász, 1991).

Hydrostratigraphically the Pre-Neogene basement has a hydraulic conductivity (K) of about 10^{-8} ms^{-1} (Bérczi and Kókai, 1976), increasing to 10^{-6} ms^{-1} in the Pre-Pannonian Aquifer. The Late Miocene-Pliocene sediments are divided into an upper and a lower series of hydrostratigraphic units by the regionally extensive Algyő Aquitard ($K \approx 10^{-8}$–10^{-7} ms^{-1}). Below the Algyő Aquitard, the turbiditic siltstones, sandstones and clays series of the Szolnok Aquifer and the shale series of the Endrőd Aquitard were deposited, with K varying between 10^{-9} and 10^{-6} ms^{-1}. The Algyő Aquitard is overlain by the Great Plain Aquifer with an assigned average K of $\approx 10^{-5}$ ms^{-1} (Tóth & Almási, 2001).

The climate of the Duna-Tisza Interfluve is moderately continental, the average annual temperature is 10–11°C and the average annual precipitation is about 550–600 mm. The annual precipitation maintains the water table within 6 to 8 m below the surface of the ridge region, and within 0 to 1 m in the valleys of the Duna and the Tisza rivers (Kuti and Kőrössy, 1989).

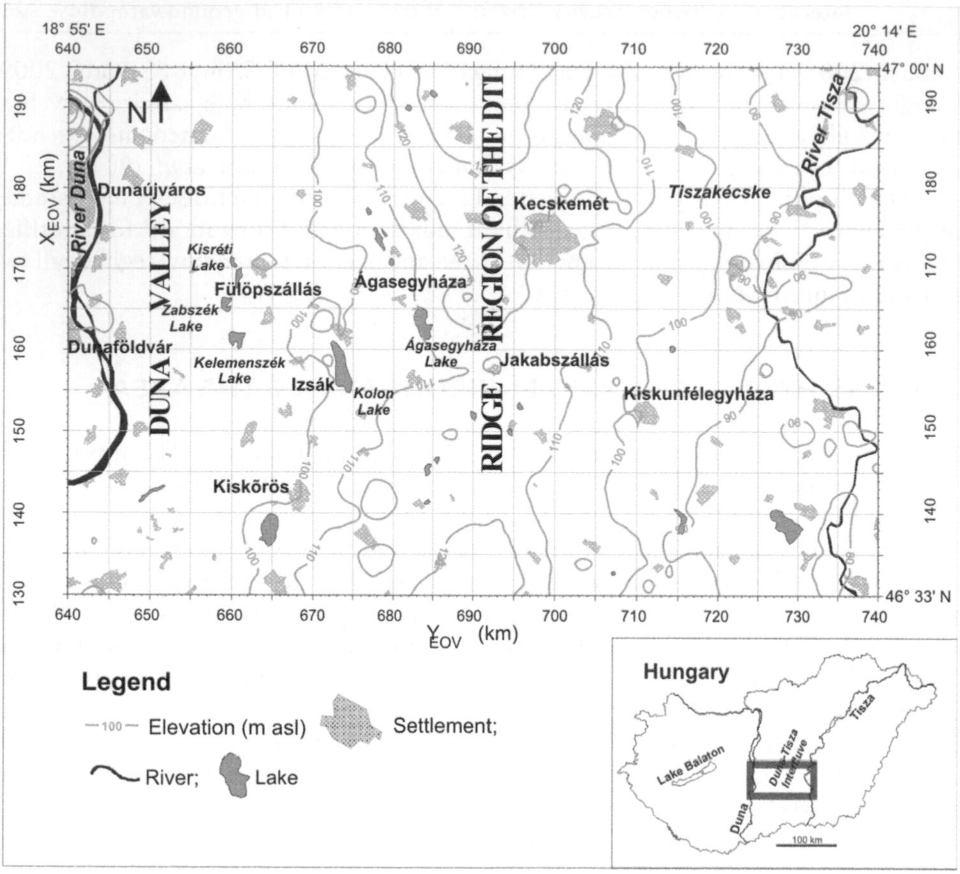

Figure 21.2 The Duna-Tisza Interfluve Region.

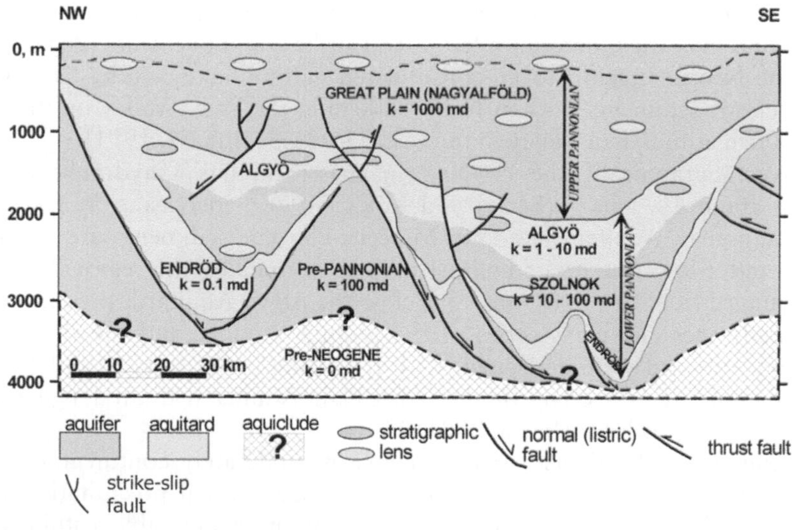

Figure 21.3 Generalised hydrostratigraphic section, Great Plain, Hungary (modified from Tóth & Almási, 2001, based on Juhász, 1991).

21.3 THE DUNA-TISZA INTERFLUVE HYDROGEOLOGICAL STUDY – REGIONAL AND LOCAL SCALE

21.3.1 The hydraulic and hydrostratigraphic context

There are two distinct groundwater flow-regimes in the area (Mádl-Szőnyi & Tóth, 2009): a locally recharged regime of gravity-driven meteoric water with its flow field adjusted to the topography-controlled water table; and the deeper regime of the ubiquitously ascending waters rising from the over-pressured deep zone (Fig. 21.4). The divide between these systems correlates fairly well with the north-south striking topographic crest of the Duna-Tisza Interfluve area. The recharge areas merge on the ridge and form a single north-south belt of about 25 to 35 km width. The gravitational discharge areas are roughly parallel with the Duna River in the west.

The area is characterised by a regionally extensive aquitard which is overlain by the Great Plain Aquifer. It is an extensive aquitard because the pore pressures are mostly near-hydrostatic above it while over-pressured underneath. However, the seismically identified strike-slip faults structures can dissect this regional aquitard and create short-cut connections between the water in the Pre-Neogen basement and the Great Plain Aquifer (Fig. 21.5).

The effect of deep overpressure for a stratified sand-shale sequence with a conduit fault was modelled by Matthäi & Roberts (1996). It was found, that in the case of a high-permeability fault the direction of fluid flow is basically upwards. In the upper hanging wall the flow is outward from the fault plane into the intersected sands. The model explains the hydraulic behaviour of strike-slip faults in the Duna-Tisza Interfluve sedimentary basin and the short cut connection with outflow from the fault plane to the sand layers.

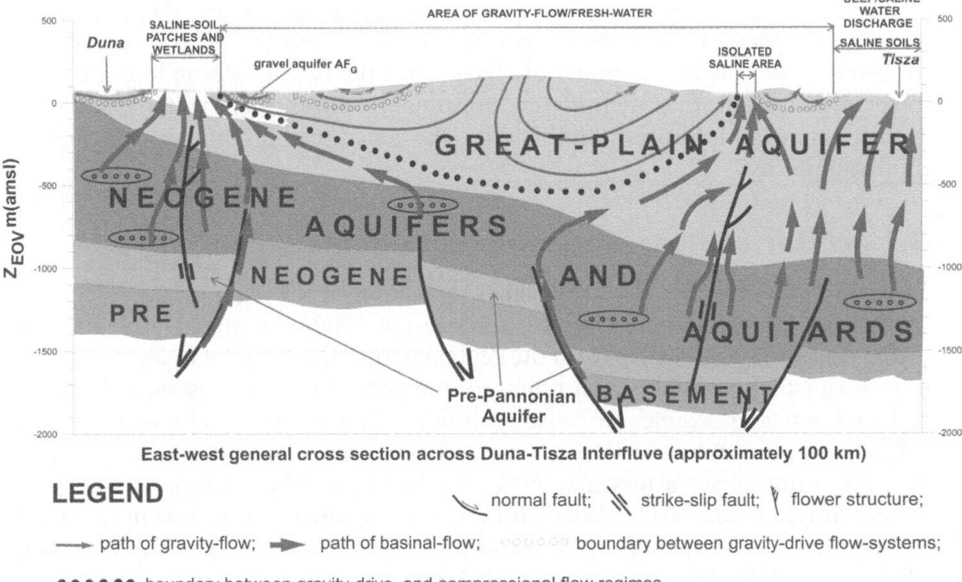

Figure 21.4 The Duna-Tisza Interfluve Hydrogeological Type Section (modified after Mádl-Szőnyi & Tóth, 2009).

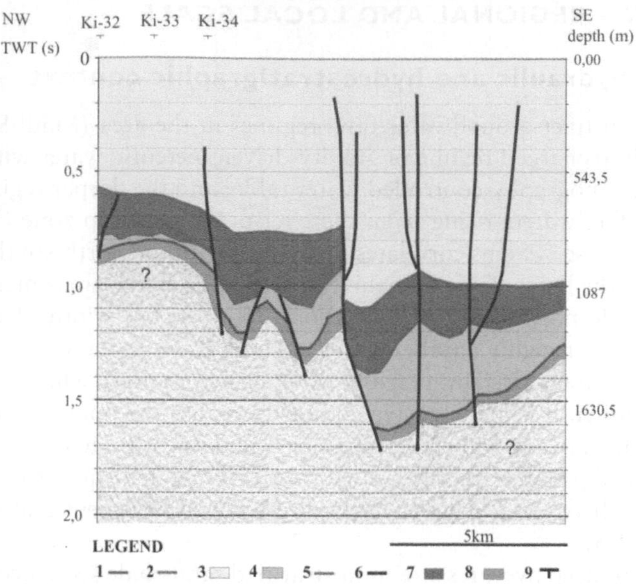

Figure 21.5 Tectonically dissected aquitard and aquifers 1: Base of Pre-Pannonian; 2: Base of Great Plain Aquifer; 3: Great Plain Aquifer; 4: Pre-Pannonian Aquifer; 5: Base of Algyő Aquitard; 6: Structural element; 7: Algyő Aquitard; 8: Pre-Neogene Basement; 9 transverse seismic profile.

The penetration depth of the gravity driven flow regime varies between 250 and 450 m. The gravitational and over-pressured flow regimes under the gravitational recharge area appear to be sharply separated, because their flow is in opposite directions. While under the gravitational discharge area the flow direction in both regimes is universally upward, thus their waters presumably mix through water-level fluctuations, dispersion, and diffusion. In this area, the meteoric and saline waters cannot be easily distinguished.

21.3.2 Water, soil and wetland salinisation

The question was whether the chemical composition of the waters in the deep basin and basement was the source of surface salinisation. Examination of the chemical and isotopic composition of waters from the central part of the Pannonian Basin sediments shows non-meteoric origin Na-Cl type groundwater above the basement highs connected to faults and sedimentary discontinuities. The isotopic composition of most Na-HCO_3 type groundwater in the basin sediments indicates a paleometeoric origin while some samples are a mixture of the Na-HCO_3 and Na-Cl type water (Simon, 2010 and Simon *et al.*, 2011). Microfiltration of the upwelling waters may leave the Cl^- below the main aquitard while water leaking through becomes depleted in Cl^- and lighter in stable isotopes (Varsányi & Ó Kovács, 2009).

The chemical composition of the waters in the Pre-Neogene basement of the Duna-Tisza Interfluve shows broad heterogenity (TDS: 3800–10000 mgl^{-1}), nevertheless the

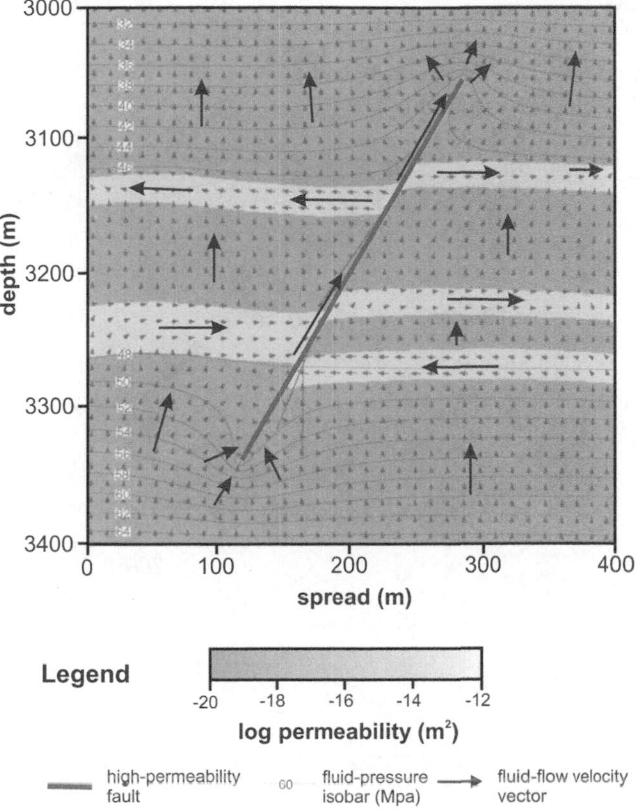

Figure 21.6 Modelled fluid-pressure isobars and fluid flow velocity vectors generated by a pressure source at the domain's base in a faulted sequence consisting high- and low-permeability strata, in the case of a conduit fault (modified from Matthäi & Roberts, 1996).

hydrogeochemical facies of these waters seems to be homogeneous. All water is Na-Cl type. The water samples from the Neogene sediments are not homogeneous and the Na-HCO_3 type water sporadically has higher Cl^- concentrations. The Cl^- concentration for approximately 15% of the samples from the Great Plain Aquifer, exceeds 100 mgl^{-1}. In all cases, these elevated Cl^- concentrations are paired with elevated TDS values. Significantly, all sample of this subgroup comes from sites of the western part of the Interfluve where the Pre-Neogene basement rises up to within 600 m below land surface (Fig. 21.4). Above the elevated basement part of the DTI the Neogene sediments contain water with relatively higher TDS and Cl^- content (Mádl-Szőnyi and Tóth, 2009). This agrees with the numerical theoretical model represented by Matthäi and Roberts (1996) and geochemical model of Varsányi and Ó Kovács (2009).

The presence and distribution of the upwelling water with elevated salinity in the near surface at the western margin of the area (Kelemenszék Lake) was proved by geophysical VES (Vertical Electric Sounding) and RMT (Radiomagnetotellurics) measurements on local scale (Fig. 21.7) (Simon, 2010). Below the surface saline tract,

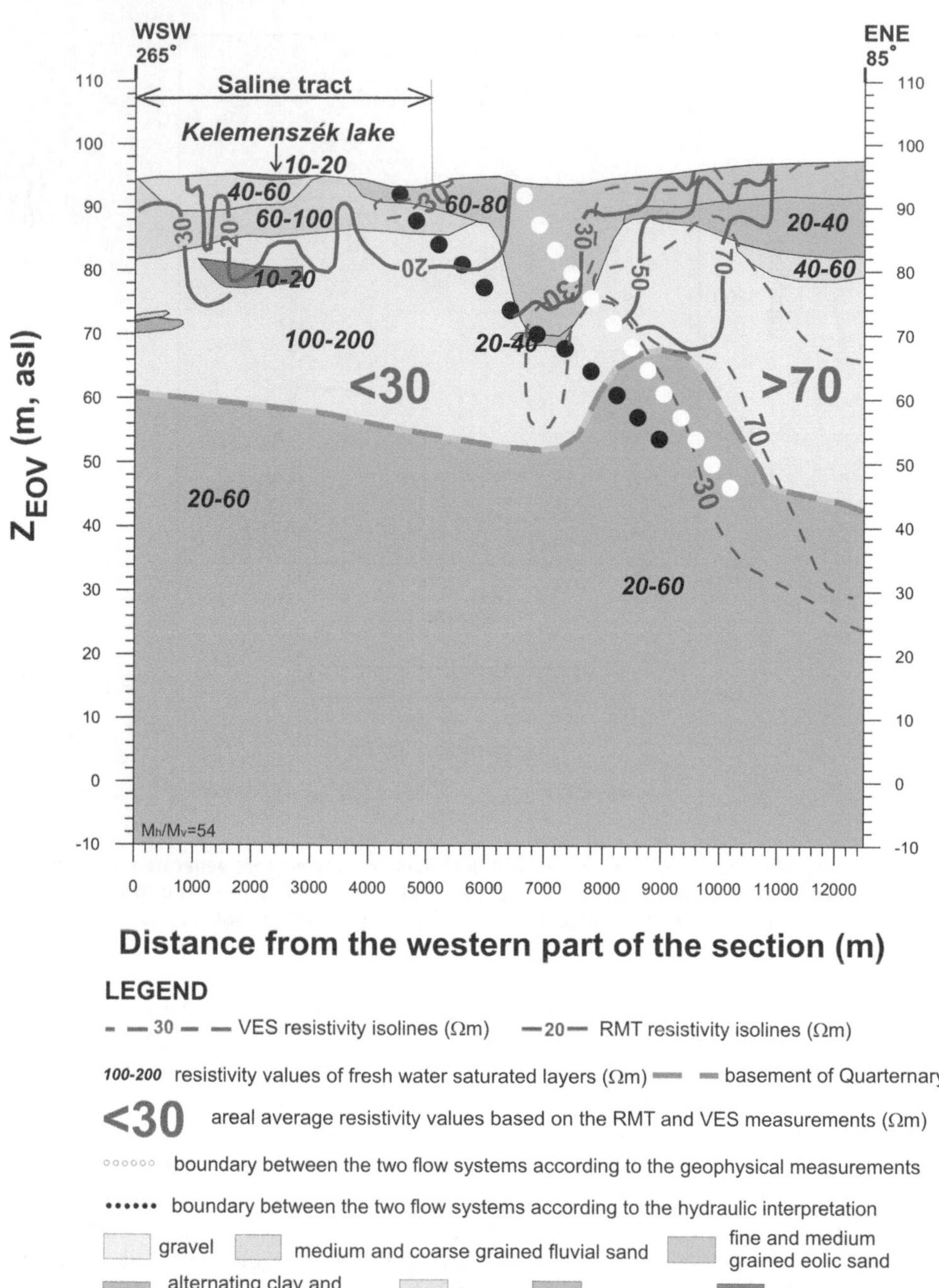

Figure 21.7 Groundwater salinisation based on VES and RMT data and hydraulic interpretation (modified from Simon *et al.*, 2011; Fig. 7c).

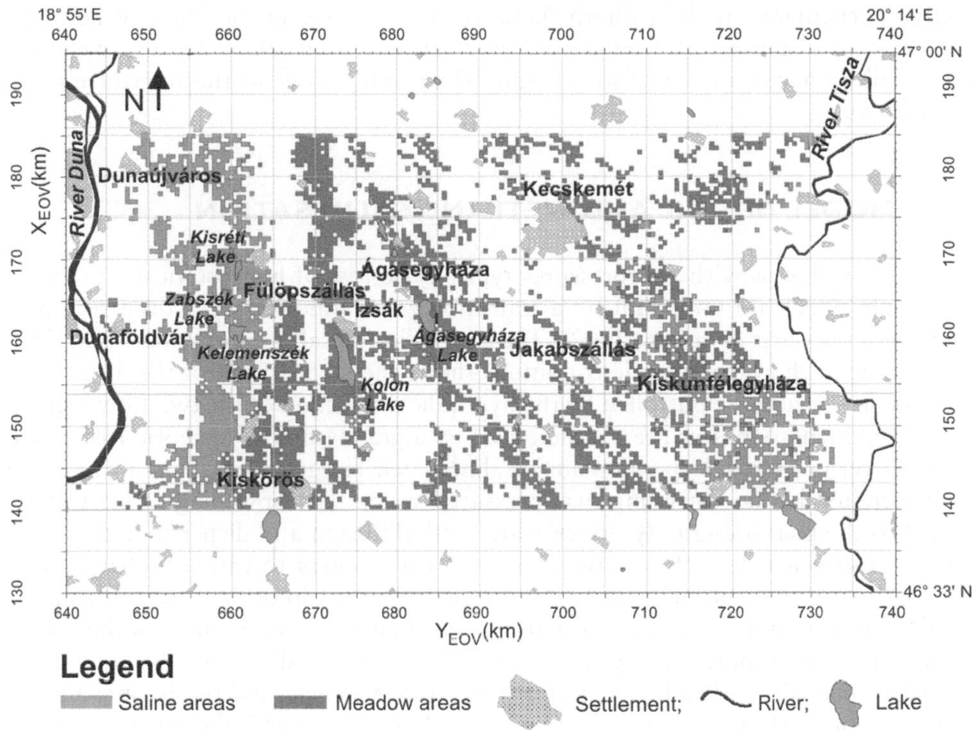

Figure 21.8 Soil and Wetland salinisation map of the Duna-Tisza Interfluve (modified after Mádl-Szőnyi & Tóth, 2009 based on Biró, 2003).

down to 70 m depth, saline water saturates the rock matrix, independently from the lithology. Eastward in the region of the gravity driven flow, fresh water of the Ca-Mg to Ca-HCO_3 type is prevalent. The boundary between the two geochemical regimes is represented by the 30 Ωm values, which depict a 1–3 km wide transition zone (Simon, 2010; Simon *et al.*, 2011).

The soil and wetland salinisation map displays the soil- and wetland salinity for the Duna-Tisza Interfluve (Mádl-Szőnyi & Tóth, 2009) (Fig. 21.8). It is characterised by regional patterns of contrast in soil types and plant-ecology. The ridge region has fresh-water meadows on its crest and both flanks, while saline soils, saline lakes and marshes, and phreatophyte and salt-tolerant vegetation dominate in the river valleys. In the eastern part of the interfluve, saline and fresh-water marshes alternate at the eastern flank of the ridge to the Tisza River. Saline soils occur on both sides of the Interfluve, generally below elevations of approximately 95 to 100 m asl on the east flank, and 90 to 95 m asl on the west flank (Fig. 21.8). The upslope boundaries of these areas run roughly parallel to the north-south oriented topographic elevation contours. Areas of salt-affected soils of irregular size and shape give way to patches of non-saline soils without any apparent regularity over the entire 25 to 30 km width of the Tisza Valley.

The transition between the saline- and non-saline soils and wetlands is relatively sharp, particularly on the western flank as, for instance, in the Duna Valley. Sporadically, and apparently irregularly, small areas of salt-affected soils, meadows, and shallow marshes occur also at higher elevations, closer to the Interfluve's crest (Fig. 21.8).

21.4 MODEL OF SOIL AND WETLAND SALINISATION

Sedimentary basins with overpressures typically consist of hydrostatically pressured sediments which are underlain by a transition zone, followed by a deep section of high fluid pressures (Deming, 2002). The hydraulic and salinisation model (Fig. 21.9) starting-up such hydrodynamic conditions shows the mechanisms of the distribution of salts on the surface controlled by the hydraulic interaction of the two flow regimes. These sedimentary basins are characterised by a gravity-driven meteoric fresh water and an over-pressured deeper domain of saline water. The gravity systems are hydraulically perched by the ascending over-pressured water. The upwelling saline water is added to the meteoric gravity driven water with diffusion and dispersion. The superjacent fresh-water 'lens' forces the ascending deep waters towards discharge areas. The pathways of ascending waters are influenced by the dip of the basement, and the variable thickness and tectonic structure of the aquitards and aquifers in the layered system. Lateral tectonic shifting caused by strike-slip faults dissecting the aquitards in the basin provides hydraulic connections between basement and the basinal aquifers based on the mechanism represented on Figure 21.6. These vertical elements can result in a 'short cut' and/or 'water exchange' connection between the basement and the basin sediments. In addition to the tectonically controlled ascent of the deep waters, cross-formational flow through the matrix of the aquitards may occur. Cl^- is a conservative element and is preserved through the upwelling and behaves as a natural

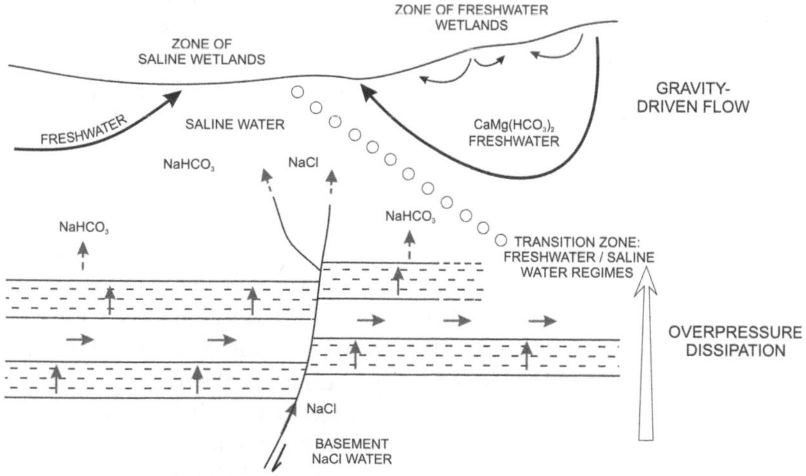

Figure 21.9 Generalised model for soil and wetland salinisation in case of gravity-driven fresh and over-pressured saline flow regimes in a stratified sedimentary basin.

tracer for the deep basement water. The elevated chloride concentration ($>100\ \mathrm{mgl^{-1}}$) of the near surface groundwater can be interpreted as an indicator of basement fluid upwelling. It proves that the basement-origin Na-Cl type water can pass through the conduit structures with direct upwelling.

The hydraulic interaction between the two systems controls the transport routes to, and distribution of the salts at the land surface. Salinity distribution at the surface concerning saline soils and wetlands can be explained by the tectonically driven cross-formational rise of deep saline waters and their mixing with fresh waters of gravity flow-systems in the near-surface sediments (Fig. 21.9).

21.5 SUMMARY AND CONCLUSION

The hydrodynamic interaction between gravity-driven and over-pressured flow is responsible for salt distribution and soil and wetland salinisation in the Duna-Tisza Interfluve. The gravitational flow-systems of meteoric fresh water are perched hydraulically upon the rising over-pressured salt waters. The hydraulic interaction between the two regimes controls the transport routes and distribution of the salts on the land surface. The salinity distribution at the surface can be explained by the tectonically driven cross-formational upwelling of deep saline waters mixed with fresh waters in near-surface sediments and by gravity flow-systems.

A flow and salinisation conceptual model displays the origin of salts and control factors of their distribution mechanism in the near-surface. The results could be generically transferred to any stratified sedimentary basin with a similar hydrogeological environment and driving forces.

ACKNOWLEDGEMENTS

The authors are pleased to acknowledge the contributions received from Gy. Pogácsás in seismic interpretation, I. Müller in VES and RMT measurements and F. Zsemle in technical assistance. The research was supported by the Hungarian Science Foundation (OTKA) grant No. T 047159 to J. Szőnyi-Mádl, and the Canadian Natural Sciences and Engineering Research Council's "Discovery grant" No. A-8504 to J. Tóth.

REFERENCES

Arany, S. (1956) *A szikes talaj és javítása [The Salt-Affected Soil and its Amelioration]. Mezõgazdasági Kiadó*, Budapest 407 [in Hungarian].

Avisar, D., Rosenthal, E., Flexer, A., Shulman, H., Ben-Avraham, Z. & Guttman, J. (2003) Salinity sources of Kefar Uriya wells in the Judea Group aquifer of Israel. Part 1 – conceptual hydrogeological model. *Journal of Hydrology*, 270, 27–38.

Balogh, J. (1840) *A magyarországi szikes vidékek [The salt-affected soil regions of Hungary].* Természettudományi Pályamunkák, A Magyar Tudós Társaság, Buda, 123 [in Hungarian].

Batelaan, O., De Smedt, F. & Triest, L. (2003) Regional groundwater discharge: phreatophyte mapping, groundwater modelling and impact analysis of land-use change. *Journal of Hydrology*, 275 (1–2), 86–108.

Bérczi, I. & Kókai, J. (1976) Hydrogeological features of some deep-basins in SE-Hungary as revealed by hydrocarbon exploration. In: Rónai, A. (ed.) *Hydrogeology of Great Sedimentary Basins. Hungarian Geological Institute*, Mémoires XI. Budapest. pp. 69–93 [in Hungarian].

Bíró, M. (2003) A Duna-Tisza köze aktuális élőhelytérképe [The Map of the Present Bio-habitat of the Duna-Tisza Interfluve]. In: Molnár Zs (ed.) *A Kiskunság száraz homoki növényzete [Sanddunes in Hungary, Kiskunság]*. Természet BÚVÁR Alapítvány Kiadó, Budapest [in Hungarian].

Brezsnyánszky, K., Haas, J., Kovács, S. & Szederkényi, T. (2000) *Geology of Hungary 2000 – Basement Geology*. Hungarian Geological Institute, Budapest. Special Publication.

Engelen, G.B. & Kloosterman, F.H. (1996) *Hydrological Systems Analysis*. Methods and Applications. Water Science and Technology Library 20, Kluwer Academic Publishers.

Erdélyi, M. (1976) Outlines of the hydrodynamics and hydrochemistry of the Pannonian Basin. *Acta Geologica Academiae Scientiarium Hungarica*, 20, 287–309.

Erdélyi, M. (1979) A Magyar medence hidrodinamikája [Hydrodynamics of the Hungarian basin] *VITUKI Közlemények*, 18, 1–82.

Haas, J., Hámor, G. & Korpás, L. (1999) Geological setting and tectonic evolution of Hungary. *Geologica Hungarica Series Geologica*, 24, 179–196.

Hayashi, M., van der Kamph, G. & Rudolph, D.L. (1998) Water and solute transfer between a prairie wetland and adjacent uplands, 2. Chloride cycle. *Journal of Hydrology*, 207, 56–67.

Hámor, G., Pogácsás, Gy & Jámbor, Á. (2001) Paleogeographic/structural evolutionary stages and related volcanism of the Carpathian–Pannonian Region. *Acta Geologica Hungarica*, 44 (2–3), 193–222.

Horváth, F. & Cloething, S.A.P.L. (1996) Stress-induced late-stage subsidence anomalies in the Pannonian Basin. *Tectonophysics*, 266, 287–300.

Ingebritsen, S., Sanford, W. & Neuzil, C. (2006) *Groundwater in Geologic Processes*. 2nd edition. Cambridge, Cambridge University Press. 536 p.

Juhász, Gy (1991) Lithostratigraphic and sedimentological framework of the Pannonian (s.l.) sedimentary sequence in the Hungarian Plain (Alföld), Eastern Hungary. *Acta Geologica Hungarica*, 34, 53–72.

Kiss, I. (1979) Vízfeltörések szerepe a szikes talajok "tarkasága" kialakításában The role of groundwater outbreaks in the "variegation" of salt-affected soils). *Botanikai Közlemények*, 66 (3), 177–184 [in Hungarian].

Kiss, I. (1990) A vízfeltörések formái és szerepük a szikes területek kialakulásában [The forms and roles of water outbreaks in the genesis of salt-affected areas]. *Hidrológiai Közlöny*, 70 (5), 281–287 [in Hungarian].

Klijn, F. & Witte, J.P.M. (1999) Eco-hydrology: Groundwater flow and site factors in plant ecology. *Hydrogeology Journal*, 7, 65–77.

Kovács Gy (1960) A szikesedés és a talajvízháztartás kapcsolata. [Relation between soil-salinisation and the groundwater budget]. *Hidrológiai Közlöny*, 40 (2), 131–139 [in Hungarian].

Kovács-Láng, E., Molnár Gy, Kröel-Dulay, S. & Barabás (1999) *Long Term Ecological Research in the Kiskunság, Hungary*. KISKUN LTER. Institute of Ecology and Botany, Hungarian Academy of Sciences, Vácrátót 64 [in Hungarian].

Kuti, L. (1977) Az agrogeológiai problémák és a talajvíz kapcsolata az Izsáki térképlap területén [Relation between agricultural problems and groundwater in the area of the Izsák map-sheet] In: Magyar Állami Földtani Intézet Évi Jelentése az 1977. évrõl, Budapest 121–130 [in Hungarian].

Kuti, L. & Kőrössy, L. (1989) Az Alföld Földtani Atlasza: Dunaújváros-Izsák [The gelogical atlas of the Hungarian Great Plain: Dunaújváros-Izsák]. Magyar Állami Földtani Intézet Budapest [in Hungarian].

Mádl-Szőnyi, J. & Tóth, J. (2009) The Duna-Tisza Interfluve Hydrogeological Type Section, Hungary. *Hydrology Journal*, 17, 961–980.

Matthäi, S.K. & Roberts, S.G. (1996) The influence of fault permeability on single-phase fluid flow near fault-sand intersections: results from steady state high-resolution models of pressure-driven fluid flow. *AAPG Bulletin*, 80 (11), 1763–1779.

Mészáros, E. (2005) *Hidrogeológiai célú szeizmikus értelmezés a Duna-völgy északkeleti részén. [Seismic interpretation for hydrogeological purposes at the NE part of the Duna Valley]*. M.Sc. Thesis, Eötvös Loránd University, Department of Applied and Environmental Geology, Budapest. 84 p. [in Hungarian].

Molnár, B. & Kuti, L. (1978) A Kiskunsági Nemzeti Park III. sz. területén található Kisréti-, Zabszék-, és Kelemenszék-tavak környékének talajvízföldtani viszonyai [Shallow groundwateronditions around Kisrét-, Zabszék-, and Kelemenszék-lakes in area No. III of the Kiskunsági National Park]. *Hidrológiai Közlöny*, 58 (8), 347–355 [in Hungarian].

Molnár, B., Iványosi-Szabó, A. & Fényes, J. (1979) A Kolon-tó kialakulása és limnogeológiai fejlõdése [The origin and limnogeological evolution of Kolon Lake]. *Hidrológiai Közlöny*, 59 (12), 549–560 [in Hungarian].

Nagymarossy, A. (1990) Paleogeographical and paleotectonical outlines of some Intracarpathian Paleogene basins. *Geologica Carpathica*, 41 (3), 259–274.

Nemčok, M., Pogácsás Gy & Pospíšil, L. (2006) Activity timing of the main tectonic systems in the Carpathian - Pannonian Region in relation to the rollback destruction of the lithosphere. In: Golonka, J. & Picha, F.J. (eds.) The Carpathians and their foreland: Geology and hydrocarbon resources . *American Association of Petroleum Geologists Memoir*, 84, 743–766.

Paine, J.G. (2003) Determining salinisation extent, identifying salinity sources, and estimating chloride mass using surface, borehole, and airborne electromagnetic induction methods. *Water Resources Research*, 39 (3), 1–10.

Pogácsás Gy, Lakatos, L., Barvitz, A., Vakarcs, G., Farkas Cs (1989) Pliocén kvarter oldalel-tolódások a Nagyalföldön. Pliocene and Quaternary strike-slipes in the Great Hungarin Plain. *Általános Földtani Szemle*, 24, 149–169 [in Hungarian].

Royden, L. & Horváth, F. (1988) The Pannonian Basin: A Study in basin evolution. *American Association of Petroleum Geologists Memoir*, 45, 394.

Rumpler, J. & Horváth, F. (1988) Some representative seismic reflection lines and structural interpretation from the Pannonian Basin. *American Association of Petroleum Geologists Memoir*, 45, 153–170.

Sigmond, E. (1923) A hidrológiai viszonyok szerepe a szikesek képződésében [The role of hydrological conditions in the genesis of saline soils]. *Hidrológiai Közlöny*, 3 (1), 5–9 [in Hungarian].

Simon Sz (2003) *Tó és felszín alatti víz közötti kölcsönhatás vizsgálata a Duna-Tisza közi Kelemen-szék tónál [Study of the interaction between lake water and groundwater at Kelemen-szék Lake of the Duna-Tisza Interfluve]*. M.Sc. thesis, Eötvös Loránd University of Sciences, Hungary [in Hungarian].

Simon Sz (2010) *Characterization of Groundwater and Lake Interaction in Saline Environment, at Kelemenszék Lake, Danube–Tisza Interfluve, Hungary*. Ph.D. Thesis. Eötvös Loránd University, Budapest. 167 p.

Simon Sz, Mádl-Szőnyi, J., Müller, I. & Pogácsás Gy (2011) Conceptual model for surface salinisation in an overpressured and a superimposed gravity flow field, Lake Kelemenszék area, Hungary. *Hydrogeology Journal*, 19 (3), 701–717.

Sulin, B.A. (1946) Воды нефтяных месторождений в системе природных вод [Oil field waters in the system of natural waters]. Современня Нефтяная Техника, Гостоптехиздат, Москва 95 [in Russian].

Tesmer, M., Möller, P., Wieland, S., Jahnke, C., Voigt, H. & Pekedeger, A. (2007) Deep reaching fluid flow in the North East German Basin: origin and processes of groundwater salinisation. *Hydrology Journal*, 15, 1291–1306.

Tessedik, S. (1804) A tiszavidéki szikes földek műveléséről, hasznosításáról [About the culture and beneficiation of saline soils in the Tisza region]. *Patriotisches Wochenblatt für Ungarn*, 27, 6.

Tóth, J. (1963) Theoretical analysis of groundwater flow in small drainage basins. *Journal of Geophysical Research*, 68 (16), 4795–4812.

Tóth, J. (1971) Groundwater discharge: A common generator of diverse geologic and morphologic phenomena. *Bulletin of the International of Scienific Hydrology*, XVI, I/3.

Tóth, J. (1984) The role of regional gravity flow in the chemical and thermal evolution of ground water. *First Canadian/American Conference on Hydrogeology, Practical Applications of Ground Water Geochemistry*. pp. 3–39.

Tóth, J. (1999) Groudwater as a geologic agent: An overview of the causes, processes, and manifestations. *Hydrogeology Journal*, 7 (1), 1–14.

Tóth, J. & Almási, I. (2001) Interpretation of observed fluid potential patterns in a deep sedimentary basin under tectonic compression: Hungarian Great Plain, Pannonian Basin. *Geofluids*, 1 (1), 11–36.

Tóth, T. (1999) Dynamics of salt accumulation in salt-affected soils. In: Kovács-Láng, E., Molnár Gy, Kröel-Dulay, S. & Barabás (eds.) *Long Term Ecological Research in the Kiskunság, Hungary*. KISKUN LTER. Institute of Ecology and Botany, Hungarian Academy of Sciences, Vácrátót 64, 407.

Treitz, P. (1924) *A sós és szikes talajok természetrajza [Nature Study of the Saline and Salt-Affected Soils]*. Stádium, Budapest 311 [in Hungarian].

Varsányi, I. & Ó Kovács, L. (2009) Origin, chemical and isotopic evolution of formation water in geopressured zones in the Pannonian Basin, Hungary. *Chemical Geology*, 264 (1–4), 187–196.

Várallyay Gy (1967) A dunavölgyi talajok sófelhalmozódási folyamatai [Salinisation processes of the processes of the Duna Valley soils]. *Agrokémia és Talajtan*, 16 (3), 327–349 [in Hungarian].

Wallick, E.I. (1981) Chemical evolution of groundwater in a drainage basin of Holocene age, east-central Alberta, Canada. *Journal of Hydrology*, 54, 245–283.

Yaalon, D.H. (1963) The origin and accumulation of salts in groundwater and in soils in Israel. *Bulletin of the Research Council of Israel*, 11G, 105–131.

Yechieli, Y. & Wood, W. (2002) Hydrogeologic processes in saline systems: Playas, sabkhas and saline lakes. *Earth-Science Reviews*, 58, 343–365.

Chapter 22

Relationship between certain phreatophytic plants and regional groundwater circulation in hard rocks of the Spanish Central System

Miguel Martín-Loeches[1] *& Javier G. Yélamos*[2]

[1]*Geology Department, Science Faculty, Alcala University, Madrid, Spain*
[2]*Geology & Geochemistry Department, Autonomous University of Madrid, Madrid, Spain*

ABSTRACT

The Spanish Central System is a hard rock (igneous and metamorphic) range bounded by faults. It is located in the middle of the Iberian Peninsula. Igneous and metamorphic rocks are normally poor aquifers, with low salinity and calcium-bicarbonate type groundwater. There are a few springs and wells with a different chemical composition that reflect deeper groundwater flow paths. Around these points Spiny rush bushes are found, a bioindicator of salty and/or alkaline soils but so far only described in soils formed over detrital formations. Thus, the presence of Spiny rush can be an indicator for the location of deep groundwater flows into hard rocks.

22.1 INTRODUCTION

The relationship between plants and groundwater and the use of certain species as bioindicators of the chemical quality of spring discharges is well known (González Bernáldez & Rey Benayas, 1992). Spiny rush (*Juncus acutus*) also known as Sharp rush is one indicator; it belongs to the Juncaceae family, and is distributed worldwide. It tolerates coastal conditions, alkaline soil, salt, no drainage and seasonal flooding. Its habitats are coastal flats, arid regions and inland saline environments such as the typical salty soils that cover the Triassic rocks of Keuper facies.

Spiny rush is not common to hard rocks, such as granites or gneisses. Soils covering the plutonic and metamorphic rocks of mountainous regions in temperate climates are often acidic or neutral, non-alkaline, and of low salinity due to the poor solubility of the silicate minerals. Rey Benayas *et al.* (1999), considered 232 species of plants in the study area, among which were eight different Juncus, but not *Juncus acutus*, although bushes of this species are quite common in the discharge areas of the regional flow system in the detrital aquifer of the Madrid basin.

In this work, effort was concentrated on showing how the presence of Spiny rush can be used as a biological indicator, giving information on the flow, residence time and origin of the water in hard rocks, a tool that has already been established by previous authors in other lithologies (Herrera, 1987; González Bernáldez *et al.*, 1989), such as the Tertiary sedimentary basin of Madrid, in the southern side of Spanish Central System. In the study area there is a good match between the presence of *Juncus acutus* and some of the chemical compositions of groundwater in hard rocks that are coming from deep flow. Hard rocks in arid or semi-arid areas are excluded (Chambel, 1999, 2006), since the presence of salty soil can also be due to high evaporation.

22.2 DESCRIPTION OF THE STUDY AREA

22.2.1 Geological setting

The study region belongs to the Spanish Central System, a sector of the Hesperian Massif. The Spanish Central System extends east north east to west south west from central Spain into Portugal, separating the Duero and Tagus Tertiary sedimentary basins. The study area extends over the southern slope of this system, approximately between the Madrid and Talavera de la Reina meridians (Fig. 22.1) covering about 3500 km^2. It comprises the Guadarrama and Gredos Mountains, with elevations ranging from 500 to 2400 masl.

It consists largely of rocks belonging to the Hercynian basement and a locally preserved thin covering of upper Cretaceous sedimentary rocks, mainly limestone. The basement rocks that make up the Spanish Central System are composed of Carboniferous age granite bodies (adamellite, coarse grained granite, leucogranite and late monzonitic granitoid) in the western sector, and high to medium grade metamorphic rocks and some older metasediments in the central sector of the Range. The eastern sector is composed mainly of low grade Ordovician metamorphic rocks, including shale and quartzite.

All these metamorphic and igneous rocks were originally formed by the Variscan orogeny and to a lesser extent, by the later Alpine orogeny. The southern limit of the Spanish Central System is marked by a very important reverse fault, known as the Southern Central System fault, a fracture that crosses the whole upper crust (Rosales *et al.*, 1977). More information about the structural characteristics of this area can be found in Warburton & Álvarez (1989), Doblas *et al.* (1994) and De Vicente *et al.* (1996).

22.2.2 Hydrogeological characteristics

Groundwater is present in numerous small and shallow independent water-table aquifers, generally less than 10 m deep. These aquifers are created by the weathering of both metamorphic and igneous rocks, and several Quaternary deposits such as talus slopes, small alluvial terraces, alluvial fans, and moraines.

Recharge occurs by direct rain and snow infiltration, and discharge takes place through numerous small springs and streams. During summer, the flow from the springs reduces and some dry up completely. The short residence time of this water reflects on the low to very low salinity (with electrical conductivity under 200 μS/cm), on pH values of 6–7.5, and on calcium (or calcium-sodium) bicarbonate hydrogeochemical facies.

Most of the wells are old, large-diameter, dug wells, generally less than 10 m deep, and the water table is close to the ground surface during the winter, but they can dry up in summer.

Some deep boreholes (up to 705 m) have flowing artesian conditions within the hard rocks. This indicates that the Guadarrama Mountain Range has a deep flow

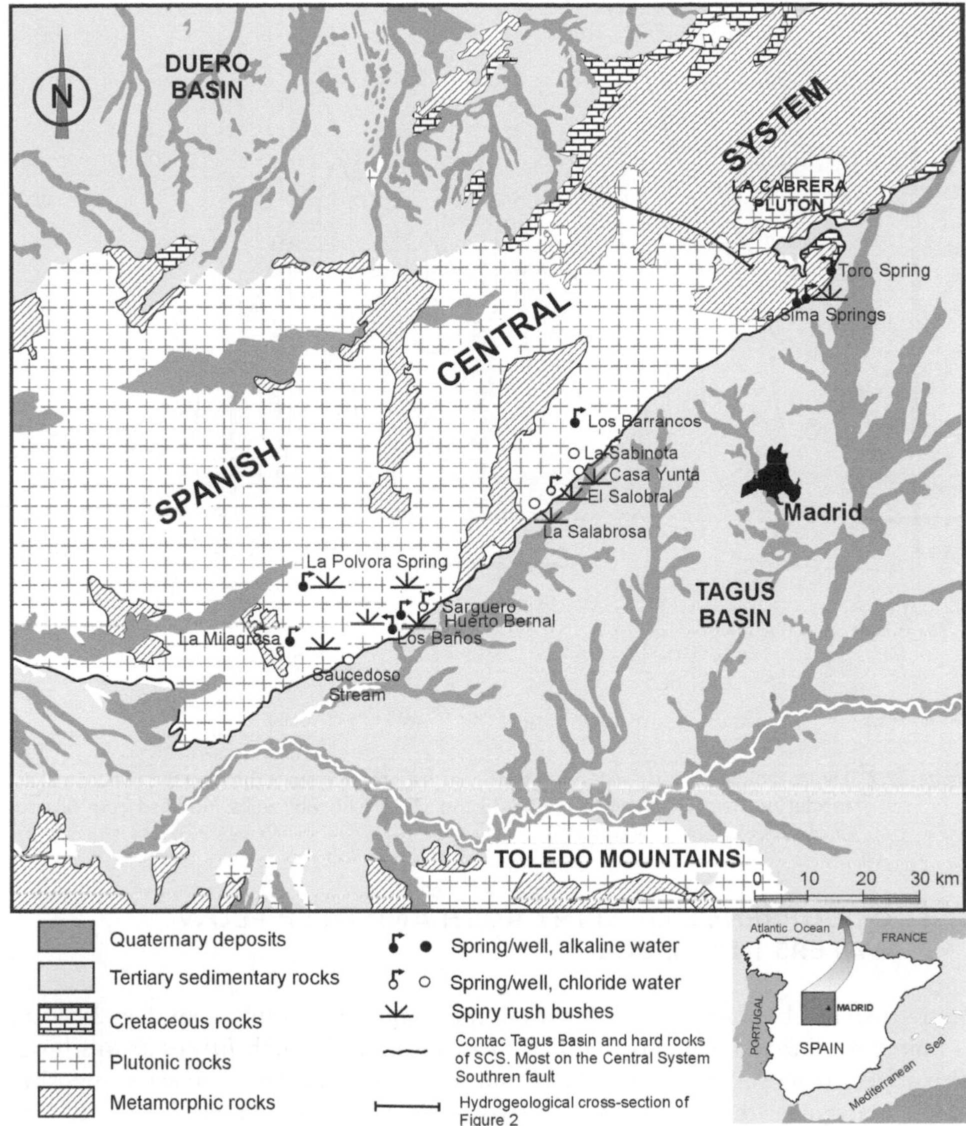

Figure 22.1 Geological setting with the location of deep water and Spiny rush on the hard rocks of the Spanish Central System.

system through fractures and joints. The hard rock massif behaves as a very deep unconfined aquifer, following the 'classic' flow model proposed by Tóth (1962, 1963). This model proposes three flow types (Fig. 22.2): a) shallow, local flows discharging into small springs b) deeper, medium flows discharging into streams and c) regional flows moving to the lowest topographic areas, perhaps discharging into the sedimentary basins of the Tagus and Duero rivers, instead of reaching the hard rock ground surface (Villarroya *et al.*, 2006).

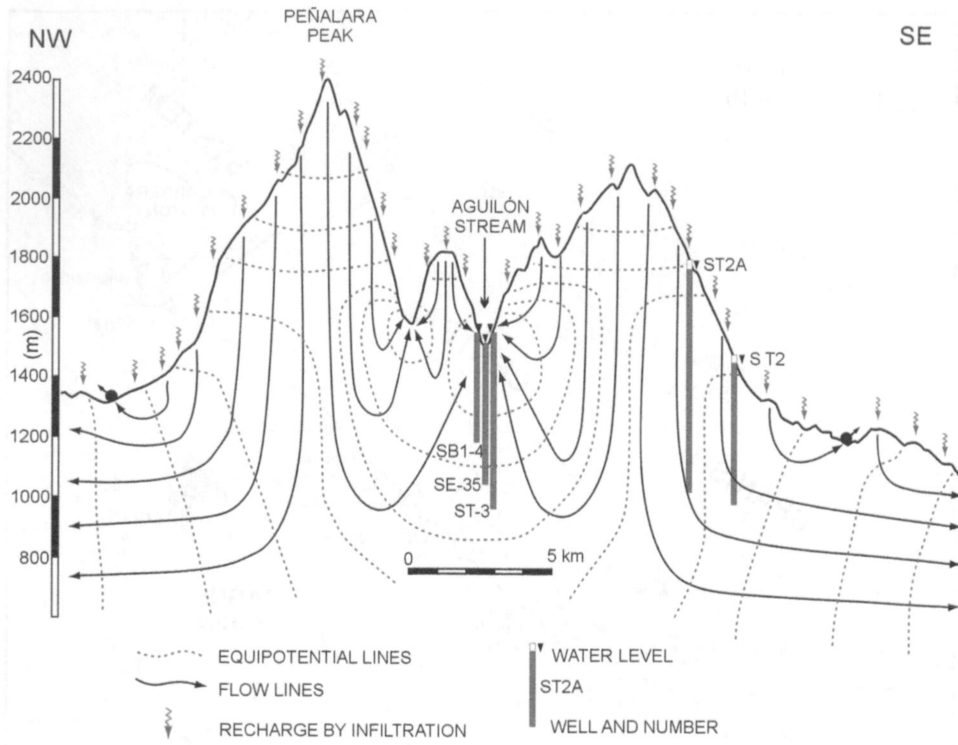

Figure 22.2 Hydrogeological cross-section of the Spanish Central System showing the conceptual flow model for the rock massif. SB1-4, SE-35 and ST-3 are flowing wells. Modified from Villarroya *et al.* (2006).

22.3 OCCURRENCE OF SPINY RUSH AND DEEP FLOW WATERS IN THE SCS

Spiny rush has been found at nine sites in the Spanish Central System (Fig. 22.1) and is linked to nine water points, most of them springs, although *Juncus acutus* bushes also appear where in damp areas there is no standing water. The chemical composition of the water near the specimens is of two types, alkaline sulphide (Table 22.1) and brackish. The chemical composition of these deep waters, and their relationship with Spiny rush, is shown in the hydrochemical map (Fig. 22.3).

22.3.1 Alkaline sulphide water

There are some groundwaters that have reduced forms of sulphur among, up to 14 ppm of H_2S, and have a very constant flow rate throughout the year (Table 22.1). The main hydrochemical features of these waters are:

- They belong to Na-HCO_3 hydrochemical facies, although in some cases they can be of Cl–Na type.

Table 22.1 Chemical composition of alkaline sulphide waters in the study area. All values in mg/l expect electrical conductivity, EC (μS/cm) and error (%).

	EC	*Cl^-*	*HCO_3^-*	*SO_4^{-2}*	*NO_3^-*	*Na^+*	*K^+*	*Mg^{2+}*	*Ca^{2+}*	*SiO_2*	*Error*
El Toro spa	1460	239	308	252	n. a.	209	3	6.8	146	n. a.	*
La Sima-1	520	100	90	62	0.0	108	5.9	0.7	11	24	1.1
La Sima-2	586	120	95	72.0	0.5	118	6.9	1.4	16.0	26	1.7
Los Barrancos	215	4	118	5.0	1.5	50	1.7	0.14	2.3	20	8
Huerto Bernal	306	25	98	4.8	<0.1	61	0.8	<0.4	5.7	36	6.5
Los Baños	546	81	127	24.2	<0.1	91	2.0	3.9	22	34.4	2.02
La Pólvora	353	55	107	3.9	2.0	80	0.6	0	8.8	19.6	5.7
Milagrosa	410	25	175	10.0	0.0	76	0.7	2	8	19.7	2.8

*Bicarbonate estimated by difference in charge balance; n. a. not analysed.

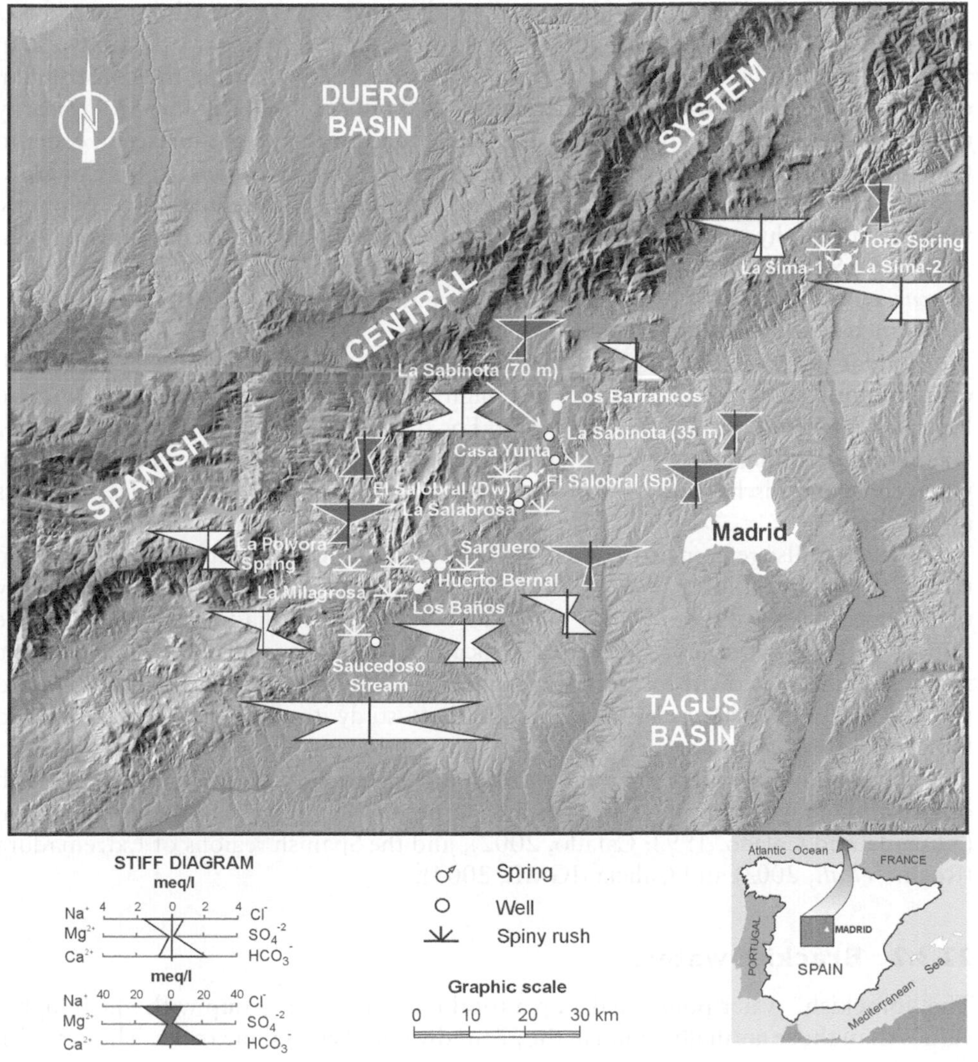

Figure 22.3 Hydrogeochemical map of deep waters on the southern side of the Spanish Central System.

- Their salinity is higher than the regional average (up to 308 ppm in Cl^-).
- They have a high concentration of F^- (up to 14 ppm), high pH, generally between 7.8 and 9, and a low concentration of Ca^{2+}.

Some of them are well known, as their waters were used in the past for therapeutic purposes. Some of them are still in use. Some of the tritium measurements of these waters have low values (0.02 ± 0.8 and 0.04 ± 0.8 TU) compared to the average 6 TU of the other springs (ENRESA, 1996), indicating that they are not recent shallow waters, but, probably, the discharge points of the deep crystalline groundwater system. Spiny rush plants are found at five points with these characteristics: La Sima-1, La Sima-2, Los Baños de Nombela, Huerto Bernal and La Pólvora but the number of specimens found around these points is not great. The size of the plants is also poor. Commonly, *Juncus acutus* appears as an isolated plant among other species or is not present. The plant is not found around El Toro, Los Barrancos or La Milagrosa springs.

Los Barrancos is a spring with naturally radioactive water, non-sulphurous, but its chemical composition and constant discharge indicate a deep flow. A possible reason for the absence of Spiny rush plants in Los Barrancos is that during the construction of a spa in the area a water tank was built and the spring water drained to a nearby creek. It is necessary for groundwater to discharge naturally on the surface in order to turn normal granite soil into the alkaline or salty soil in which *Juncus acutus* can grow.

The springs at La Sima and El Toro are a special case. Both sites have sulphuric waters but their hydrochemical facies is sodium chloride, with different salinity values: 520–586 μS/cm in the two springs at La Sima and 1450 μS/cm at El Toro. There are *Juncus acutus* around the two springs at La Sima, which are the only sulphurous waters in the metamorphic rocks of the SCS.

Similar to Los Barrancos, there was a Spa at El Toro from the middle of the19th century until 1940 (Sánchez, 1992) although its therapeutic properties were known long before then (Limón Montero, 1697). For therapeutic purposes, the spring water was drained through pipes to a swimming pool. It is possible that before the Spa, when the water discharged naturally over the ground, there were *Juncus acutus* plants at the site.

There are also references to other two historical sulphurous waters, both located in the more eastern batholith of the study area: La Cabrera. The first is in the municipality of Bustarviejo, but Prado (1864) indicates that the spring had disappeared by the middle of the 19th century. The second is located in the Mangirón area. Nowadays the spring discharges under the waters of the main reservoir that supplies the city of Madrid (El Atazar dam). In these cases it is impossible to study its chemical composition and look for *Juncus acutus*.

These alkaline waters are very similar and have many points in common with other sulphide waters in other areas of the Hesperian Massif, such as those in Portugal (Almeida and Calado, 1993; Calado, 2002), and the Spanish regions of Extremadura (Rosino *et al.*, 2003) and Galicia (IGME, 2004).

22.3.2 Brackish waters

The "brackish" water points make up a total mineralisation group with much higher values than those normally found in the groundwater of crystalline rocks. The brackish

Table 22.2 Chemical composition of "brackish" waters in the study area. All values in mg/l except electrical conductivity, EC (μS/cm) and error (%).

	EC	*Cl⁻*	*HCO_3^-*	*SO_4^{-2}*	*NO_3^-*	*Na^+*	*K^+*	*Mg^{2+}*	*Ca^{2+}*	*SiO_2*	*Error*
Sabinota (35 m)	2060	598	66	58	31.2	240	2.5	10	160	11.7	0.60
Sabinota (70 m)	2870	855	18	55	33.2	385	0.6	2.0	185	24.5	2.2
Casa Yunta	1145	79	181	35	43.4	46	0.2	7.0	80	51.9	0
Salobral, spring	2770	891	154	54	1.5	441	9.5	12.4	194	21.6	5
Salobral, dug well	1670	440	411	102	7	114	7.9	37.6	306	34	10
La Salabrosa	2690	1050	13	105	0.0	495	31	0.6	254	25	9
El Sarguero	4050	1250	170	0	16.0	610	11.3	20.0	210	16	2.2
Saucedoso	1400	194	449	77	4.6	176	24	25.0	90	21	2.21

waters have electrical conductivities between 2000 and 4000 μs/cm, whereas most groundwater points in these rocks have values which seldom reach 200 μS/cm. They appear to be related in some way with the meridional fault of the Central System Range.

There are six areas and in all of them Spiny rush appears in profusion, both in terms of numbers of plants and the surface area they cover. Table 22.2 shows the hydrochemical results obtained at the points where sampling was possible. The chemical quality is illustrated by Stiff diagrams (Fig. 22.3).

22.3.2.1 The Sabinota, Casa Yunta and the small gorge in the Perales river valley

The 'Casa Yunta' and 'Sabinota' are boreholes where the water has the characteristics:

1 High Cl^-/HCO_3^- ratio, with a value of 82.5.
2 High concentration of Cl^- (700 mg/l).
3 Important bromide content, comparable to sea water. The Br/Cl ratio from one of the analysed samples is 0.028, while the sea water ratio is 0.0034.
4 One of the lowest contents of tritium of the hard rock region, with 2.2 ± 0.4 TU.

According to periodic electrical conductivity measurements, the value in these boreholes remains constant whatever the sampling season.

Near these two wells, the Perales river valley cuts abruptly through the granite, forming a small gorge where the deepest sector of the crystalline valley is located. Here, numerous specimens of *Juncus acutus* are found. They always appear where there are fractures. The presence of Spiny rush does not seem to depend on the surface water of the Perales river, as the plants are found up to 5 m above the stream level (Fig. 22.4).

The sector where the *Juncus* appear corresponds to the part of the stream that flows from 554 to 548 masl through crystalline rocks. *Juncus acutus* is not present in the lowest 50 m before the river reaches the limit of the Madrid detrital basin, or in the adjacent parts of the sedimentary basin. It seems evident, taking into account the permeability barrier that the crystalline rocks represent, that if the presence of *Juncus*

Figure 22.4 Juncus acutus growing in the fractures of the granite in the Perales river valley.

acutus were due to a discharge of regional saline water from the basin, its distribution would not be interrupted southward and would appear above the sedimentary rocks near the contact.

Groundwater from the wells bored in the sedimentary rocks near these sites does not show clear signs of any significant hydrogeochemical evolution.

The *Juncus acutus* plants in the Perales valley indicate a discharge area of brackish waters quite similar to the water from the boreholes of Casa Yunta and La Sabinota.

22.3.2.2 El Salobral and La Salabrosa

La Salabrosa is an area represented by a 3 m deep dug well and a dry spring. The water from the well has a strong H_2S smell and electrical conductivities ranging from 2600 to 3200 μS/cm (Yélamos, 1991; Martín-Loeches, 1995). The water is of the Ca-Mg-Cl type, with high Ca/SO_4 and Ca/HCO_3 ratios. They were first reported by IGME (1983).

The well is dug in the weathered zone of the granite; 400 m away is a dry spring called 'La Salabrosa' (the name refers to the salty character of the water).

In the *El Salobral* area (first named by Gálvez & Yordana, 1941) there are three water points with TDS higher than 1300 ppm, a well and two springs, one of which is dry at present. In the detrital basin, at 300 m from the contact with the granite, there is a borehole which penetrates granite at a depth of 112 m, and which has a similar chemical composition.

22.3.2.3 *El Sarguero*

A detailed description of this point is included in Martín-Loeches (1995), Martín-Loeches *et al.* (1995) and Martín-Loeches *et al.* (1997). El Sarguero is a spring near the village of Aldeaencabo de Escalona, in the province of Toledo.

The water comes out at the confluent point of a small creek and a quartz dyke. It is located 120 m from the contact between the crystalline rocks and the Tertiary Madrid Basin.

It has always been difficult to collect representative samples of water from this point, as it does not yield enough water to allow a continuous flow along the stream and the thick vegetation makes it difficult to reach the point where the water emerges. The water in the stream is stagnant, creating small pools where samples can be collected. These factors probably contribute to the fact that it has measured conductivities ranging between 2000 and 4000 μS/cm, depending on the evaporation rate.

Among the chemical characteristics of this site, is a very low sulphate content of less than 2.1 mg/l, and a high bromide content of 29 mg/l. ENRESA (1996) presents a tritium value for this site of 5 TU.

The geological and ecological features of the location are significant. The relationship between the spring and the quartz dyke is evident and there are many signs of important tectonic activity throughout the area covered by the spring, up to the limit between the crystalline massif and the Tagus basin. From an ecological perspective, it is important to point out the presence of numerous specimens of *Juncus acutus* along the stream that extend from the spring up to the geologic contact of the range with the basin, and the existence of two points with the same type of vegetation following the dyke at 20 and 40 m above the main spring itself (Fig. 22.5).

22.3.2.4 Saucedoso stream

There is Spiny rush near the Saucedoso stream, and a well close to the plants shows Na-HCO_3 type groundwater. Despite this, Saucedoso is not geologically linked to the other brackish water sites. The presence of granites southward from this point prevents salty water coming from the regional flow of the Tagus tertiary depression.

100 km westward is the westernmost water point, a brackish spring over granites in the foothills of the Gredos range (Vicente, 1986; Vicente & Sastre, 1987), and the existence of three other brackish points in the Campo Arañuelo sedimentary basin, also strongly related to the granites.

Most of the general geochemical characteristics of the brackish waters in the studied sector of the Spanish Central System are similar to Campo Arañuelo, as are the geological and ecological aspects linked to them.

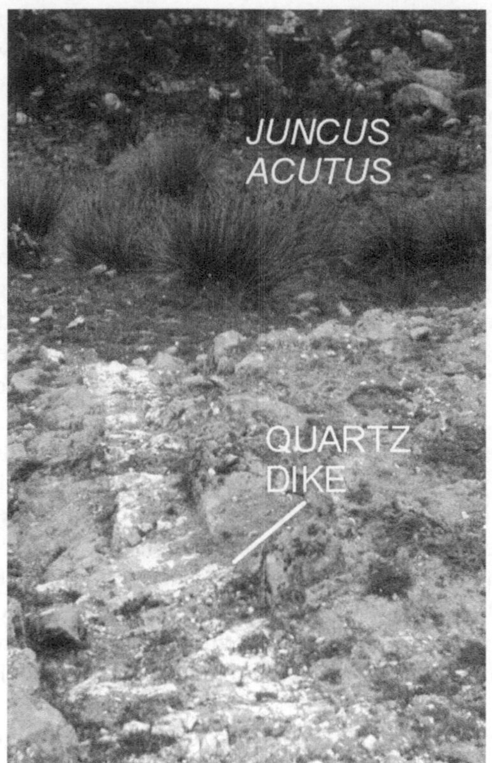

Figure 22.5 Spiny rush profusely distributed around a quartz dyke, 20 m above where the water of El Sarguero spring discharges.

22.4 DISCUSSION

The origin of the two groundwater types is not clear. It seems that alkaline sulphide waters are linked to the medium or long term evolution of infiltrating rainwater in the hard rock bodies; there are still some uncertainties regarding the presence of sulphur species, which are rare in the hard rock of the Spanish Central System. The origin of the brackish waters is more difficult to establish in this model, because of the scarcity of Cl in these rocks. Brackish waters in the Spanish Central System are different from the discharge waters of the Tertiary detrital basin of the Tagus, due to their low SO_4^{2-} and their high Ca^{2+} concentrations. Furthermore, their position in hard rocks makes it difficult from a hydrodynamic point of view to be directly affected by waters from the Tertiary rocks. The isotopic composition of this water indicates likely mixture with surface water, due to the difficulties in water sampling.

Although the chemical make-up of the alkaline and brackish waters of the Spanish Central System have evolved independently, both of them appear to be of deep circulation in the hard rocks, with a longer residence time in the subsoil than waters of low salinity and Ca-HCO_3 facies. Taking into account the range of sites where Spiny rush appears in the hard rocks of the Spanish Central System, it is evident that this plant is an indicator of geochemical and hydrodynamic areas.

In the low flow rate springs and streams in the hard rocks of the Spanish Central System, the most common plant indicating moisture is by far *Scirpus holoschoenus*. Like *Juncus acutus*, this plant is also a Juncacea, but it only indicates humidity, it is not a bioindicator of salty or alkaline soil. Even so, both juncaceas can be found at the same site.

According the observations in the Spanish Central System, the greater the groundwater salinity associated with the soil, the more frequent is the presence of *Juncus acutus* and scarcer that the presence of *Scirpus holoschoenus*, up to its complete absence.

It is easy to distinguish between the two species: *Juncus acutus* has darker tones and it has prickles (the origin of its other common name: "sharp" rush) and its florescence is very different to that of *Scirpus holoschoenus* (Fig. 22.6). As it is a perennial plant, it is possible to use it as a tool for searching for deep water flows all year round.

Figure 22.6 Florescence of the two juncacea on the hard rock of the Spanish Central System. On the left side, the *Scirpus holoschoenus* and on the right side a bush of *Juncus acutus*.

22.5 CONCLUSION

The main conclusion reached from this study is that there is a good relationship between Spiny rush and alkaline and brackish groundwater in hard rock mountainous regions in temperate climates, making *Juncus acutus* a good bioindicator of the presence of these types of water in that kind of rocks.

Although not all the deep circulation springs and wells can be linked to the presence of Spiny rush, the opposite is clearly the case: if Spiny rush is present, the hydrochemical composition of nearby wells is always alkaline or salty, which can be indicative of the emergence of water that has evolved in the massif through deep or large flows. It is not known whether the same relationship occurs in the numerous hard rock massifs of temperate climate mountainous areas in other regions.

ACKNOWLEDGEMENTS

The authors wish to thank Bruce Dudley Misstear and Antonio Chambel Gonçalves for his detailed and constructive reviews, which significantly improved the final version of the paper. The comments of another anonymous reviewer were also appreciated.

REFERENCES

Almeida, C. & Calado, C. (1993) Chemical components of deep origin in sulphide waters of the Portuguese sector of the Hesperian Massif. *Hydrogeology of Hard Rocks, XXIVth IAH International Congress*, 1, 377–388.

Calado, C. (2002) *Ocorrência de águas sulfurosas alcalinas do Maciço Hespérico [The occurrence of alkaline sulfurous waters in the Hesperian Massif]*. Ph.D. Thesis, University of Lisboa, Portugal.

Chambel, P.A. (1999) *Hydrogeologia do Concelho de Mertola [Hydrogeology of Mértola municipality]*. Ph.D. Thesis, University of Évora, Portugal. 380 pp.

Chambel, P.A. (2006) Groundwater in semi-arid Mediterranean areas: desertification, soil salinization and ecosystems. In: Baba *et al.* (eds.) *NATO Science Series, IV. Earth and Environmental Sciences*, 70, 47–58.

De Vicente, G., Giner, J.L., Muñoz-Martín, A., González-Casado, J.M. & Lindo, R. (1996) Determination of present-day stress tensor and neotectonic interval in the Spanish Central System and Madrid Basin, central Spain. *Tectonophysics*, 266, 405–424.

Doblas, M., Lopez-Ruiz, J., Oyarzun, R., Mahecha, V., Sanchez Moya, Y., Hoyos, M., Cebriá, J.M., Capote, R., Hernandez Henrile, J.L., Lillo, J., Lunar, R., Ramos, A. & Sopeña, A. (1994) Extensional Tectonics in the Central Iberian Peninsula during the variscan to alpine transition. *Tectonophysics*, 238, 95–116.

ENRESA (1996) El Berrocal Project: Characterization and validation of natural radionuclide migration processes under real conditions on the fissured granitic environment. Empresa Nacional de Residuos Radiactivos S A, Volume II, Hydrogeochemistry, Topical Reports, 557 pp.

Gálvez, A. & Jordana, L. (1941) *Geologic Map of Spain at scale, 1:50,000 Sheet nº 558. "Villaviciosa de Odón"*. Madrid, Instituto Geológico y Minero de España.

González Bernáldez, F., Montes, C., Besteiro, A.G., Herrera, P. & Pérez, C. (1989) *Los humedales del acuífero de Madrid [The wetlands of Madrid aquifer]*. Canal de Isabel II. Madrid, 92 pp.

González Bernáldez, F. & Rey Benayas, J.M. (1992) Geochemical relationships between groundwater and wetland soils and their effects on vegetation in central Spain. *Geoderma*, 55 (3–4), 273–288.

Herrera, P. (1987) *Aspectos ecológicos de las aguas subterráneas en la facies arcósica de la Cuenca de Madrid [Ecological aspects of the groundwater of the arcosic facies of Madrid basin]*. Ph.D. Thesis, University of Alcalá de Henares, Spain.

IGME (1983) *Hydrogeological map of Spain at scale, 1:50,000 Sheet nº 558, "Villaviciosa de Odón"*. Instituto Geológico y Minero de España.

IGME (2004) *Estudio hidrogeológico y de las condiciones de captación para mejorar los recursos hidrominerales de la Comunidad Autónoma de Galicia [Hydrogeological study and conditions of captation to improve hydro-mineral resources in the Autonomous Community of Galicia]*. Volume 4, Study of relation between physical and chemical composition and mineral waters and the geological setting of Galicia. Instituto Geológico y Minero de España, Internal report. 78 pp.

Limón Montero, A. (1697) *Espejo cristalino de las aguas de España. Hermoseado y embellecido en el marco de las diversidad de fuentes y baños [Picture of Spanish waters. To embellish in the context of diversity of springs and baths]*. García Fernández, Alcalá de Henares. Facsimile Edition in 1979 by Instituto Geológico y Minero. 472 pp.

Martín-Loeches, M.(1995) *Hidrogeología de las rocas ígneas y metamórficas de un sector de la Cuenca hidrográfica del río Alberche y su relación con las aguas subterráneas de la cuenca de Madrid [Hydrogeology of the igneous and metamorphic rocks in a sector of the Alberche river basin and its relationship with the groundwater of the Madrid basin]*. Ph.D. Thesis, University of Alcalá de Henares, Spain.

Martín-Loeches, M., Sastre, A. & Almeida, C. (1995) Hidrogeoquímica de las aguas subterráneas del sector suroccidental de la cuenca paleozoica del río Alberche [Hydrogeochemistry of the groundwater of the S-W sector of the Paleozoic basin of Alberche river]. *Hidrogeología y Recursos Hidráulicos*, 20, 133–147.

Martín-Loeches, M., Muñoz, I., Sastre, A. & Vicente, R. (1997) Brackish groundwater manifestations in the meridional limit of Spanish Central System. Hydrogeology of hard rocks, Ed AIH-GE, 1, 267–290.

Prado, C. (1864) *Descripción física y geológica de la provincial de Madrid [Physical and geological description of Madrid province]*. Facsimile Edition in 1975 by Colegio de Ingenieros de Caminos, Canales y Puertos, 352 pp.

Rey Benayas, J.M., Scheiner, S.M., García Sánchez-Colomer, M. & Levassor, C. (1999) Commonness and rarity: theory and application of a new model to Mediterranean montane grasslands *Conservation Ecology*, 3 (1), 5. http://www.consecol.org/vol3/iss1/art5/.

Rosales, F., Carbó, A. & Cadavid, S. (1977) Transversal gravimétrica sobre el Sistema Central e implicaciones corticales [Gravimetric transversal over the Spanish Central System and cortical implications]. *Bol. Geol. Min.*, 88 (6), 99–105.

Rosino, J., Martín-Loeches, M., Álvarez, I. & Galán, J.I. (2003) *Recursos minerales de Extremadura: las aguas minerales [Mineral resources of Extremadura: the Mineral Waters]*. Consejería de Economía, Industria y Comercio, Junta de Extremadura, Mérida, Spain. 303 pp.

Sánchez, J. (1992) *Guía de establecimientos balnearios de España [Spa Resort directory of Spain]*. Ministerio de Obras Públicas y Transportes, Spain. 357 pp.

Tóth, J. (1962) A theory of groundwater motion in small drainage basins in Central Alberta, Canada. *Journal of Geophysical Research*, 67 (11), 4372–4387.

Tóth, J. (1963) A theoretical analysis of groundwater flow in small drainage basins. *Journal of Geophysical Research*, 68 (16), 4795–4812.

Vicente, R. (1986) *Hidrogeología regional de la depresión de Campo Arañuelo [Regional hydrogeology of Campo Arañuelo basin]*. PhD Thesis, University of Alcalá de Henares, Spain.

Vicente, R. & Sastre, A. (1987) Contribución al conocimiento de la hidrogeología regional del Campo Arañuelo (provincias de Toledo y Cáceres) [Contribution to the regional hydrogeological knowledge of Campo Arañuelo (Toledo and Cáceres provinces)]. *Hidrogeología y Recursos Hidráulicos*, 7, 665–675.

Villarroya, F., Yélamos, J.G., Molina, M.A. & Sanz, E. (2006) Hydrogeology of igneous and metamorphic rocks in the Guadarrama Sierra tunnel of the Madrid-Segovia high velocity railway (Spain). In: Chambel (ed.) *Proceedings of the 2nd Workshop of the IAH Iberian Regional Working Group on Hard Rock Hydrogeology, AIH-GP*. pp. 189–200.

Warburton, J. & Álvarez, C. (1989) *A thrust tectonic interpretation of the Guadarrama Mountains,Spanish Central System. Asociación de Geólogos y Geofísicos del Petróleo.* Libro homenaje a Rafael Soler, Madrid. pp. 147–155.

Yélamos, J.G. (1991) *Hidrogeología de las rocas plutónicas y metamórficas en la vertiente meridional de la Sierra de Guadarrama [Hydrogeology of plutonic and metamorphic rocks in the south side of Guadarrama Sierra].* Ph.D. Thesis. Autonomous University of Madrid, Spain, 334 pp.

Chapter 23

Surface/groundwater interactions: Identifying spatial controls on water quality and quantity in a lowland UK Chalk catchment

Nicholas J.K. Howden[1], *Howard S. Wheater*[2], *Denis W. Peach*[3] & *Adrian P. Butler*[4]

[1]*Queen's School of Engineering, University of Bristol, University Walk, Bristol, UK*
[2]*Global Institute for Water Security, University of Saskatchewan, National Hydrology Research Centre, Saskatoon, Canada*
[3]*British Geological Survey, Environmental Science Centre Keyworth, Nottingham, UK*
[4]*EWRE, Department of Civil and Environmental Engineering, Imperial College London, London*

ABSTRACT

This chapter focuses on the Piddle catchment in Dorset in southern England, where surface/groundwater interactions play a key role in determining the spatial distributions of water quality and availability in the catchment river systems and, hence, a fundamental control on the supply of water to dependent ecosystems. The hydrogeological controls on surface/groundwater interactions are identified and the spatial variation in surface water chemistry at individual sites and along the length of the river is discussed. It is found that, whilst there are numerous spring sources in the lower part of the Piddle catchment which contribute large volumes of water to river flow, the most significant spring sources are those in the Upper catchment. This is because there are very few sources, but these exert the main control on the quantity of water in the Upper Piddle and, consequently, control the water quality throughout the year.

23.1 INTRODUCTION

The interactions between surface water and groundwater are complex. Sophocleous (2002) provides a comprehensive review of the subject and emphasises the importance of a sound hydrogeological framework within which to understand the influence of factors such as climate, landform, geology, and biota. This is needed for effective management of water resources.

Groundwater has a major influence on the surface water flow regime in lowland catchments which overly a permeable geology such as the Chalk, supporting dry-weather flows which are often of significant environmental and amenity value. However 42% of river habitats suffer from the effects of groundwater abstraction and urbanisation (NERC, 1999). In addition, there are increasing regional and national concerns regarding water quality and its effects on the ecology and amenity status of southern lowland UK rivers (Neal *et al.*, 2000a, 2000b). This is in part due to a lack of sufficient nutrient dilution under low flow conditions, in addition to problems encountered with nutrient inputs from sewage sources, exacerbated by an increasing population density in the southern lowland areas. These various concerns are compounded by effects of climate variability in the UK (Marsh and Sanderson, 1997), and scenarios of climate change show major impacts on summer low flows in the southern and south-eastern lowland areas, which experience the lowest rainfall and highest evaporation rates in the country.

River/aquifer interactions have always been recognised as important in Chalk streams, which are characterised by stable flows and support a high diversity and density of river biota, including in some streams an economically-valuable salmonid fishery (Sear *et al.*, 1999). Ecological studies of groundwater rivers have tended to be dominated by research into Chalk systems (Berrie, 1992), but it is only relatively recently that researchers have begun to focus upon the importance of river-aquifer interactions, in the context of the impact of groundwater abstractions to improve understanding of the role that groundwater plays in the heterogeneity of floodplain and river hydrochemistry and ecology (Neal *et al.*, 2000b, 2000c, 2000d; Petts & Amoros, 1995). For example, ecological concepts of 'site' and site conditions may be determined and maintained by upward seepage, essential for the survival of various relatively rare plant species and communities and may be altered easily by (mis-)management (Klijin & Whitte, 1999).

This chapter uses geological, hydrogeological and hydrochemical data to investigate the linkage between different hydrogeological units, and river water chemistry in the Piddle catchment in west Dorset, UK. Following from Howden *et al.* (2004), the impact of surface/groundwater interactions upon the surface water chemistry is considered, and the relative importance of different spring sources in determining overall river water quality is discussed.

23.2 CATCHMENT

23.2.1 General

The River Piddle drains a lowland, permeable, heavily-abstracted Chalk catchment in West Dorset, UK. It is relatively small and rises at around 150 m above sea level from Chalk springs in the North Dorset Downs, and then flows 40 km roughly south-east to a common estuary with the river Frome, before discharging into the English Channel at Poole Harbour (see Fig. 23.1). Annual average rainfall is between 900 and 1000 mm, with river flows of around 400 mm and actual evapotranspiration of 500 mm. There are two principal tributaries, the Devil's Brook and the Milborn/Bere Stream, which are also Chalk streams, the whole catchment comprising an area of high amenity and ecological value (Stevens, 1999). The catchment is heavily utilised for water abstraction, with a total maximum daily licensed quantity of 197 Ml day^{-1}. Of this, 43.5% is abstracted from groundwater and reflects the importance of the Chalk aquifer as a source of supply. In 1993 the River Piddle was included in the list of the top 40 rivers in England and Wales suffering from unacceptably low flows (NRA, 1993).

23.2.2 Geology and hydrogeology

The Piddle is located on the western-most outcrop of the southern Chalk province, as described in Howden *et al.* (2004), and a summary of the geological succession

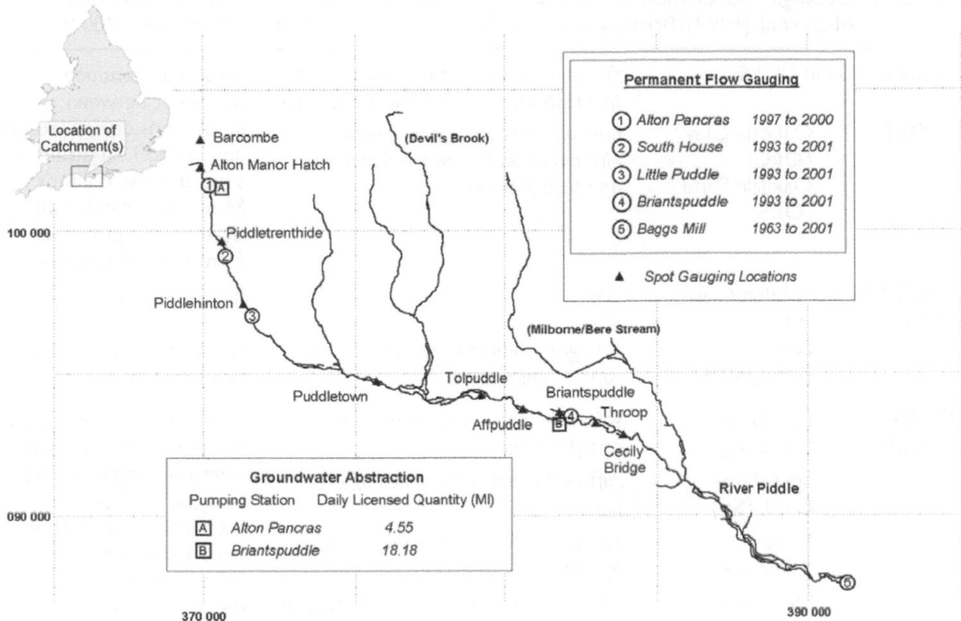

Figure 23.1 Map of the River Piddle catchment.

and hydrogeological characteristics of the principal formations is given in Table 23.1. The surface geology of the Piddle catchment comprises three distinct geological units: the headwaters of the Piddle and its tributaries rise from Jurassic limestones and mudstones or the Greensands of the Lower Cretaceous, very rapidly giving way to chalklands for the remainder of the upper and middle reaches. The lower reaches flow over the Palaeogene deposits of the Wareham basin before discharging into Poole Harbour (Newell *et al.*, 2002).

The Chalk strata in the Piddle catchment are extensively deformed. A number of small folds and faults are associated with the Purbeck Monocline. This strikes east-west and lies to the southern edge of the Frome catchment and dominates the geological structure of the area (Allen *et al.*, 1997). To the southeast of the catchment, the Chalk is overlain by Palaeogene strata which confine the Chalk and form an overlying aquifer of interbedded sands, gravels and clays of the London and West Park Farm formations.

The Chalk is frequently described as a dual or double porosity aquifer. The matrix pores provide porosity and little permeability and the fractures have limited storage but form permeable pathways. In reality the hydraulics of the Chalk aquifer are more complicated than this as the aquifer properties vary vertically, laterally, with lithostratigraphy and geological structure and, perhaps most importantly, with geological and geomorphological history. The matrix makes little contribution to aquifer transmissivity (Alexander, 1981; Wellings, 1984); as stored water is held in the matrix pores by capillary and molecular forces, very limited movement occurs under the influence of gravity due to the small pore size (Price *et al.*, 1977). A comprehensive description of the catchment hydrogeology is provided in Howden (2004) and Howden *et al.* (2004).

Table 23.1 Geological succession in the Piddle catchment using generalised strata descriptions from Allen et al. (1997), Bristow et al. (1998), Jones et al. (2000) and Newell et al. (2002).

Branksome Sand (**BrkS**)		*fine to very coarse grained sand with thin lenticular clays in cyclical sequence*	Hydraulic conductivity may exist between formations due to lateral and vertical variations in sand and clay content. Sandy beds can be well developed. Springs issue from the formations
POOLE	Oakdale Clay (**OkC**) Oakdale Sand (**OkS**)	*medium to coarse grained sand, alternating with brown, red-brown and pale grey clays*	
LONDON CLAY	London Clay (**LC**)	*clay*	Impermeable
	West Park Farm (**WPF**)	*clay, grey, commonly red-stained and sands, locally pebbly*	Confined groundwater head evident in sands
UPPER CHALK	Portsdown Chalk (**P**)	*soft white chalk with marl seams and flint bands*	Groundwater flow takes place in confined Chalk fractures which may have become enlarged by dissolution. Flint bands, hard and soft grounds may act as catalysts for such dissolution. The extent of fracture development depends upon depth of burial, although the zone of active groundwater movement is generally within 60 m of the surface (Buckley et al., 1998)
	Spetisbury Chalk (**Sp**)	*soft white chalk with bands of large flints*	
	Tarrant Chalk (**T**)	*soft white chalk with regular bands of flints*	
	Newhaven Chalk (**N**)	*soft to medium hard chalk with regular marl seams and regular flint bands*	
	Seaford Chalk (**S**)	*soft, pure, smooth white chalk, with regular bands of flints, some large*	
	Lewes Nodular Chalk (**LN**)	*hard to very hard nodular chalk and hardgrounds with interbedded soft to hard gritty chalks; regular bands of nodular flint; particular hardground of the Chalk Rock at the base of the strata*	
MIDDLE CHALK	New Pit Chalk (**NP**)	*massively bedded pure white to pale green chalks with regularly spaced marl seams and marly wisps, sparse flints towards the top*	
	Holywell Nodular Chalk (**HN**)	*medium hard to very hard nodular chalk and regular flaser marls, very shelly and gritty towards the top; particular hardground of the Melbourn Rock towards the base of the strata*	
LOWER CHALK	Zig Zag Chalk (**Z**)	*a soft to medium hard, greyish, blocky chalk with marl-limestone rhythms in the lowest part*	Low permeability caused by high marl content
	West Melbury Marly Chalk (**W**)	*rhythmically bedded soft off-white to grey marls, grey marly chalk and hard to very hard grey and reddish-brown limestone bands*	Strata absent in the catchment
Upper Greensand (**UGS**)		*glauconitic sands topped by a layer of calcareous sandstone*	Permeable sandstone strata, with confined heads in some places

23.3 DATA

Data required to study the surface/groundwater interactions in the River Piddle include locations of spring sources, details of lithostratigraphy and geological structure along the river valley corridor, flow gauging along the river (permanent flow gauging and accretion flow profiles from spot flow measurements) and hydrogeological descriptions of strata underlying the river valley corridor (Howden *et al.*, 2004). These were gained from the following sources: underlying lithology and structure of the river channel was obtained directly from the geological map (Newell *et al.*, 2002); flow gauging data was obtained for the five permanent streamflow gauging stations maintained by the Environment Agency and from spot flow measurements taken by Wessex Water Authority and the National Rivers Authority from 1960 to 1990; spring sources in the catchment were obtained by visual inspection, obtained directly from the OS 1:25 000 and 1:10 000 Maps (Ordnance Survey with dates of map editions) and from the geological map of the area (Newell *et al.*, 2002); topography of the river channel, approximated from 5 m contours on the OS 1:25 000 Map; boreholes from the BGS archive were used to characterise the general hydrogeology of the valley corridor (summarised in Table 23.1).

Further, water chemistry data from the Environment Agency's Harmonised Monitoring Scheme (HMS) (1977 to 2000) was analysed to illustrate the variations in different chemical components at specified sites along the river over the period 1977 to 2000.

23.4 METHOD

Conventionally a cross section would be used to illustrate the hydrogeology of the river valley corridor. However in this instance the angle of dip of the Chalk strata (around 2 to 3°) is very close to the slope of the topography, the geological structure affecting the river channel is complex and supporting data are limited. Therefore, a topographic schematic section of the Piddle (shown in Fig. 23.2) is used here, linked to an illustration of the lithostratigraphy and geological structure affecting the river channel and the location of spring sources along the river, together with flow accretion profiles.

Analyses of the river water chemistry is presented in the form of box and whisker plots, together with a schematic showing the location of sampling sites with respect to the geology. In these plots, the extent of the boxes indicates the interquartile range and whiskers show 5th and 95th percentiles. Values above the 95th or below the 5th percentiles are individually plotted. A horizontal line across the boxes indicates median concentration, and a further horizontal line, joined from site-to-site, indicates mean concentration. (Note: The distance between sampling sites is not to scale – sampling sites are represented on the geology plot by a circle).

23.5 RESULTS

The results are presented in: Figure 23.2 showing river bed geology, flow accretion profiles from spot gaugings and major spring locations; Figures 23.3, 23.4 and 23.5, box and whisker plots showing chemical variations at, and between, sites along the river.

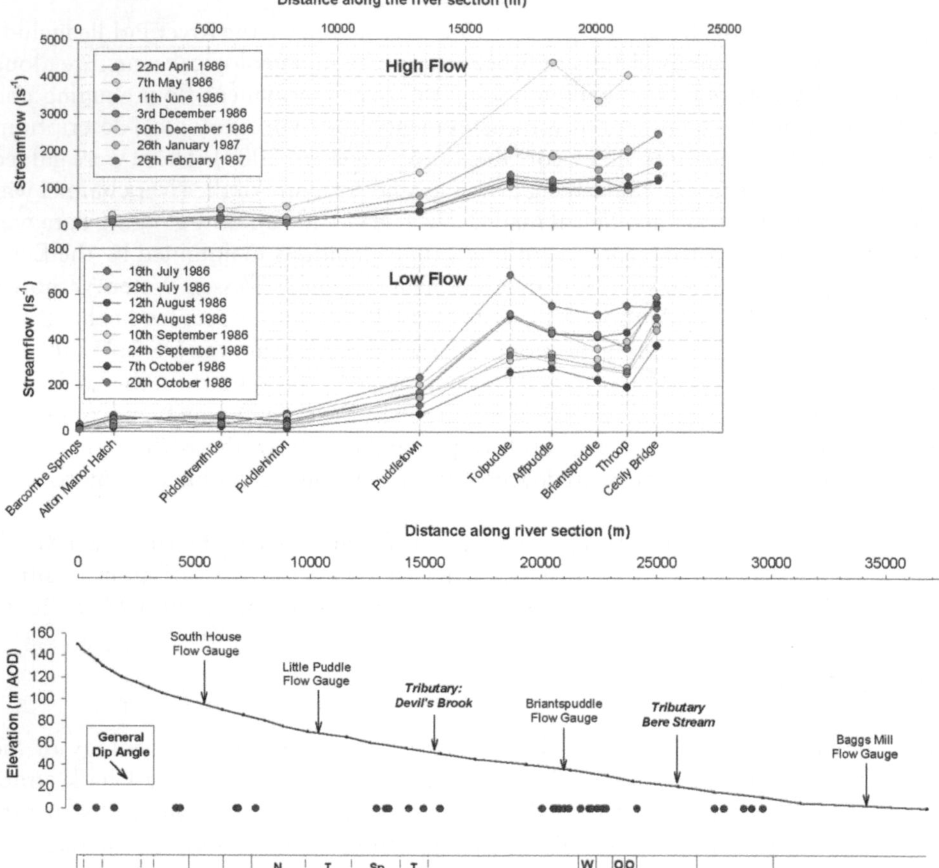

Figure 23.2 Schematic section of the River Piddle showing topography, stream bed geology and geological structure, significant spring sources (blue dots), tributaries and flow accretion curves derived from spot gaugings.

From the original conclusions of Howden *et al.* (2004), surface/groundwater interactions in the river Piddle valley corridor take two forms: spring flow and stream leakage. Both of these processes are linked to the lithology and geological structure underlying the river valley. The river drains a series of discrete springs, between which streamflow depletion may occur. The main hydrogeological controls on the generation of spring flow (see Table 23.2 for a summary) are as follows: Upper Greensand, confined groundwater head in the Upper Greensand overlain by Zig Zag Chalk, or faulting of an Upper Greensand surface outcrop against less permeable strata; Holywell Nodular Chalk, gravity drainage at the Holywell Nodular/Zig Zag Chalk boundary or faulting of the Holywell Nodular Chalk against less permeable strata; Portsdown Chalk, the surface outcrop of marl and flint beds; Portsdown Chalk/West Park Farm

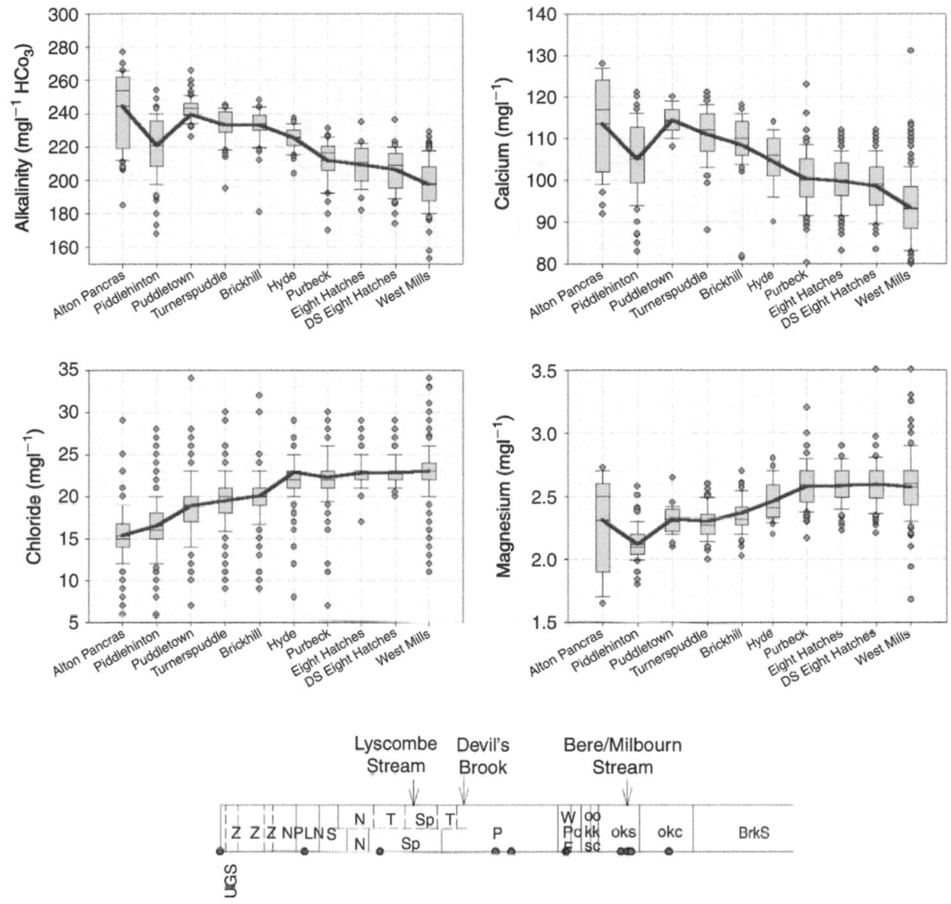

Figure 23.3 Major Ions. Box and whisker plots of major ion determinands at sites along the River Piddle. Concentrations in mg l^{-1} of alkalinity (HCO_3), calcium (Ca^{2+}), chloride (Cl^-) and magnesium (Mg^{2+}).

Boundary, the West Park Farm strata are a confining layer overlying the Portsdown Chalk, creating some artesian springs associated with this confinement; and Poole Formation, interbedded sands and clays of the Poole formation enable spring lines to be formed due to groundwater head in sand layers confined by overlying clays, or drainage from the base of sands overlying clay layers.

There is very little data to characterise the relative flows from different spring sources. However, from observations in the catchment it is generally the spring sources in the Upper Piddle that are the most significant, because there are very few sources in the Piddle headwaters, and these are, for the majority of the year, the sole mechanism for streamflow generation in the Upper Piddle catchment. Whilst the springs controlled by the Chalk/Palaeogene boundary, and by the boundaries between sands and clays within the Palaeogene are more numerous and most likely supply a greater volume of

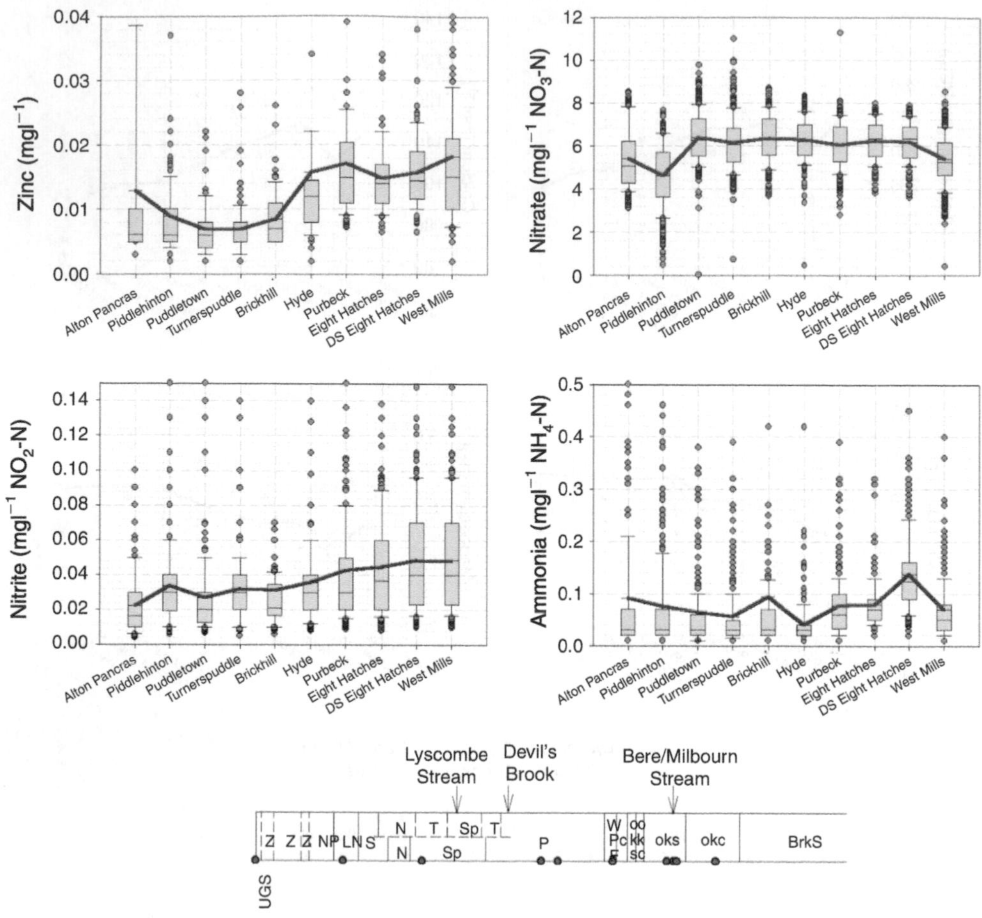

Figure 23.4 Nitrogen and Zinc. Box and whisker plots of major ion determinands at sites along the River Piddle. Concentrations in mg l^{-1} of zinc (Zn), nitrate (NO_3-N), nitrite (NO_2-N) and ammonia (NH_4-N).

water to the river, their overall impact upon the flow regime is less critical than the few sources in the Upper catchment around Alton Pancras.

Upper Greensand, Chalk and Palaeogene groundwater sources contribute to streamflow in the river Piddle, but outcrops of Upper Greensand only make small contributions to headwater springs. Chalk springs provide baseflow in the upper and middle reaches of the river Piddle. This Chalk-dominated streamflow is diluted in the lower Piddle by inputs from the Palaeogene groundwaters and springs to the river channel.

Hydrochemical determinands measured at sites from source to outfall reflect these source areas and the mixing zones between them:

- average alkalinity, calcium and water hardness decrease along the length of the river;

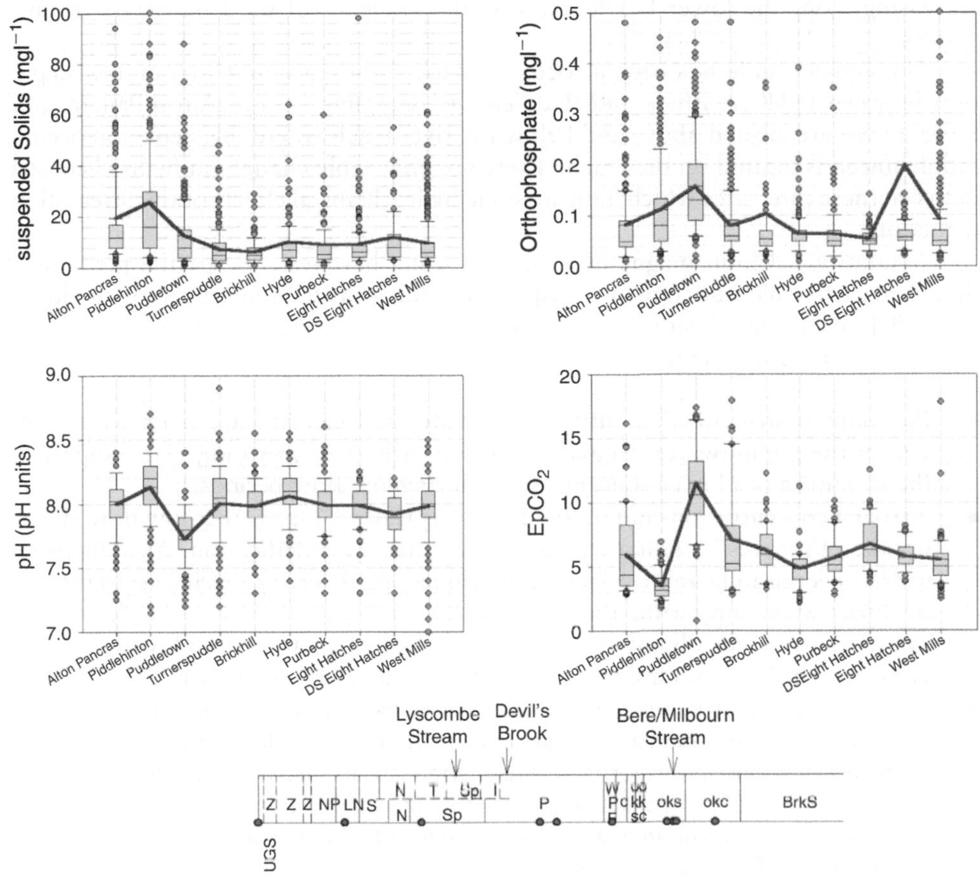

Figure 23.5 Suspended Solids, Orthophosphate or Soluble Reactive Phosphate (SRP), pH and Dissolved Carbon Dioxide ($EpCO_2$). Box and whisker plots at sites along the River Piddle. Concentrations of suspended solids and orthophosphate (PO_4^{3-}-P) in mg l^{-1}.

Table 23.2 Geological controls on spring sources in the Piddle catchment.

	Boundary	*Outcrop*	*Fault*	*Confinement*	*Total*
Upper Greensand	–	–	3	–	3
Chalk	–	5	7	2	14
Palaeogene	16	2	–	–	18
Total	16	7	10	2	35

- average magnesium, chloride and nitrite concentrations increase along the length of the river;
- soluble reactive phosphorus (SRP) concentrations increase along the upper Piddle, before decreasing along the middle and lower Piddle.

Zinc concentrations decrease from the headwaters of the upper Piddle, before increasing along the lower Piddle marking the influence of the Palaeogene groundwaters.

The Upper Piddle has been a focus of much attention due to streamflow depletion between Piddletrenthide and Puddletown. Sampling sites at Alton Pancras and Puddletown are located above and below the river reach where this depletion occurs. Piddlehinton is located on the reach. There are noticeable changes in hydrochemistry between these three sites which indicate some hydrochemical effects of this streamflow depletion.

Alkalinity, calcium, magnesium, nitrate, water hardness, the partial pressure of dissolved carbon dioxide ($EpCO_2$), suspended solids and SRP concentrations are lower at Piddlehinton, and the site also records an increase in average nitrite and pH levels. These differences are due to:

- the main sources of alkalinity (bicarbonate) and calcium are from weathering within the groundwater (Chalk) aquifer. A lack of fresh spring water supplying the stream causes lower concentrations during low flow periods;
- nitrogen concentrations fall to low levels (typically $<1\,mg\,l^{-1}$) at low flow, due to uptake of nitrates by aquatic plants which causes denitrification. Also, increases in nitrite levels indicates the lack of dilution along this river reach (due to the lack of spring water supporting the river flow);
- average suspended solids concentrations are higher, an artefact of low streamflow volumes which are not able to transport the sediments downstream;
- elevated average SRP concentrations are most likely due to a combination of two factors. Firstly the intensity of farming in the Upper Piddle catchment leads to more intensive phosphorus-rich fertiliser applications. Secondly, strong association between phosphorus and suspended sediments due to microparticulates passing through the filters: the increase in suspended solids concentrations has the effect of increasing SRP concentrations;
- average carbon dioxide partial pressures are much lower at the Piddlehinton site, and pH levels are consistently high, due to degassing of dissolved $EpCO_2$.

At Puddletown, average concentrations of alkalinity ($239\,mg\,HCO_3^-\,l^{-1}$), $EpCO_2$ (11.5), calcium ($114\,mg\,l^{-1}$), nitrate ($6.36\,mg\,NO_3\text{-}N\,l^{-1}$) and SRP ($0.16\,mg\,PO_4\text{-}P\,l^{-1}$) are the highest and pH (7.73) the lowest average concentrations recorded along the length of the river respectively. These data are showing the combined effect of perennial springs and artesian boreholes at Waterston and some sewage effluents from Piddlehinton.

Hydrochemical variations are largest at sites in the upper Piddle, and could raise water quality concerns, particularly during the summer periods and extreme low flow events.

23.6 CONCLUSIONS

River water quality of the Piddle is high and predominantly calcium bicarbonate bearing, the source of which is weathering processes within the Chalk aquifer. There are hydrochemical variations along the length of the river and at particular sites due to the heterogeneity of surface/groundwater interactions in the river valley corridor, and changing sources of groundwater baseflow from Upper Greensand, Chalk and Palaeogene springs. Along the River Piddle there are five distinct areas: three source areas (Upper Greensand, Chalk and Palaeogene) with two mixing zones between the three.

Streamflow depletion in the upper Piddle causes a number of hydrochemical variations which were identified at a particular site. Where similar streamflow depletion occurs elsewhere in the catchment similar hydrochemical effects may occur. It is noteable that, whilst the most significant numbers and flow volumes of spring sources occur at the Chalk/Palaeogene boundary, and from lithological boundaries within the Palaeogene, the few springs around Alton Pancras are the most significant in maintaining water quality along reaches of the Upper Piddle.

The Chalk hydrogeology dictates that the use of groundwater, whether for public supply or otherwise, will entail abstraction from a preferential flow horizon in the Chalk. The nature of surface/groundwater interactions in the Piddle suggests that springs contributing to streamflow are driven by heads in such flow horizons, and that groundwater abstraction for any purpose will inevitably cause streamflow depletion.

The water quality of the River Piddle is high and, due to sufficient volumes of baseflow to enable dilution of nutrient and anthropogenic effluent discharges to the river, are much less vulnerable to nutrient enrichment and low water quality than the smaller tributaries, where flow volumes are low and the rivers exhibit flow depletion during the summer. However, it is important that water resource management ensures groundwater abstractions in the Upper Piddle catchment do not cause springflow depletion to the point where water quality could be adversely affected.

ACKNOWLEDGEMENTS

This research has been carried out under a NERC non-thematic PhD studentship with CASE support from the British Geological Survey. The first author is indebted to the EA hydrometry team at Blandford Forum for supplying data, Peter Orton, David Simmonds and John Winters for assistance with fieldwork, and to Nick Robins for his support and advice.

REFERENCES

Alexander, L.S. (1981) *The hydrogeology of the Chalk of south Dorset*. Ph.D. thesis, University of Bristol, UK.

Allen, D.J., Brewerton, L.J., Coleby, L.M., Gibbs, B.R., Lewis, M.A., MacDonald A.M., Wagstaff, S.J. & Williams, A.T. (1997) *The Physical Properties of Major Aquifers in England and Wales*. BGS Technical Report WD/97/34 (Environment Agency R&D Publication 8 equivalent).

Berrie, A.D. (1992) The chalk stream environment. *Hydrobiologia*, 248, 3–9.

Bristow, C.R., Mortimore, R.N. & Wood, C.J. (1998) Lithostratigraphy for mapping the Chalk of southern England. *Proceedings of the Geologists Association*, 109, 293–315.

Howden, N.J.K. (2004) *Hydrogeological Controls on Surface/Groundwater Interactions in a Lowland Permeable Chalk Catchment: Implications for Water Chemistry and Numerical Modelling*. Ph.D. Thesis, University of London, London.

Howden, N.J.K., Wheater, H.S., Peach, D.W. & Butler, A.P. (2004) Hydrogeological controls on surface/groundwater interactions in a lowland permeable Chalk catchment. *In Hydrology: Science & Practice for the 21st Century: Volume II*. The Netherlands, British Hydrological Society.

Jones, H.K. *et al.* (2000) *The Physical Properties of Minor Aquifers in England and Wales*. British Geological Survey Technical Report.

Klijin, F. & Witte, J-P.M. (1999) Eco-hydrology: Groundwater flow and site factors in plant ecology. *Hydrogeology Journal*, 7, 65–77.

Marsh, T.J. & Sanderson, F.J. (1997) A review of hydrological conditions throughout the period of the LOIS monitoring programme. *Science of the Total Environment*, 194/5, 59–70.

Neal, C., Neal, M., Wickham, H. & Harrow, M. (2000a) The water quality of a tributary of the Thames, the Pang, southern England. *Science of the Total Environment*, 251/252, 459–475.

Neal, C., Jarvie, H.P., Howarth, S.M. *et al.* (2000b) The water quality of the river Kennet: initial observations on a lowland chalk stream impacted by sewage inputs and phosphorus remediation. *Science of the Total Environment*, 251/252, 477–495.

Neal, C., Williams, R.J., Neal, M. *et al.* (2000c) The water quality of the river Thames at a rural site downstream of Oxford. *Science of the Total Environment*, 251/252, 441–457.

Neal, C., Neal, M. & Wickham, H. (2000d) Phosphate measurements in natural waters: two examples of analytical problems associated with silica interference using phosphomolybdic acid methodologies. *Science of the Total Environment*, 251/252, 511–522.

NERC (1999) Proposal for a NERC Thematic Programme in Lowland Catchment Research (LOCAR) 1999–2004. http://www.nerc.ac.uk.

Newell, A. *et al.* (2002) *The Geological Framework of the Frome-Piddle Catchment*. Commissioned report CR/02/197 Integrated Geoscience Surveys – Southern Britain, BGS.

NRA (1993) *Low Flows and Water Resources. Facts from the Top 40 Low Flow Rivers in England and Wales*. Bristol, UK, National Rivers Authority. 30 pp.

Petts, G.E. & Amoros, C.A.L. (1995) *Fluvial Hydrosystems*. Edward Arnold. 355 pp.

Price, M., Robertson, A.S. & Foster, S.D.D. (1977) *Chalk Permeability – A Study of Vertical Variation using Water Injection Tests and Borehole Logging*. Water Services, Institute of Geological Sciences.

Sear, D.A., Armitage, P.D. & Dawson, F.H. (1999) Groundwater dominated rivers. *Hydrological Processes*, 13, 255–276.

Sophocleous, M. (2002) Interactions between groundwater and surface water: the state of the science. *Hydrogeology Journal*, 10, 52–67.

Stevens, A. (1999) Impacts of groundwater abstraction on the trout fishery of the river Piddle, Dorset; and an approach to their alleviation. *Hydrological Processes*, 13, 487–496.

Wellings, S.R. (1984) Recharge of the Upper Chalk at a site in Hampshire, England: Water balance and unsaturated flow. *Journal of Hydrology*, 69, 259–273.

Chapter 24

Modelling stream-groundwater interactions in the Querença-Silves Aquifer System

José Paulo Monteiro[1], *Luís Ribeiro*[2], *Edite Reis*[3], *João Martins*[4], *José Matos Silva*[5] & *Núria Salvador*[1]

[1]*Geo-Systems Centre/CVRM, CTA, University of the Algarve, Faro, Portugal*
[2]*Geo-Systems Centre/CVRM, Instituto Superior Técnico, Lisbon, Portugal*
[3]*Algarve River Basin Administration, Faro, Portugal*
[4]*EDZ – Environmental Consulting, Lisbon, Portugal*
[5]*Catholic University of Portugal, School of Engineering, Lisbon, Portugal*

ABSTRACT

Streamflow in the Central Algarve, during dry seasons, is primarily base-flow. The stream network is also affected by water losses along influent reaches in wet seasons, across streambeds and karstic swallow holes. The representation of these streams in a finite element groundwater flow model allowed an investigation of the factors controlling the spatial distribution of the stream-groundwater interactions. After calibration of transmissivity, the implemented model provided a reliable representation of the equipotential surfaces of the aquifer, accommodating the values of the regional scale water balance. The subsequent introduction of the stream network in the model allowed the identification and systematisation of the parameters and variables needed to improve the conceptualisation of the river-aquifer relations at the local scale. The use of models combined with collected data contributes to the optimisation of monitoring networks and promotes the understanding of the boundary conditions at the regional and local scale.

24.1 INTRODUCTION

The study of the balance of the Querença-Silves Aquifer System located in the south of Portugal is of great regional importance as its waters support an important part of the Algarve urban water supply system, further enhanced during the severe drought that affected Portugal in 2004 and 2005. A detailed analysis of water use in historic terms, in that crisis period, was described by Monteiro *et al.* (2006), Monteiro & Ribeiro (2006) and Monteiro *et al.* (2007). Available data show that, during the drought period, the extractions in QSAS were the most intense in its entire historic period of water exploitation. According to the values presented (op. cit), the extraction volumes for urban supply in municipalities located in the area of QSAS in this period were: 4.6×10^6 m^3/year (Silves); 1.9×10^6 m^3/year (Lagoa); 3.5×10^6 m^3/year (Albufeira) and 0.4×10^6 m^3/year (Loulé). Additionally, in this same period, *Águas do Algarve* extracted 11.0×10^6 m^3/year in the *Vale da Vila* well field, for urban supply in other areas of the Algarve. Finally, as the average of extractions for irrigation in the area of the Querença-Silves Aquifer System was around 31.24 × 10^6 m^3/year (Nunes *et al.*, 2006), it is estimated that the total extractions in the Querença-Silves Aquifer System, in that period was around 52.7 × 10^6 m^3/year.

More recently, Salvador *et al.* (2012a) and Salvador *et al.* (2012b) highlighted the importance of the river-aquifer interactions in the area of the Querença-Silves Aquifer System, as well as the contribution of allogenic recharge in terms of the water balance. Allogenic recharge (Field *et al.*, 1999) is derived from runoff of neighbouring or overlying non-karst rocks that drains into a karst aquifer. Earlier estimated values have been exclusively based on recharge derived from precipitation directly onto the karst landscape (autogenic recharge).

Allogenic recharge is important in the case of the Querença-Silves Aquifer System, as well as in other Early Jurassic karst aquifers in Algarve, due to their northern contact with the low permeability Carboniferous turbiditic sequence of shale and greywacke present to the north of the sedimentary Mesocenozoic basin in this region. These carboniferous rocks, popularly referred to as Serra, constitute the upland section of the basin of the Algarve streams, where the drainage density d is high (more than 3.5 km/km^2, with areas where values are higher than 6.5 km/km^2 (Almeida, 1985)), due to the low infiltration rates. The mid-sections of the streams are located in Early and Middle Jurassic karstic rocks, where infiltration rates are very high and, therefore, d is between 0 and 2 km/km^2 (as in the case of the Querença-Silves Aquifer System). To the south in the coastal strip, the infiltration rates are less important than in the karstic outcrops, because of the low permeability of the Upper Jurassic and Cretaceous rocks and because Miocene carbonate and detritic aquifers are covered by low permeability Pliocene to Quaternary sediments. Thus d values in these areas are near 3.5 km/km^2 (Almeida, 1985). Due to short length of the watercourses between the mountains constituting the northern limit of the region (Serra) and the Atlantic Ocean to the south, the concentration time of streams in the Algarve river basin is always less than 24 hours. The most important water courses in the Mesocenozoic basin of the Algarve receive baseflow in effluent reaches hydraulically connected with the carbonate Jurassic aquifers. Therefore, they are active during periods spanning from a few months to more than one drought year, in some effluent reaches as shown in the recent period of 2004–2005.

Due to the complexity of the hydrogeological factors controlling stream-aquifer interactions along reaches in areas with these very different hydrogeological properties, a traditional rainfall-runoff model cannot take all the relevant factors into account, to simulate streamflow at the watershed scale in a realistic way (Sophocleous & Perkins, 2000; Sophocleous, 2002; Arnold & Fohrer, 2005). This is especially true for the reaches present in the area of the Jurassic karstic rocks. In these areas, the generation of streamflow is not necessarily the final destiny of overland flow, after the fulfilment of the processes of interception and infiltration. The generation of stream-flow in the area where outcrops of Jurassic carbonate formations are present is almost entirely derived from baseflow, resulting from the recharge of karstic aquifers supported by these rocks.

The discharges of rivers are the result of the hydrogeological control of streamflow in the several influent and effluent stream reaches along the Querença-Silves Aquifer System area. Taking these conditions into account, a regional scale finite element flow model was used to investigate the river-aquifer interactions, namely to evaluate the reliability of the use of a regional groundwater flow model to quantify the volume of water transfers between the Querença-Silves Aquifer System and the stream network. The proposed model differs from conventional groundwater flow models in that the simulated terms of the natural water balance include the successive quantification of local-scale river-aquifer transfers occurring in the successive effluent and influent streams. Conventional large-scale groundwater flow models only take into account the total water input (recharge) and output (outflow in the main discharge areas) for the natural water balance of the aquifer at the regional scale. Temporary residence of water in streams inside the area of the aquifer corresponds usually to secondary terms of the overall water balance. However, in the cases where part of the river flow is generated upstream of the aquifers these secondary terms can became relevant and affect the overall water balance. Local simulation of interactions between surface water and groundwater can be an innovative way to integrate observed data from river flow gauging and piezometric monitoring. This allows the quantification of allogenic recharge in the water balance, which can affect the water budget at the regional scale. The model presented in this paper sets different kinds of boundary conditions to improve the calculation of the local-scale groundwater-surface water interactions in order to improve the integration of these secondary terms into the regional scale balance. In addition to the improved accuracy of the water balance estimate at the regional and local scale, the understanding of the local-scale river-aquifer transfers is also very important to determine the dependence of baseflow on the reaches of the stream network which are in hydraulic contact with the Querença-Silves Aquifer System. As these water courses are Mediterranean groundwater dependent ecosystems (GDEs), their ecological sustainability must be taken into account when defining river basin management plans and measures to protect the water wells used for public water supply.

24.2 BACKGROUND (GEOLOGICAL AND HYDROGEOLOGIC SETTING)

The starting point for the characterisation of the river-aquifer transfers at local scale is provided by earlier studies on the Querença-Silves Aquifer System, including the contributions for the definition of a conceptual model of the system during the last three decades (Almeida, 1985; Andrade, 1989; Monteiro *et al.*, 2006; Stigter *et al.*, 2009; Stigter *et al.*, *in press*; Hugman *et al.*, 2012; Salvador *et al.*, 2012a,b). This aquifer is formed by Jurassic (Lias-Dogger) carbonate sedimentary rocks covering an irregular area of 324 km^2 that spans from the River Arade in thye west to Querença, to the north of the Algibre fault, which is the main onshore thrust in the Algarve Basin. A map of the Querença-Silves Aquifer System and the stream network in the area is shown in Figure 24.1. A representative cross-section of the lithologies present in the central area of the aquifer is shown in Figure 24.2.

The Algibre thrust, which constitutes the southern limit of the Querença-Silves Aquifer System, is a boundary that divides the Algarve Basin into two distinct morphological domains: the northern domain is limited by the Algibre thrust fault scarp to the south and by the cuestas cut in the Triassic-Hettangian rocks of the São Bartolomeu de Messines area to the north (Terrinha, 1998). In this domain the country rock consists

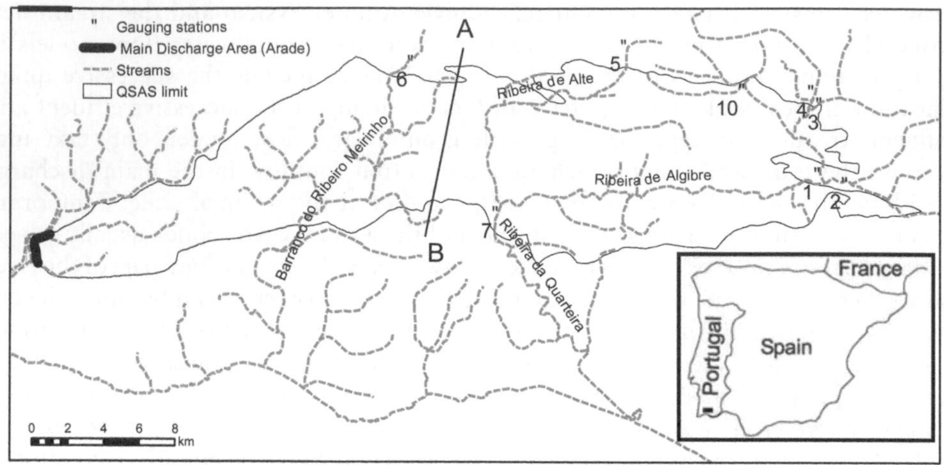

Figure 24.1 Stream network of the central Algarve in relation with the Querença-Silves Aquifer System and location of the gauging stations where stream discharge is monitored.

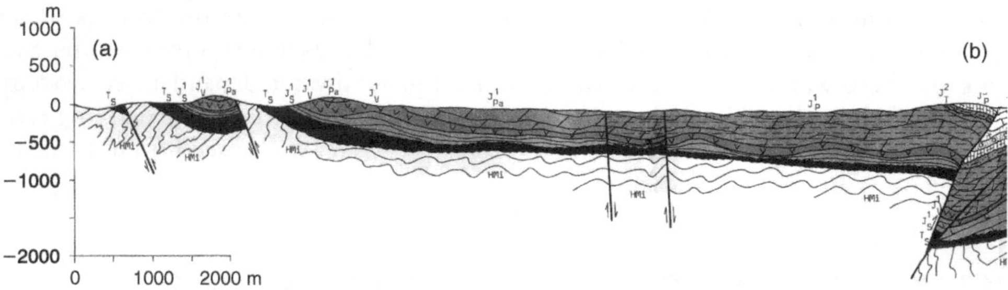

Figure 24.2 Geological cross-section of the Jurassic lithologies supporting the Querença-Silves Aquifer System (J^1_{Pa} – Sinemurian and J^1_P – Carixian to Toarcian carbonate rocks) and the surrounding formations: Carboniferous basement (HMi), Triassic red sandstones (T_s), Hettangian evaporites (J^1_s) and volcano-sedimentary series (J^1_v). Adapted from Manuppella, 1992. Points A and B are located on the map of Figure 24.1.

mainly of karstified Lower Jurassic dolomites with little bedding or other sedimentary structures evident. The morphology is hilly, the ground is rocky and scattered poljes are the site of soil accumulation and hence the best place for agriculture in this region. The cross section (Fig. 24.2) is located along the points A and B in (Fig. 24.1) and shows the geometry of the Jurassic lithologies supporting the QSAS (J^1_{Pa} – Sinemurian and J^1_P – Carixian to Toarcian carbonate rocks) and the lithologies defining its limits.

Due to the hydrological and hydrogeological conditions, the streams in hydraulic connection with the Querença-Silves Aquifer System include the three types of reaches present at the regional scale in the Algarve Region: (1) upstream of the aquifer, in the low permeability Paleozoic rocks, where a rainfall-runoff surface water provides a reasonable approximation for description of surface water flow; (2) in the area of the Querença-Silves Aquifer System, where overland flow toward streams is negligible, due to the high infiltration rates; streamflow is strongly dependent on hydraulic connection with groundwater in several influent and effluent stream reaches and (3) the downstream reaches where streamflow toward the sea occurs over areas with variable infiltration rates and, in some cases, contributing to water transferences toward other aquifer systems located downstream of the Querença-Silves Aquifer System. In these cases effluent reaches in the area of aquifer constitute secondary outflow areas, in addition to the main discharge area where boundary conditions are imposed in the models implemented to simulate the groundwater flow pattern at the regional scale.

The monitoring network used to analyse the local-scale balance of river aquifer transferences was installed by the water basin authorities (currently *Administração de Região Hidrográfica do Algarve – ARH Algarve*) and consists of the gauging stations for sequential measurements of stream discharge (Fig. 24.1) (Reis *et al.*, 2007). These gauging stations were installed to allow the quantification of water entering in the streams located in the northern area of the Querença-Silves Aquifer System (upstream of the aquifer) and downstream, near the southern limits of the system. Based on their location it is expected that these monitoring points allow the estimation of the local balance of each stream. The gauging stations (Fig. 24.1) started only in 2005 and the available datasets are limited. Field work was identified influent and effluent reaches of the Ribeira do Algibre stream (Fig. 24.1), although difficult to identify in the field, particularly the influent points, which are only active during short periods after storms. Effluent points are more obvious, as they are active during long periods in spring and summer, when baseflow is the only component of active streamflow. Field campaigns are ongoing in order to map all the possible influent and effluent areas in the stream network in the area of the Querença-Silves Aquifer System. The georeferenced points already identified for the Ribeira do Algibre are mapped in Figure 24.3 and represented in profile in Figure 24.4.

The integration of the hydrographs (examples are shown in Fig. 24.5) obtained in the gauges located in points 1, 2 and 7 (Fig. 24.1) allows the quantification of a total discharge of 23.3×10^6 m^3 (sum of gauges 1 and 2) and 25.1×10^6 m^3, (gauge 7) for the 2005/2006 hydrological year. These values show an outflow volume in the order of 1.8×10^6 m^3, transferred from the Querença-Silves Aquifer System toward the reach of the Algibre Stream between these gauging points. For the 2006/2007 hydrological year, the total transfers were 14.8×10^6 m^3 (sum of gauges 1 and 2, upstream) and 13.0×10^6 m^3, (gauge 1, downstream). Thus, in the 2006/2007 hydrological year a volume of recharge in the order of 1.8×10^6 m^3 occurred from the reach between these gauges toward the Querença-Silves Aquifer System. These differences show that the Algibre River can export water from the aquifer, or import water derived from runoff generated upstream, on the Paleozoic rocks of the Serra region. The estimated average volume of annual recharge associated with the Algibre Stream is in a range between 4.2×10^6 m^3/year and 16.4×10^6 m^3/year (including contributions from non-gauged tributary streams), depending on the applied methodology. The monitoring network for stream flow is new and not all gauging stations have data. For example, there are some gaps in the records of gauging station 2 and the actual datasets only provide a short and incomplete record. In these conditions, recent studies dedicated to the investigation of river-aquifer interactions in the Querença-Silves Aquifer System presented calculations of stream flow generated upstream of the aquifer in the Serra

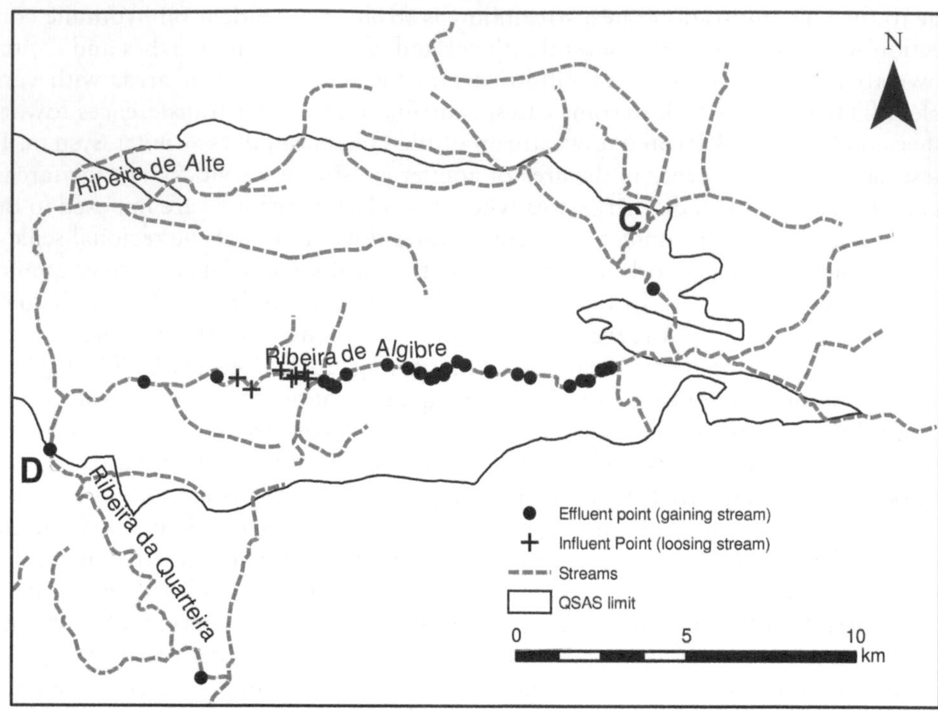

Figure 24.3 Map of the Ribeira de Algibre Stream, showing reaches with different river-aquifer relations, according to the lithologies: influent (+) and effluent (•) reaches identified in the field.

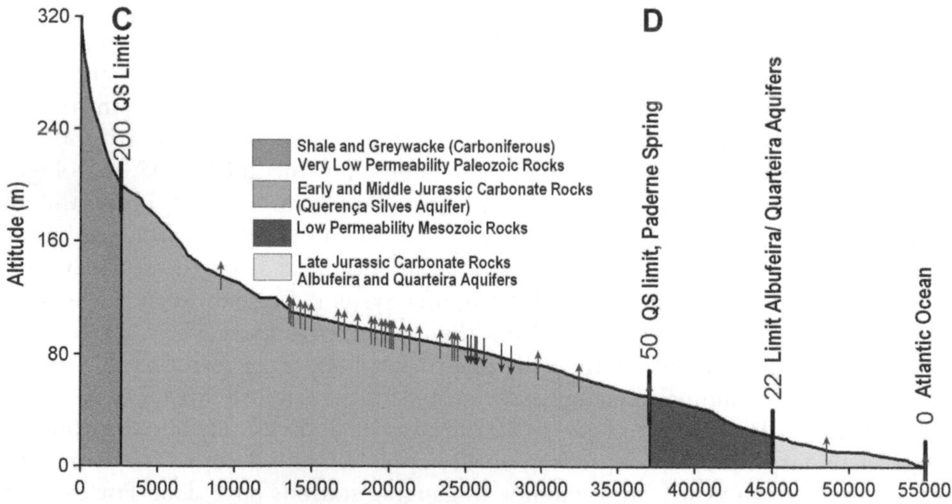

Figure 24.4 Schematic cross-section of the Ribeira de Algibre Stream, showing reaches with different river-aquifer relations, according to the lithologies. Arrows represent influent (head down) and effluent (head up) reaches identified in the field. Distances are in metres. Location of points C and D are located, respectively, at the upstream and downstream limits of the Querença-Silves Aquifer System intercepted by the Algibre stream shown in Figure 24.3.

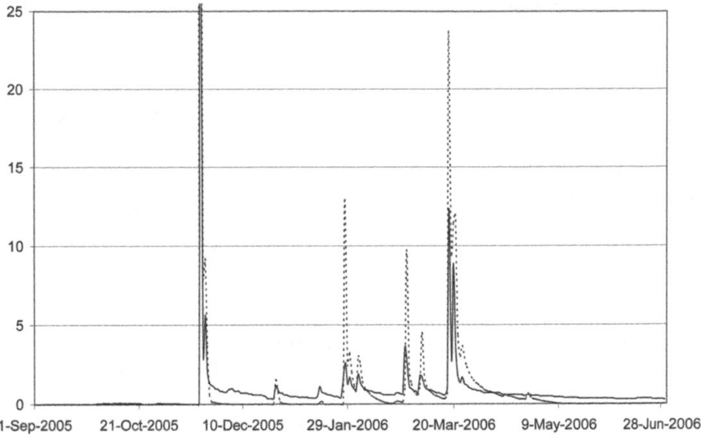

Figure 24.5 Hydrographs of the Algibre Stream in gauging station 1 (solid line) and 7 (dashed line). The hydrographs are truncated to keep the figure legible. Maximum discharge is 36 m^3/s for gauge 7 and 102 m^3/s for gauge 1 in November 2005.

Region (based on precipitation, evapotranspiration and infiltration values) and values of baseflow, obtained by the use of hydrograph separation techniques, for the case of streamflow datasets corresponding to stream gauges located near the downstream limits of the Querença-Silves Aquifer System (Salvador *et al.*, 2012a,b). It is expected that the increase of historical data in stream gauges and thus the increase of accuracy of estimation of water volumes involved in recharge and baseflow could be used to obtain information about the variables necessary for the calibration of parameters, allowing the simulation of surface water-groundwater interactions, using a numerical groundwater flow model.

24.3 METHODS

24.3.1 Inverse calibration of the groundwater flow model

First attempts to synthesise the hydraulic behaviour of the Querença-Silves Aquifer System at the regional scale using numerical models were exclusively based on previously available data regarding: (1) geometry, (2) boundary conditions, (3) recharge and discharge rates (4), spatial distribution and temporal evolution of state variables and (5) hydraulic parameters obtained in pumping tests performed at the well scale. However, it is well known that hydraulic parameters obtained from pumping tests in individual boreholes are not adequate representations of aquifers. Therefore, the simulations presented in this work are based on a synthetic bi-dimensional numerical representation of the aquifer, where the conductive parameter transmissivity (T) was estimated by inverse modelling for zones where the behaviour of piezometers allows a reasonable fit of field data using a single value of transmissivity. The results obtained allowed an important improvement of the reliability of the simulation of the regional

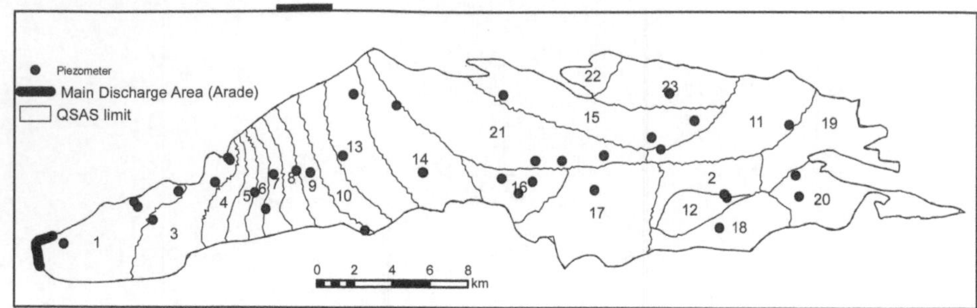

Figure 24.6 Map with zonation established for estimation of transmissivity (T) and location of the points where piezometric data are collected in the field for model calibration.

flow pattern observed at the Querença-Silves Aquifer System, particularly regarding a better characterisation of the spatial distribution of the hydraulic head.

The defined conceptual flow model was translated to a finite element mesh with 11 663 nodes and 22 409 triangular finite elements. The physical principles at the basis of the simulation of the hydraulic behaviour of the Querença-Silves Aquifer System are expressed by:

$$S\frac{\partial h}{\partial t} + \mathrm{div}(-[T]\,\overrightarrow{\mathrm{grad}}\,h) = Q \tag{24.1}$$

where T is transmissivity $[L^2T^{-1}]$, h is the hydraulic head [L], Q is a volumetric flux per unit volume $[L^3T^{-1}L^{-3}]$, representing sources and/or sinks and S is the storage coefficient [–]. In steady state conditions the variables are time-independent. In this case equation 1 is reduced:

$$\mathrm{div}(-[T]\,\overrightarrow{\mathrm{grad}}\,h) = Q \tag{24.2}$$

The direct solution of equations 1 and 2 was implemented using a standard finite-element model based on the Galerkin method of weighted residuals The software used was a continuously improved version of the open source Fortran code FEN (Kiraly, 1985), as well as FEFLOW, a commercial software for finite element groundwater flow modelling (Diersh, 2002). The simulations were systematically performed using both codes, which provided the same results. The exceptions happen only in the case of human errors and, therefore, the duplication of calculations using different software is a quality control test. The calibration of transmissivity was performed by inverse modelling, using the Gauss-Marquardt-Levenberg method, implemented in the nonlinear parameter estimation software PEST (Doherty, 2002). Additionally to the zonation of recharge areas, sub-areas were also defined for the characterisation of T (Fig. 24.6). Individual values of T were defined for 23 zones where the behaviour of piezometers allows a reasonable fit of field data using a single value for this parameter. The optimisation of the results was based on 1000 model runs performed by inverse modelling.

Table 24.1 Calculated values of transmissivity for each zone defined in the Querença-Silves Aquifer System.

Zone	*T [m²/s]*	*Zone*	*T [m²/s]*
t1	4.00E+00	t12	2.21E−03
t2	1.58E+00	t13	3.08E+00
t3	4.00E+00	t14	4.00E+00
t4	3.11E+00	t15	5.00E−04
t5	2.73E+00	t16	4.00E+00
t6	2.00E+00	t17	4.04E−02
t7	1.50E+00	t18	7.69E−04
t8	2.64E−01	t19	3.63E−02
t9	1.55E+00	t20	1.92E−02
t10	1.53E+00	t21	2.43E−03
t11	2.33E−03	t22	5.00E−04
		t23	5.00E−04

Values of T obtained from the inverse calibration in each zone are listed in Table 24.1.

24.3.2 Introduction of boundary conditions for identification of influent and effluent reaches of streams

The calibration of the finite element model was based on a conceptual flow model, for which the local scale interactions between groundwater and the stream network were ignored. In the first variants of the model, boundary conditions were restricted to the hydraulic connection of the Querença-Silves Aquifer System with the estuary of the Arade River at its western limit. This representation of the area was adequate for the analysis of water management at the aquifer scale. Transfers to surface water bodies were only considered at the boundaries where water is 'definitively exported' from the aquifer. Thus, the analysis of water balance in these simulations does not quantify water exported and imported in successive effluent and influent reaches of the streams within the aquifer boundaries. The regional-scale representation of aquifer balance was assumed prior to the work of Reis *et al.* (2007), Salvador *et al.* (2012a) and Salvador *et al.* (2012b), because it was assumed that these local scale transfers would not significantly affect the overall water balance of the aquifer at the regional scale.

Taking into account the knowledge provided by the description of the position of the effluent and influent reaches in the Algibre River (Figs. 24.3 and 24.4), the second phase of the implementation of the model consisted of the inclusion of the entire stream network present in the Querença-Silves Aquifer System, with the aim of simulating the local-scale river-aquifer transfers. The study of water balances at the local scale of the most important water courses in the Querença-Silves Aquifer System area is important because these constitute temporary Mediterranean GDEs. Initial estimates of these transfers assumed that the groundwater system is hydraulically connected with streams along the entire length of each reach. The simulation of the relations between surface water and groundwater was implemented by the imposition of specified heads (type 1 or Dirichlet boundary condition) at the location of each node of the model corresponding to the geometry of the stream network. The analysis of the stream network in the the Querença-Silves Aquifer System shows that the water courses with relevant stream-aquifer transfers are the Algibre, Alte and Meirinho streams and some tributaries of the left bank of the Arade River. Therefore boundary conditions were imposed on these water courses and not on other minor streams, in which no field evidence was found for the occurrence of surface-groundwater transfers.

24.4 RESULTS

24.4.1 Simulation of the regional flow pattern and the long-term steady state water balance

The regional-scale balance of the Querença-Silves Aquifer System was estimated based on infiltration rates proposed by Vieira & Monteiro (2003) for the different areas where carbonate rocks occur as outcrops or covered by different types of sedimentary deposits. The climate series used for the average long-term steady state balance consists of a dataset calculated for the period 1959/60–1990/91 (Nicolau *et al.*, 2000; Nicolau, 2002). According to values calculated by Nicolau (2002) for Portugal, using methodologies discussed in Nicolau *et al.* (2000), the precipitation in the area of the Querença-Silves Aquifer System in the 1959/60–1990/91 period is 653 mm/year. These values were obtained using an orthogonal grid with a resolution of 1×1 km and calculated by Kriging with elevation as external drift. This method proved to be the best-suited option among different auxiliary variables and resolution for the characterisation of the physiographic factors affecting the spatial distribution of rainfall in Portugal.

For the areas where carbonate rocks are covered by sedimentary deposits, recharge rates were estimated on the basis of the convergence of calculated values of potential and actual evapotranspiration, using the Coutagne, Turc and Thornthwaite methods and the transient water balance in the soil for different values of field capacity (Almeida, 1985; Andrade 1989). In the case of the outcrops of carbonate rocks, the recharge rates were estimated using the Kessler method (Kessler, 1965).These techniques were extensively applied in the southern area of Portugal and showed a good adaptation to the particular climatic conditions of the region (Vieira & Monteiro, 2003). A synthetic representation of data related to these pre-processing features, used to characterise the conceptual flow model, is presented in Figure 24.7.

According to the spatially distributed values of precipitation and the recharge rates (Fig. 24.7), the estimated average recharge for the Querença-Silves Aquifer System is of 93×10^6 m^3/year. The study of the water balance of the Querença-Silves Aquifer System is of regional importance as it supplies part of the Algarve urban water supply as well as agricultural use. Available data show that the maximum value of extraction occurred in the hydrological year 2004/2005. Extractions in the Querença-Silves Aquifer System in that period were in the order of 52.7×10^6 m^3/year (Nunes *et al.*, 2006).

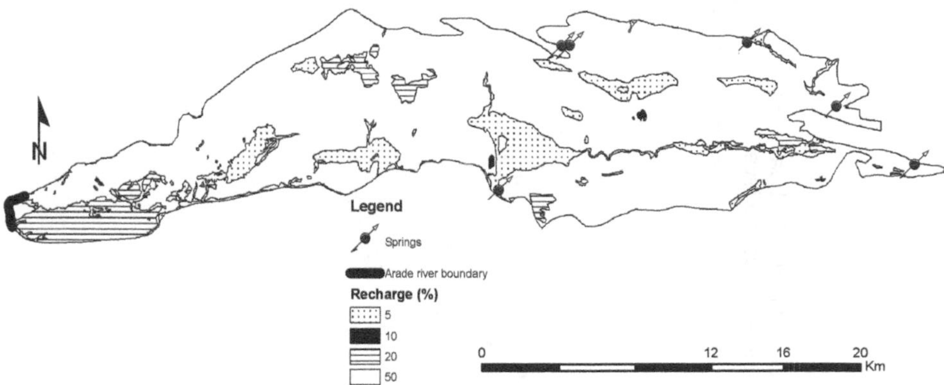

Figure 24.7 Spatial distribution of recharge expressed as percentage of precipitation and boundary conditions. Points represent springs and the bold line, at the western limit, represents the main aquifer discharge area, corresponding to a reach of the Arade River with highly effective hydraulic connection with the Querença-Silves Aquifer System.

Table 24.2 Hydraulic heads monitored in piezometers in the field, calculated by the model, and calculated errors (measurements in meters).

Obs. Pt.	*Measured*	*Model*	*Error*	*|Error|*	*Obs. Pt.*	*Measured*	*Model*	*Error*	*|Error|*
137	267.19	228.29	38.90	38.90	7905	16.40	31.61	−15.21	15.21
527	124.50	154.99	−30.49	30.49	8596	16.33	11.88	4.45	4.45
868	111.81	121.75	−9.94	9.94	8928	13.35	10.54	2.81	2.81
1101	112.34	92.00	20.34	20.34	9034	13.10	10.65	2.45	2.45
2127	40.21	59.24	−19.03	19.03	9418	12.97	10.88	2.09	2.09
2710	128.94	132.79	−3.85	3.85	9572	11.25	10.38	0.87	0.87
3059	86.47	87.40	−0.93	0.93	9823	10.75	10.13	0.62	0.62
3372	87.85	87.86	−0.01	0.01	10001	11.00	9.98	1.02	1.02
3619	43.87	43.89	−0.02	0.02	10050	10.70	10.23	0.47	0.47
3677	26.00	26.39	−0.39	0.39	10195	10.32	9.71	0.61	0.61
4416	115.81	110.67	5.14	5.14	10211	10.31	9.56	0.75	0.75
4599	104.06	104.19	−0.13	0.13	10364	10.29	9.42	0.87	0.87
4788	23.33	22.09	1.24	1.24	10618	7.02	7.33	−0.31	0.31
5065	78.25	69.71	8.54	8.54	10875	5.20	6.03	−0.83	0.83
5276	106.31	111.69	−5.38	5.38	10965	5.15	5.48	−0.33	0.33
6425	20.53	29.73	−9.20	9.20	10984	4.10	5.40	−1.30	1.30
7767	14.49	13.06	1.43	1.43	11551	2.57	3.68	−1.11	1.11
					Average				5.62
					Min.	2.57	3.68		0.0078
					Max	267.19	228.29		38.90

The steady state calibration of the model, using the zonation of transmissivity characterised in Figure 24.6 and Table 24.1, accommodating the regional-scale calculated water balance, allowed an accurate description of the flow pattern in the Querença-Silves Aquifer System, as revealed by the errors of the simulated hydraulic heads at the wells of the piezometric monitoring network (Table 24.2 and Fig. 24.8).

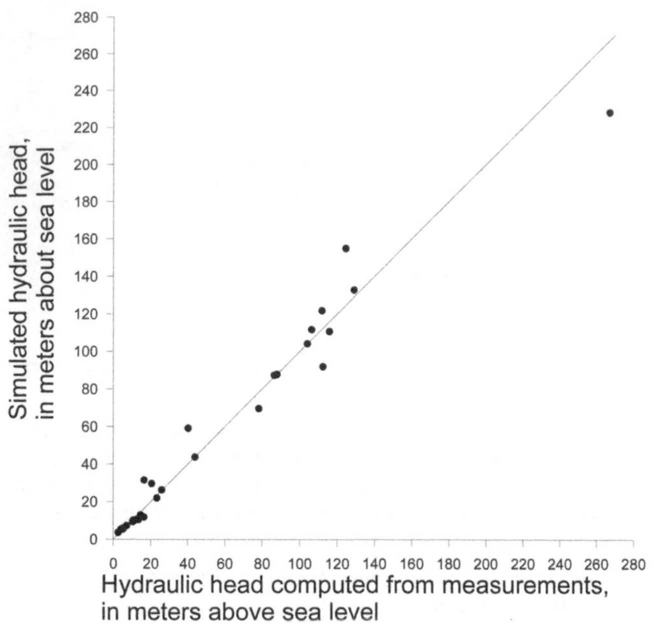

Figure 24.8 Relation between hydraulic head computed from observation points and simulated using the numerical model.

24.4.2 Simulation of transient outflow in the main discharge area of the QSAS

The initial objective of the simulation of the regional scale balance of the Querença-Silves Aquifer System was to evaluate the risks related to the occurrence of gradient inversions at the western aquifer boundary (the main discharge area of the Querença-Silves Aquifer System) and, thus, assess the risk of salt-water intrusion. Periods when estuary water enters the aquifer were confirmed by field observations. However, these transfers are not responsible for degradation of the water quality, as they result from the diurnal influence of the tide (Stigter *et al.*, 2011). Available datasets of a specific monitoring network installed in this area include the monitoring points of *ARH Algarve* for the entire aquifer (Fig. 24.6). Additionally, four continuous groundwater monitoring points equipped with CTD divers, collecting electrical conductivity (EC), temperature and water depth data in groundwater wells located within the fresh/saltwater interface were installed in 2005. The monitoring of this area since the severe drought that affected the region in 2004 and 2005 shows that the movement of the interface depends mainly on the variation of regional aquifer recharge and abstraction rates for irrigation and public water supply. Locally, groundwater heads are also influenced by the tidal fluctuations in the Arade estuary.

The analysis of the datasets presented by Stigter *et al.* (2011) show that during the severe drought of 2005, salinities of groundwater discharging from the springs

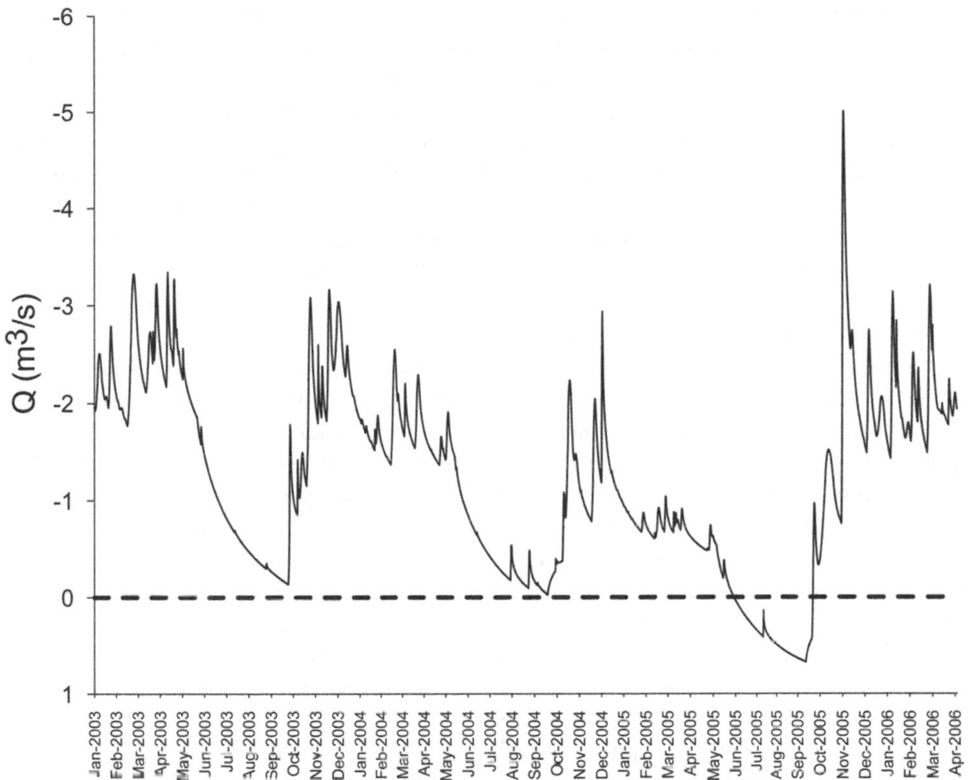

Figure 24.9 Hydrograph of transferences between the Querença-Silves Aquifer System and the Arade River simulated using daily values of recharge for a period between 1 January 2003 and 24 April 2006.

toward the Arade River were extremely high. This was due to low freshwater discharges and intense mixing with seawater entering the estuary (attaining EC values in the order of 30 000 μS/cm). However, datasets also shows a completely different picture at the end of the rainy season of 2008, when discharges from the aquifer were much higher, providing large amounts of freshwater to the Arade discharge area, with maximum values of EC between 5000 and 10 000 μS/cm (Stigter *et al.*, 2011). The field measurements presented by Stigter *et al.* (2011) for 2009 and 2010 showed intermediate situations. These data reflect the high resilience of the Querença-Silves Aquifer System during periods of drought and intense aquifer exploitation, rapidly returning to normal conditions in wet and even normal hydrological years.

In order to simulate the transient water transfers between the aquifer and the reach of the Arade River, which constitutes the main discharge area of the Querença-Silves Aquifer System, a transient variant of the regional finite element was implemented. The hydrograph (Fig. 24.9) shows the simulated transfers between the aquifer and the Arade River in the period between January 2003 and April 2006, including the drought period of 2004–2005. Negative discharges in this period represent outflow from the aquifer and positive values represent water volumes entering the aquifer, as result of gradient inversions between the Querença-Silves Aquifer System and the Arade River. This simulation was performed with a single value of the storage coefficient (S) of 5×10^{-2} [–] for the entire aquifer system flow domain. The gradient inversions are generated by the influence of the pumping rates of 193 wells, some of which pump all year (for water supply – water wells used for irrigation are only active between the last week of May and the end of September). Part of the seasonal period of irrigation, in which extraction in water wells increases in a period without recharge, is responsible for the simulated inversion of hydraulic gradients (Fig. 24.7). This inversion would cause saltwater from the estuary of the Arade River to enter the aquifer, but this has not been observed. The reliability of these modelled inversions of gradient is, of course, strongly dependent on the value of S.

Several variants of spatial distribution of S were tested in order to obtain the best fit of the temporal evolution of hydraulic head monitored in the available piezometric monitoring network. The different variants of the model, regarding the values and spatial distribution of S are presented in Stigter *et al.* (2011) and Hugman *et al.* (2012) and are not discussed further here. In general terms Hugman *et al.* (2012) were able to calibrate the Querença-Silves Aquifer System flow domain defining eight separate zones, the location of which was intersected with the areas of equal T (Fig. 24.6). The results from the distribution of S, varying between 7.5×10^{-2} and 1.0×10^{-3}, showed a satisfactory fit with observed head time series for most of the observation wells. The calibration of S allowed the elimination of gradient inversion (not detected in the experimental data). The transient model reflects both the spatial distribution and temporal evolution data for hydraulic head in piezometers, as well as the evolution of water quality. However, the spatial distribution of hydraulic parameters characterising the flow domain of the Querença-Silves Aquifer System was obtained with an overall water balance based on the autogenic recharge (derived from precipitation directly onto the karst landscape) as the only source of water input. Therefore, a risk exists associated with the actual estimates of T and S, which neglect the importance of allogenic recharge as a relevant term of the Querença-Silves Aquifer System balance. In these conditions, despite the actual acceptable accuracy of the transient simulations, it necessary to investigate if the improvement of the local-scale balance of the streams in hydraulic connection with this aquifer system could be responsible for an improvement of reliability, particularly in the case of extreme climatic events. The combined results of monitoring and modelling show that the resilience of the Querença-Silves Aquifer System during drought periods could be related to the additional allogenic recharge, derived from runoff generated upstream of the aquifer, rather than just the hydraulic properties.

24.4.3 Simulation of river-aquifer transferences at the local scale

The calibration of the numerical model was based on a conceptual model, for which the local-scale interactions between groundwater and the stream network were ignored. Therefore, the analysis of the water balance in these simulations excludes the quantification of water exported and imported in successive effluent and influent reaches of the rivers. However, the results presented in Salvador *et al.* (2012a) and Salvador

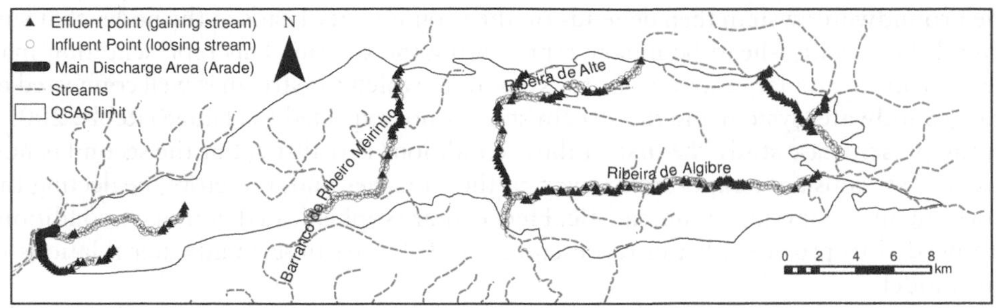

Figure 24.10 Stream network in the area of the model and the identification of the simulated distribution of influent (triangles) and effluent (circles) reaches of the rivers.

et al. (2012b) indicate that these local-scale transfers are relevant to the overall water balance of the aquifer.

As expected, the representation of the hydraulic river-aquifer relations using a type 1 boundary condition fails in the reaches of the rivers with elevations that are too high to be in hydraulic connection with the Querença-Silves Aquifer System. In these cases, transfers from streams to the aquifer are overestimated. The identification of the error associated with the use of type 1 boundary conditions is possible, to some extent, by the visual identification of these elevated reaches. This exercise depends on the accuracy of the available piezometric datasets to characterise the spatial distribution of hydraulic head in the Querença-Silves Aquifer System. Despite the limitations of a simulation of river-aquifer interactions using a type 1 boundary condition, the variants of the model using this conceptualisation are useful as they provide an image of the possible locations of influent and effluent reaches of watercourses. The representation of the location of the effluent and influent reaches, corresponding to the nodes with negative and positive flux values in the finite element mesh are shown in Figure 24.10. The analysis of the distribution of positive and negative values present a general pattern that is compatible with the effluent and influent reaches identified in the field (Figs. 24.3 and 24.4). It was observed that the annual average volumes of transfers between the Ribeira do Algibre and the Querença-Silves Aquifer System are of the same order of the volume of transfers estimated using data from the stream gauges and analysis of climatic and hydrologic variables for the Ribeira do Algibre (Salvador *et al.*, 2012a; Salvador *et al.*, 2012b).

In addition to the selection of the reaches to be considered as hydraulically connected with the Querença-Silves Aquifer System, there are other sources of error when specified heads are used as boundary conditions for the simulation of river-aquifer transferences. The most significant source is related to the assumption that the stage of the stream does not vary with water transfers between the river and the aquifer (Reilly, 2001). When a stream is represented as a specified-head boundary, nodes in the model where the stream is located are simulated with a head that is constant and set at the stage of the stream. It implies that there is no head loss between the stream and the groundwater system and that the flow of groundwater into or from the stream will not affect the stage of the stream. The amount of water flowing between the stream and the groundwater system then depends on the groundwater heads in the nodes that surround the specified-head boundary representing the stream. This representation may be appropriate for large streams or for systems in which the stream is well connected to the groundwater system and the stream stage is not expected to change (Reilly, 2001). In the present case study the first of these conditions is realistic, but the second is not, because streams are temporary in most of their reaches and, therefore, neglecting the lowering in stream stage is unrealistic. Hence, more sophisticated boundary conditions are needed to provide a better representation of the stream-groundwater relations in this model.

The ongoing research for the improvement of the actual state of knowledge of the stream-aquifer interactions in the Querença-Silves Aquifer System is based on further development of the conceptualisation of the boundary conditions representing these interactions. The reliability of the simulations depends on the combined use of models and integrated planning of monitoring of hydraulic head in piezometers and stream discharge. Most models are not well equipped to deal with mechanisms operating at the groundwater-surface water interface (Sophocleous, 2002). However the different kinds of boundary conditions available for groundwater flow models allow the definition of strategies to investigate the more adequate conceptualisation of different types of stream-groundwater interactions. According to Reilly (2001), depending on each particular case a stream may be represented as: (1) a specified-head boundary (also known as a Type 1 or Dirichlet boundary); (2) a specified-flow boundary (also known as a Type 2 or Neumann boundary); (3) a head-dependent or 'leaky' boundary (also known as a Type 3 or Cauchy boundary) and (4) non-linear variations of the 'leaky' boundary. The available information for the present case study allows the identification of situations where the use of different kinds of boundary conditions improves the implemented model. The actual state of development of the implemented models allows the incorporation of all these kinds of boundary conditions, according to the conveniences of the available field information. As the gauging stations (Fig. 24.1) started operating only in 2005, the available datasets of stream discharge are limited. However, it is expected that during the following years monitoring data will start to include drought periods as well as wet and average hydrological years. This will allow adapting the models according to the best technical solutions for dealing with the available monitoring data.

Using the notation proposed by Diersh (2002), the application of Type 3 or Cauchy boundary conditions consists of the use of a fluid transfer coefficient Φ_h (leakage parameter), whose value can be different when affecting transfers from the aquifer to the river ($\Phi_h^{out} =$ for $h^R \leq h$) or from the river toward the aquifer ($\Phi_h^{in} =$ for $h^R \geq h$), where h^R are the prescribed boundary values of hydraulic head h. Therefore, if the influent and effluent reaches are mapped the values of this parameter can be adjusted in order to control the intensity of transferences.

Another aspect resulting from the observation of the available stream monitoring datasets is the importance of the contribution of water input in the Querença-Silves Aquifer System. This is a result of concentrated allogenic recharge resulting from runoff generated in the upland section of the Algarve streams, where the drainage density is high, due to the low infiltration rates in the thick turbiditic sequence of shale and greywacke present upstream of the Querença-Silves Aquifer System.

In the cases where the values of concentrated allogenic recharge are known the best way to simulate the water input is by using specified-flow boundaries. This is possible when data are available to quantify the volume of water input in these points. This representation may be appropriate for streams that are disconnected from the ground-water system (Reilly, 2001), such as streams at high altitudes that lose their water as they enter the valley deposits from the mountains or, as in the present case, when they enter in high permeability lithologies flowing from areas with low recharge rates.

24.5 CONCLUSIONS

The approach used to simulate stream-aquifer hydraulic relations was based on a sequence of operations started by the implementation of a bi-dimensional numerical representation of the Querença-Silves Aquifer System. The conductive parameter transmissivity (T) was estimated by inverse modelling for 23 zones, using the Gauss-Marquardt-Levenberg method, implemented in a non-linear parameter estimation software, coupled to a standard finite-element model based on the Galerkin method of weighted residuals. The introduction of the stream network in this model allowed the identification and systematisation of the parameters and variables needed to improve the conceptualisation of the river-aquifer relations.

The study of water balances at the local scale of the more important water courses in the Querença-Silves Aquifer System is important because they constitute temporary Mediterranean GDEs. The only way to calibrate and validate a model with the characteristics described for this case study is by designing a specific monitoring network, coupling, at least, piezometric and stream discharge data, respectively, in the aquifer and at selected points of the stream network. Despite the absence of a monitoring network specifically designed with this objective, the Querença-Silves Aquifer System has been more or less continuously monitored for piezometric data in 34 observation points in the last 20 years. More recently, the discharge of some of the more important rivers and streams started to be gauged in specific locations, allows control of the variables at stake at a reasonable number of points. The numerical experiments were designed to deal with the available information and to plan the design of future monitoring networks. This will allow a deeper analysis of the relations between rivers and aquifers in complex karst environments in the future, whereby regional groundwater and surface water flows will be coupled in a complex regional pattern.

REFERENCES

Almeida, C. (1985) *Hidrogeologia do Algarve Central [in Portuguese]. Hydrogeology of Central Algarve.* Ph.D., Universidade de Lisboa. Dep. Geologia. FCUL, Portugal. 180 pp.

Andrade, G. (1989) *Contribuição para o Estudo da Unidade Hidrogeológica Tôr-Silves [in Portuguese]. Contribution for the Study of the Tôr-Silves Hydrogeologic Unit.* Diss. para Obt. do Grau de Mestre em Geologia Económica e Aplicada. Departamento de Geologia da FCUL, Lisboa, 1989. 179 pp.

Arnold, J.G. & Fohrer, N. (2005). SWAT2000: Current Capabilities and Research Opportunities in Applied Watershed Modeling. *Hydrological Processes*, 19 (3), 563–572.

Kessler, H. (1965) Water balance investigations in the karstic regions of Hungary. Act. Congress AIHS – UNESCO, 73, 91–105.

Diersh, H.J. (2002). *Feflow. Finite Element Subsurface Flow & Transport Simulation System. Reference Manual. Physical Basis of Modeling*. Berlin, Germany, WASY Institute for Water Resources Planning and Systems Research Ltd. 278 pp.

Doherty, J. (2002) PEST *Model-Independent Parameter Estimation*. 4th edition. Australia, Watermark Numerical Computing. 279 pp.

Field, M.S., Kraemer, S.R. & Palmer, A.N. (1999) *A Lexicon of Cave and Karst Terminology with Special Reference to Environmental Karst Hydrology*. Washington, DC, USA, Environmental Protection Agency, National Center for Environmental Assessment. 201 pp.

Hugman, R., Stigter, T.Y., Monteiro, J.P. & Nunes, L. (2012) Influence of aquifer properties and the spatial and temporal distribution of recharge and abstraction on sustainable yields in semi-arid regions. *Hydrological Processes*, 26, 2791–2801.

Kiraly, L. (1985) *FEM301, A Three Dimensional Model for Groundwater Flow Simulation*. NAGRA Technical report 84-89. 96 p.

Manuppella (coordinator, 1992) *Carta Geológica de Portugal na escala 1/100 000 [in Portuguese]. Geologic Map of Portugal at the Scale, 1, 100 000*. Lisboa, Serviços Geológicos de Portugal.

Monteiro, J.P., Vieira, J., Nunes, L. & Younes, F. (2006) *Inverse Calibration of a Regional Flow Model for the Querença-Silves Aquifer System (Algarve-Portugal). Integrated Water Resources Management and Challenges of the Sustainable Development*. International Association of Hydrogeologists, IAH, Marrakech. pp. 44, doc. Elect. CD-ROM, 6 pp.

Monteiro, J.P. & Ribeiro, L. (2006) *Modelação Matemática de Cenários de Exploração do Sistema Aquífero Querença-Silves. Calibração e Simulações de Escoamento. Relatório Técnico [in Portuguese]. Mathematical Modelling and Scenarious of Exploitation of The Querença-Silves Aquifer System. Calibration and Flow Simulations*. Technical Report. Instituto da Água (INAG). 26 pp.

Monteiro, J.P., Ribeiro, L. & Martins, J. (2007) *Modelação Matemática do Sistema Aquífero Querença-Silves. Validação e Análise de Cenários. Relatório Técnico [in Portuguese]. Mathematical Modelling of The Querença-Silves Aquifer System. Validation and Analysis of Scenarious*. Technical Report. Instituto da Água (INAG). 52 pp.

Nicolau, R., Ribeiro, L., Rodrigues, R., Pereira, H. & Camara, A. (2000) Mapping the spatial distribution of rainfall in Portugal. In: Kleingeld, W.J. and Krige, D.G. (eds.). *Geostats 2000*, South Africa. 11 pp.

Nicolau, R. (2002) *Modelação e Mapeamento da Distribuição Espacial da Precipitação – Uma Aplicação a Portugal Continental. [in Portuguese]. Modeling and Mapping Spatial distribution of Precipitation – The Portuguese Case Study*. Ph.D., FCT, Universidade Nova de Lisboa, Lisboa (Portugal), 356 p.

Nunes, G., Monteiro, J.P. & Martins, J. (2006) Quantificação do Consumo de Água Subterrânea na Agricultura por Métodos Indirectos. [in Portuguese]. Quantification of Groundwater Use in Agriculture by Indirect methods – Teledetection. *Detecção Remota. IX Encontro de Utilizadores de Informação Geográfica (ESIG). 15–17 de Novembro, Tagus Park, Oeiras*. Doc. Electrónico em CD-ROM. 15 pp.

Reilly, T.E. (2001) *System and Boundary Conceptualization in Ground-Water Flow Simulation*. Techniques of water-Resources Investigations of the United States Geological Survey. Book 3, Applications of Hydraulics. Reston, Virginia. 30 pp.

Reis, E., Gago, C., Borges, G., Matos, M., Cláudio, A., Mendes, E., Silva, A., Serafim, J., Rodrigues, A. & Correia, S. (2007) *Contribuição para o Cálculo do Balanço Hídrico dos Principais Sistemas Aquíferos do Algarve [in Portuguese]. Contribution to the Calculation of the Water Balance of the Main Aquifer systems in the Algarve*. Technical Report. Ministério do Ambiente, do Ordenamento do Território e do Desenvolvimento Regional, Comissão de Coordenação e Desenvolvimento Regional do Algarve. 41 pp.

Salvador, N., Monteiro, J.P., Hugman, R., Stigter, T. & Reis, E. (2012a) Quantifying and modelling the contribution of streams that recharge the Querenča-Silves aquifer in the south of Portugal. *Natural Hazards and Earth System Sciences*, 12, 3217–3227.

Salvador, N., Oliveira, M., Reis, E., Oliveira, L., Lobo-Ferreira, J.P. & Monteiro, J.P. (2012b) *Contribuição Para a Quantificação das Relações Rio-Aquífero no Sistema Aquífero Querença-Silves [in Portuguese]. Contribution for the Quantification of River–Aquifer Interactions in the Querença-Silves Aquifer System.* Associação Portuguesa de Recursos Hídricos (APRH). 11° Congresso da Água, Porto, 15 pp.

Sophocleous, M.A. (2002) Interactions between groundwater and surface water: the state of the science. *Hydrogeology Journal*, 10, 52–67.

Sophocleous, M.A. & Perkins, S.P. (2000) *Methodology and Application of Combined Watershed and Ground-Water Models in Kansas. Journal of Hydrology*, 236, 185–201.

Stigter, T.Y., Monteiro, J.P., Nunes, L.M., Vieira, J., Cunha, M.C., Ribeiro, L., Nascimento, J. & Lucas, H. (2009). Screening of sustainable groundwater sources for integration into a regional drought-prone water supply system. *Hydrology and Earth System Sciences*, 13, 1–15

Stigter, T., Monteiro, J.P., Nunes, L., Ribeiro, L. & Hugman, R. (2012) Regional spatial–temporal assessment of groundwater exploitation sustainability in the south of Portugal. In: Maloszewski, P.,Witczak, S., Malina, G. (eds.) *Groundwater Quality Sustainability.* International Association of Hydrogeology Selected Papers, 17, 153–162.

Stigter, T., Hugman, R., Monteiro, J.P, Ribeiro, L., Samper, J., Pisani, B., Yanmei, L., Fakir, Y. & Mandour, A. (2011) *Assessing and Managing the Impact of Climate Change on Coastal Groundwater Resources and Dependent Ecosystems. Work Package 3 – Groundwater Monitoring and Modelling. Climate Impact Research Coordination for a Larger Europe – Mediterranean Integrated Coastal Zones And Water Management.* Final Report. 88 pp.

Terrinha, P. (1998) *Structural Geology and Tectonic Evolution of the Algarve Basin, South Portugal.* Thesis submitted for the Degree of Doctor of Philosophy at the University of London, 430 pp.

Vieira, J. & Monteiro, J.P. (2003) *Atribuição de Propriedades a Redes Não Estruturadas de Elementos Finitos Triangulares (Aplicação ao Cálculo da Recarga de Sistemas Aquíferos do Algarve) [in Portuguese]. Attribution of Properties to Unstructured Triangular Finite Elements (Application for the Calculation of Recharge of Aquifer System of The Algarve).* As Águas Subterrâneas no Sul da Península Ibérica. Assoc. Intern. Hidrog. APRH publ. pp. 183–192.

Author index

Barton, A.B. 69
Bellin, A. 125
Bense, V. 81
Berens, V. 253
Bonte, M. 81
Brodie, R.S. 183
Butler, A.P. 295

Carvalho Dill, A. 153
Cendón, D. 57
Clifton, C.A. 1
Condesso de Melo, M.T. 221
Cook, P.G. 1
Cox, J.W. 69
Crowe, A.S. 237
Culver, D.C. 47
Custodio, E. 169

Dahl, M. 95
Dahlhaus, P.G. 69
Davies, M. 197
D'Elia, M.P. 137

Ellery, W.N. 9
Evans, L.R. 113
Evans, R.S. 1

Fitzpatrick, A.D. 253
Froend, R. 197, 207

Geris, J. 81
Green, R.T. 183

Hatch, M.A. 253
Herczeg, A.L. 69
Higueras, H. 169
Hinsby, K. 95
Holland, K.L. 253
Hollins, S. 57
Howden, N.J.K. 295
Howe, P. 1
Hughes, C. 57

Irvine, E. 1

Jolly, I.D. 253

Kooi, H. 81
Kotze, D. 9

Lewandowski, J. 23
Loomes, R. 207
Lorentz, S.A. 9
Lozano, E. 169

Mádl-Szőnyi, J. 267
Malta, E.-j. 153
Manzano, M. 169
Martin, M. 197
Martins, J. 307
Martín-Loeches, M. 281
Marzadri, A. 125
Matos Silva, J. 307
McEwan, K.L. 253
Meredith, K. 57
Milne, J. 237
Monteiro, J.P. 307
Munday, T.J. 253

Ngetar, S.N. 9
Nützmann, G. 23

Paris, Marta del C. 137
Peach, D.W. 295
Perez, M.A. 137
Pipan, T. 47
Post, V.E.A. 81
Pretorius, J.J. 9

Reis, E. 307
Ribeiro, L. 307
Riddell, E.S. 9

Salvador, N. 307
Santos, R. 153
Sena, C. 221
Simon, K.S. 47
Simon, S. 267
Sommer, B. 207
Souter, N.J. 253
Stigter, T.Y. 153
Stone, D. 57

Tóth, J. 267
Tujchneider, O.C. 137

van Dijk, H.J.A.A. 81

Wheater, H.S. 295
White, M.G. 253
Williams, R.M. 183

Yélamos, J.G. 281
Yesertener, C. 33
Youngs, J. 113

Subject index

Page numbers in *italic* denote figures. Page numbers in **bold** denote tables.

Aakaer Stream catchment area
chemical status *108*, 109–110
GSI response units 105, 107–109, 110
nitrates 107, *108*
Acacia caven 142, *143*
Acacia stenophylla 254, 263–264
agriculture
and degradation
Pateira de Fermentelos lagoon 224
Sand River 10
groundwater use
Alstonville Plateau 183, 185
Doñana aquifer system 172–173
Ria Formosa basin 157–158
Santa Fe province 138–139, 142
Aguas Santafesinas, groundwater extraction 142
Águeda River 222, *223*, 229
algae, Ria Formosa lagoon, nutrient sources 153–165
Algarve
Querença-Silves Aquifer system 307–323
Ria Formosa lagoon 153–165
Algibre Stream *310*, 311, *312*, 313, 321
Algibre thrust fault 309
Algyõ aquitard 269, *272*
Alstonville Plateau basalt aquifer 183, *184*
GDEs 192–195, *194*
river baseflow systems 193, *194*, 195
terrestrial vegetation 193, *194*
wetland seepage and springs 192–193, *194*
groundwater discharge 189, 191, 193
groundwater dynamics 189–191
groundwater extraction 190–191
groundwater flow model 191–192
groundwater levels 189–191
groundwater management 185–187
Water Sharing Plan (2003) 186
hydrogeological conceptual model 188–189
rainfall 186, 189, *190*
recharge 186–187, 189, 191
Amberley Beach 239
E. coli 245, 246, *247*, 248
groundwater 241, *242*, 243
ammonium, hyporheic zone 125–126, 128–134
Ammophila breviligulata 242
aquifers
basalt, Alstonville Plateau 183–195
Chalk 297
Darling River catchment 59, 64
Doñana 169–180
Duna-Tisza Interfluve 269, 271, *272*
Gnangara Groundwater Mound 33, 35, 44, 197, *199*, 207–210
landscape type GSI 97
Pilbara 116, 121–123
Querença-Silves Aquifer system 307–323
riparian area 98–104
denitrification capacity 104–105
Santa Fe province 139–140
Spanish Central System 282–283
aquitards
Darling River system 64
Duna-Tisza Interfluve 269, 271, *272*
Arade River 315, 318, 319–320
Ardea alba 142, *143*
Argentina, groundwater recharge 137–150
argiudol 141
arsenic, Pateira de Fermentelos lagoon 232

Artemisia campestris 242
Ashfield Township Beach 239
  E. coli 245, 246, 248
  groundwater 241, *242*, 243
Australia
  Barwon-Darling River 57–59
    stable isotopes 59–66
  Corangamite wetlands 69–78
  groundwater dependent ecosystems
    Alstonville Plateau 183–195
    EWR toolbox 1–7
  groundwater management policy 1, 2, 185–186
  Pilbara dewatering re-injection 113–123
  River Murray floodplain management 253–265
  Swan Coastal Plain
    Banksia woodland 197–205
    wetland vegetation 207–218
  Water reform framework 1, 185
  Yanchep Caves 33–45
Aveiro Cretaceous Groundwater Body 222

bacteria, nitrifying and denitrifying 126
Balm Beach 238, *243*
  E. coli 245, *247*, 248
  groundwater *240*, 241, *242*, 243
Banksia woodland 197–205
  Banksia attenuata 200
    drawdown experiment 201–205
  Banksia ilicifolia 200
    drawdown experiment 201–205
  Banksia menziesii 200
Barwon-Darling River 57–59
  cyanobacteria bloom 58, 65–66
  drought 57–58
  groundwater 63–64
  groundwater-surface water exchange 64–66
  rainfall 61
  run-of-river 64, *65*
  salinisation 64–65
  stable isotope studies 59–60, 62, *63*
  stream flow 61
  water abstraction 58, 59
basalt, Alstonville Plateau 183, 188–189
beach grass, American 242, 249, 250
beaches, Lake Huron
  groundwater *240*, 241, *242*, 243
    E. coli 237, 239–241, 244–250
    wet and dry ecosystems 242–244, *247*, 248–250
    human impact 244, 246, 249, 250
black bean 193
black box 254, 263–264
Bookpurnong
  "Living Murray" project 254, 255, 256–265
    salinisation 256–263
    vegetation response 263–265
  salt interception schemes 256, 257
boreholes
  Alstonville Plateau basalt aquifer *184*, 185, 189, *190*, 191, *194*, 195
  brackish water, Spanish Central System 287–289, *290*
  Lake Huron beaches 239
  Pateira de Fermentelos lagoon 224, *225*, *226*
  Pilbara mine dewatering 117, 122
  Pinjar Borefield *199*, 200
  Yanchep groundwater flow model 38–39, 44
Bourke, Barwon-Darling River *58*, 61, 62, 63
Brewarrina, Barwon-Darling River *58*, 61, 62, 63
Brockman Iron Formation 115, 116
bulrush 193, 243
Burtundy, Barwon-Darling River *58*, 60, 62
bushfires 213

calcite saturation, Yanchep Caves 37, 45
calcium
  Pateira de Fermentelos 228, 230, *231*
  River Piddle 304
Canada, Lake Huron beaches 237–250
carbon dioxide, partial pressure
  Pateira de Fermentelos lagoon 228, 230–231
  River Piddle 304
carbon, organic
  benthic *48*, 49, *51*
  dissolved (DOC) 47–55
  flux in caves 47–55
  particulate (POC) 47–49, *51*
Carex spp. 193, 217, 243
Casa Yunta borehole 287–288
Castanospermum australe 193
cat-tail 243

caves
  organic carbon flux 47–55
  see also Yanchep Caves; Crystal Cave
Central Graben *see* Roer Valley Graben
Cértima River 222, *223*
  ion concentration 227–229
chalk streams
  groundwater-surface water interaction 295–296
    River Piddle 299–305
chloride
  Bookpurnong trials 260
  Duna-Tisza Interfluve 272–273, 276–277
  Pateira de Fermentelos lagoon 228, 230, *231*
  River Spree 29–30, 31
  Santa Fe province groundwater 142–143, **144**, **145**, 146–147, 149
Chowilla Floodplain 254
chromium, Pateira de Fermentelos lagoon 233
Clark's Floodplain
  Bookpurnong "Living Murray" project 254, *255*, 256–265
  hydrogeology 254, *256*
clay-plugs, Manalana sub-catchment 10, 18, 20
clogging, River Spree GSI 28, 30
club-rush, round-headed 291
coastal groundwater discharge (CGD) 154
Colocasia esculenta 13
common reed 223, 243
contact types, GSIs 99, *100*
cooba 254, 263–264
Coogee Swamp *34*, 35
  MODFLOW simulation 38
Coolabah open woodland 116–117
Coonambidgal Clay Formation 254, *256*
copepods, and DOC, Postojna-Planina Cave System 50, 53, 54–55
Corangamite wetlands 69, *70*, 71
  ion concentration 73, *74*, 75–78
  lake classification 76–78
  rainfall 71
  salinity and stable isotopes 71–78
Črna jama 49, *50*, 52–53, *54*
Crystal Cave *34*
  groundwater *36*, 37, 40, 41–42
Ctenomys yolandae 142, *143*
cyanobacteria, Barwon-Darling River 58, 65–66
Darling River
  stable isotopes 57–66
  see also Barwon-Darling River
Darling River Drainage Basin *58*, *59*
  groundwater 63–66
  run-of-river 64, *65*
denitrification 104–105, 110–111, 126, 128–134
Denmark, Aakaer Stream catchment area 105–111
deuterium excess 76, *77*
dewatering, Pilbara 113
  re-injection 114–115, 117–123
    modelling 121–122
    monitoring 119, *120–121*
diffuse flow 101, *102*, 103–104
direct flow 102, 103–104
discharge
  Alstonville Plateau basalt aquifer 189, 191, 193
  Querença-Silves Aquifer system 318–320
  see also coastal groundwater discharge (CGD)
Doñana aquifer system 169–180
  eolian mantle 171, 175, *176*
  geology and hydrogeology 170–173
  groundwater flow 171–172
    chemistry 173–174, 176–177
    isotopic signature 174–175, 176–177
    and wetlands 175–180
  groundwater level 170, 178, 179
  rainfall 170, 171, 173
  recharge 171
  salinity 174, 175, 177, 180
  sediments 170–171
  surface water (lagoons) *174*, 175–177, 179–180
  vegetation 169–170, 178–180
  wetlands, role of groundwater 175–180
Dorvan-Cleyzieu basin, organic carbon flux 47
drainage flow 102, 104
drought, Barwon-Darling River 57–58
Duna-Tisza Interfluve *270*
  groundwater flow regimes 271–272
  groundwater salinisation 272–275
  hydrogeological environment 269
  soil and wetland salinisation 275–277
  tectonic evolution 269

ecological water requirements (EWRs), assessment toolbox, Australia 1–7
effluent 155, 158
  see also septic tanks; wastewater treatment
Eicchornia crassipes 223–224
El Salobral spring and well **287**, 288–289
El Sarguero spring **287**, 289, *290*
El Toro spring **285**, 286
electrical resistivity tomography (ERT), Manalana sub-catchment 13, 18, 20
Elymus lanceolatus 242
Endrod aquitard 269
eolian mantle, Doñana aquifer system 171, 175, *176*, *177*
epicormic growth, river red gums 263
epikarst 47, 48–49
  drips, Postojna-Planina Cave System 49–55
epilithon *48*, 49
erosion gullies, Sand River catchment 10, 11, *12*, 17
Escherichia coli, Lake Huron beaches groundwater 237, 239–241, 244–250
Esperanza
  groundwater recharge 147–148
  groundwater requirement 138–139, 142
eucalyptus, Doñana aquifer system 169, 170, 178–179
Eucalyptus camaldulensis 254, 263–265
Eucalyptus largiflorens 254, 263–264
Eucalyptus rudis 216
European Union
  Groundwater Directive (GWD) 95–96
  Water Framework Directive (WFD) 95–96
evaporation
  Barwon-Darling River 57, 63
  Corangamite wetlands 73, 76, 78
evapotranspiration
  Doñana Coastal Aquifer 170, 179
  River Piddle 296
  Swan Coastal Plain 208
  see also transpiration
exchange processes
  Barwon-Darling River 64–66
  River Spree 23–31
  see also groundwater-surface water interaction
exfiltration, River Spree GSI 25, *27*

faults
  Algibre thrust fault 309
  and groundwater flow 81–82
    Duna-Tisza Interfluve 271, *272*, 276
  Lower Rhine Graben 83
fertilisers, nitrogen effluent 155, 158, 304
floodplains, River Murray 253–254

geese, *E. coli* groundwater contamination 247, 248, 249, 250
Georgian Bay *see* Lake Huron, beaches
Glen Villa, Barwon-Darling River *58*, 63, 64, 65
Global Network for Isotopes in Rivers (GNIR) 60
Gnangara Groundwater Mound 33, 35, 44, 197, *199*, 207–210
  groundwater levels 209–210
  wetlands 210
Graspeel, temperature profile *84*, 85, *86*, 88, *89*
gravel-bed rivers, hyporheic zone 125
  flow modelling 126–128
  nitrates 126, 128–132
  nitrogen cycle modelling 128–132
Great Hungarian Plain
  groundwater flow 268–269
  hydrostratigraphy 269, *270*
Great Lakes *see* Lake Huron, beaches
Great Lakes Wheat Grass 242
Great Plain Aquifer 269, 270, 271, *272*, 273
groundwater
  artificial maintenance of, Yanchep Caves 33–45
  Lake Huron beaches, *E. coli* 237, 239–241, 244–250
  Santa Fe province
    chemical analysis 142–143, **144**, **145**
    water levels 142–143, **144**, 145, 147–148
  and wetlands, Doñana aquifer system 175–180
  see also discharge; human impact; recharge
groundwater body types 100–101
groundwater dependent ecosystems (GDEs)
  Australia
    basalt aquifers, Alstonville Plateau 192–195
    EWR toolbox 1–7
    Pilbara 113–114, 116–117, 123
  fault-induced seepage, Netherlands 82–91
  Santa Fe province 142
groundwater drawdown, response of *Banksia* woodland 197–205

groundwater flow
  Duna-Tisza Interfluve 271–272, 276–277
  impact of faults 81–83, 271, *272*, 276
  Lake Huron beaches *240*, 241
  River Spree 24, 25, 28
  Yanchep Caves 34, 35
groundwater-surface water interaction
  Barwon-Darling River 64–66
  chalk streams 295–296
    River Piddle catchment 299–305
  Querença-Silves Aquifer system 308–309, 320–323
  response units (GSI RUs) 105–106, 110–111
    Aakaer Stream catchment area 107–109, 110
  River Spree 23, 25–31
  temperature profiling 81–91
  typology 96–104, 110–111
    landscape type 97, *98*, 102–103
    riparian flow path *98*, 101–104, 110–111
    riparian hydrogeological 98–101, 103, 105, 111
Guadalquivir River 169, 170
gulls, *E. coli* groundwater contamination 247, 248, 249
Gum Creek, streamflow 191

Hamersley Group 116
headwaters, wetland, Sand River catchment 9–21
Holywell Nodular Chalk **298**, 300
Hottonia palustris 82
Huerto Bernal spring **285**, 286
human impact
  Alstonville Plateau basalt aquifers 183, 185, 190–191
  Barwon-Darling River 58, 59
  Corangamite Wetlands 71
  Doñana aquifer system 170, 172–173, 174
  Pateira de Fermentelos 224
  Querença-Silves Aquifer system 307–308
  Ria Formosa 153–154
  Santa Fe province 138–139, 140, 141–142
  Swan Coastal Plain 197, 198, 208–209
Hungary, Duna-Tisza Interfluve 267–277
hydrogen, stable isotopes
  Barwon-Darling River 59, 60, 62, *63*, 64
  Corangamite wetlands 73, *74*
hyporheic zone
  gravel-bed rivers 125
    flow modelling 126–128
    nitrates 126, 128–134
    nitrogen cycle modelling 128–132
    Transient Storage Model 126
  lagoons 221

infiltration, River Spree 25, *27*, 30
iron
  Pateira de Fermentelos 232, *233*
  River Spree 29, 31
  wijst seepage zones 82
irrigation
  Bookpurnong 254
  Doñana aquifer system 170, 171
  water requirements, Ria Formosa basin 157–158
isotopes *see* stable isotopes
Ituzaingó Formation 139, **140**

Jackson Park Beach 238
  E. coli 245, 246, *247*, 248
  groundwater *240*, 241, *242*, **243**
  wet ecosystem 243, *244*
Juncus acutus 288, *290*, *291*
  Spanish Central System 281, *283*, 284–292
    alkaline sulphide water 284, **285**, 286, 290–292
    brackish water 286–289, 290–292

karst aquifer
  Gnangara Groundwater Mound 35
  Querença-Silves Aquifer system 308
karst basins
  organic carbon flux 47–55
    conceptual model 48–49
Kaweide, temperature profile *84*, 85–86, 88, *89*
Kelemenszék Lake, salinisation *270*, 273, *274*, *275*
Klein Drakensberg Escarpment 10, 11
Knokerd, temperature profile *84*, 85, *86*

La Milagrosa spring **285**, 286
La Pólvora spring **285**, 286
La Rocina creek 171, *172*, *178*, 179
La Salabrosa spring and well **287**, 288–289

La Sima spring **285**, 286
lagoons
  Doñana aquifer system *174*, 175–177, 179–180
  Pateira de Fermentelos 221–234
lake classification, Corangamite wetlands 76–78
Lake Colac **72**, 78
Lake Corangamite 71, *72*
Lake Huron, beaches
  groundwater *E.coli* 237, 239–241, 244–250
  groundwater flow *240*, 241
Lake Jandabup 211, **212**, 213, 214–215
Lake Joondalup *209*, **212**, 213
Lake Mariginiup 211, **212**, 213, 214, 218
Lake Nowergup 211, **212**, 215–216, 217
Lake Wilgarup 211, **212**, 213, *215*, 216–218
Lake Yonderup **212**, 213, 214
Lamington Volcanics 188
landscape type GSI 97, *98*, 102–103
lawns, Great Lakes beaches 244, 249, 250
lead, Pateira de Fermentelos lagoon 232–233
Lexia wetlands 211, **212**, 213, 214, *215*, 216–218
Lismore Basalt 183, 188
livestock, groundwater use 138–139, 142
Llanura Pampeana 138
Loch McNess *34*, *35*, **212**, 213
  MODFLOW simulation 38, 43, 45
London Chalk Formation 297, **298**
Los Baños spring **285**, 286
Los Barrancos spring **285**, 286
Lower Rhine Graben 83, *84*

magnesium, Pateira de Fermentelos 228, 230, *231*
Manalana sub-catchment 10–11
  ERT survey 13, 18, 20
  hydrological monitoring 13
  land use 12–13
  longitudinal profile 14
  phreatic surface 10, 13, 14–20
  rainfall 12, 15–16
  soils 11, 19
  vegetation 12–13
manganese, Pateira de Fermentelos lagoon 232–233
Marandoo iron ore mine
  dewatering re-injection programme 114–123
  GDE 116–117, 123
  hydrogeology 115–117, 122–123
Marom Creek, streamflow 191
Marra Mamba Iron Formation 115, 116, *117*
Melaleuca rhaphiophylla 193, *194*, 216
Menindee, Barwon-Darling River *58*, 61, 62, 63
mining, Pilbara 113
  dewatering re-injection 114–123
MODFLOW groundwater flow simulation
  Alstonville Plateau basalt aquifer 191–192
  Yanchep National Park 37–45
    calibration and validation 40
    limitations 44
    Yanchep Caves 40–45
Monoman Sand Formation 254, *256*
Mungindi, Barwon-Darling River *58*, 60
Murray River system, stable isotope studies 60
Murray-Darling Basin 57, *58*, 59
  "Living Murray" project 254, 255, 256–265
Myriophyllum sp. 223

Narrabri Subsystem 59
Netherlands, GSI, temperature profiling 81–91
New South Wales
  GDEs, Alstonville Plateau 183–195
  groundwater management policy 185–186
nickel, Pateira de Fermentelos lagoon 232–233
nitrate
  Aakaer Stream catchment 107, *108*, 109–110
  groundwater quality standard 109–110
  hyporheic zone 126, 128–134
  Pateira de Fermentelos lagoon 228, 229, 232
  Ria Formosa lagoon 153–155, 158
  River Piddle *302*, 304
  River Spree 28–29
nitrate reduction capacities 97, 104
Nitrate Vulnerable Zones, Ria Formosa lagoon *154*, 155
nitrification, hyporheic zone 126, 128–134
nitrite, River Piddle *302*, 303, 304
nitrogen cycle, hyporheic zone 126, 128–132

ODEINT software 88
overland flow 101, *102*, 103–104
overpressure, Duna-Tisza Interfluve 271–272, 276
oxbow, River Spree 23–24, 25–30
oxygen
River Spree 28–29
stable isotopes
Barwon-Darling River 59, 60, 62, *63*, 64–65
Corangamite wetlands 73, *74*

Pampa Formation 139, **140**
Pannonian Basin, tectonics 269
Paraná Formation 139
Pateira de Fermentelos lagoon 221–234, *223*
geology *223*, *225*, *226*
hydrochemistry 226–234
major elements 227–229
minor and trace elements 230–233
monitoring 224–226
Principal Component Analysis 230–233
vegetation 223–224
Peel Block 83, *84*
Peel Boundary Fault Zone 83, *84*, 87, 90
Peelse Huis, temperature profile *84*, 86
Perales river valley, *Juncus acutus* 287–288
phosphate, River Spree 30–31
phosphorus, soluble reactive (SRP)
River Piddle 303, 304
River Spree 29
Phragmites australis 223, 243
Phragmites mauritianus 12
phreatic surface, Manalana sub-catchment 10, 13, 14–20
phreatophytes
Lake Huron beaches 243, *244*, 248–249
Netherlands 82
Spanish Central System 281–292
Swan Coastal Plain 197–205
Pilbara
mine dewatering 113
re-injection 114–115, 117–123
modelling 121–122
monitoring 119, *120–121*
rainfall 113, **114**, 119, 122
pine plantations, groundwater decline 39, 208, 213
Pinjar Borefield *199*, 200
Pippidinny Swamp 35
MODFLOW simulation 38
Pivka jama 49, *50*, 51, *53*
Pivka River 49
Planinska jama 49
platypus, Alstonville Plateau streams 193, *194*
Poole Formation **298**, 301
poplar 223
Populus sp. 223
Portsdown Chalk **298**, 300
Portugal
Pateira de Fermentelos lagoon, hydrochemistry 221–234
Querença-Silves Aquifer system 307–323
Ria Formosa lagoon, algal nutrient sources 153–165
Postojna-Planina Cave System 49–55, *50*
Postojnska jama 49, *50*, 51, *52*
Principal Component Analysis, Pateira de Fermentelos lagoon 230
Prunus pumila 242

Querença-Silves Aquifer system 307–323
discharge, transient outflow 318–320
GDEs 309, 315
geology 309–310
groundwater flow model 309, 313–315
GSI 308–309, 320–323
local-scale water transfer 320–323
rainfall 316
recharge 308, 313, 316, *317*
regional-scale water balance 316–317
streamflow 308, 310–313

radio frequency-electromagnetic surveys, Ria Formosa lagoon 156–157
Rafaela, groundwater requirement 138–139, 142
rainfall
Alstonville Plateau basalt aquifer 186, 189, *190*
Barwon-Darling River 61
Corangamite wetlands 71
Doñana aquifer system 170, 171, 173
Pilbara 113, **114**, 119, 122
Querença-Silves Aquifer system 316
Sand River catchment 10, 12, 15–16
Santa Fe province 140–141, 142–143, 146, 148–149
Swan Coastal Plain 208, *209*

rainforest, Alstonville Plateau 193, *194*
re-injection
  dewatering discharge 114–115, 117–123
    modelling 121–122
    monitoring 119, *120–121*
re-watering, Yanchep Caves 33–34
recharge
  allogenic, Querença-Silves Aquifer system 308, 313, 316, *317*, 322–323
  Alstonville Plateau basalt aquifer 186–187, 189, 191
  Santa Fe province 137–150
    estimation 142–147
  Yanchep National Park 39
    artificial 40–43, 45
red gum *see* river red gum
redox potential
  Pateira de Fermentelos lagoon 230–231
  River Spree 28–29
Ria Formosa lagoon 155
  geophysical study 159, *160*
  green algal blooms, nutrient sources 153–165
  groundwater and nutrient discharge 160–164
  nitrate balance 158
  Nitrate Vulnerable Zones *154*, 155
  radio frequency-electromagnetic surveys 156–157
  salinity 163–164
  water balance 157
riparian area aquifers 97, 98–105
riparian flow path type GSI *98*, 101–102, 103–104
  denitrification 104–105, 110–111
riparian hydrogeological type GSI 98–101, 103, 105, 111
  contact types 99, *100*
  groundwater body types 100–101
riparian zone, lagoons 221
River Murray, floodplain management 253–265
River Piddle, chalk catchment
  geology 296–297, **298**, **303**
  GSI 299–305
  hydrochemistry *301*, 302–304
  hydrogeology 300–302
  spring flow 300–302, 304–305
  streamflow 302
    depletion 304–305
river red gum 254, 263–265
River Spree
  geology 24
  groundwater flow 24, 25, 28
  GSI 25–31
    chemical data 24–25, 28–31
    clogging 28, 30
    infiltration/exfiltration 25, *27*
  water level 25, *26*, *27*, 28
Roer Valley Graben 83, *84*
runoff, Santa Fe province 145–146

Sabinota borehole 287–288
Salado River basin 138–139
  geology and hydrogeology 139–140
salinisation
  Barwon-Darling River 64–65
  Duna-Tisza Interfluve
    groundwater 272–275
    soil and wetlands 275–277
  and groundwater flow 267–268
  River Murray floodplains 253, 259–263
salinity, Corangamite wetlands 73–77
salt interception schemes, Bookpurnong 256, 257, 258
sand cherry 242
Sand River wetland
  degradation 10
  rehabilitation 10
  water table dynamics 9–21
  see also Manalana sub-catchment
Santa Fe province
  flora and fauna 142, *143*
  groundwater
    chemical analysis 142–143, **144**, **145**
    water levels 142–143, **144**, 145, 147–148
  groundwater dependent ecosystems (GDEs) 142
  groundwater recharge 137–150
    estimation 142–147
  rainfall 140–141, 142–143, 146, 148–149
  Salado River basin 138–139
    geology and hydrogeology 139–140
  soil 141
  water balance 141
Sauble Beach 239
  E. coli 245, *247*, 248
  groundwater 241, *242*, **243**
  wet ecosystem 243
Saucedoso stream **287**, 289
Scirpus holoschoenus 291

Scirpus validus 243
 see also bulrush
sedge 193, 217, 243
seepage
 Alstonville Plateau basalt aquifer 192–193, *194*
 fault-induced 82–91
  modelling 87–91
septic tanks
 E. coli groundwater contamination 246, *247*, 249
 nitrogen effluent 155, 158
sewage
 River Piddle 295, 304
 see also septic tanks; wastewater treatment
sharp rush *see Juncus acutus*
Slovenia, Postojna-Planina Cave System 49–55, *50*
sodium, Pateira de Fermentelos 228, 230, *231*
South Africa, Sand River, water table dynamics 9–21
Spain, Doñana aquifer system 169–180
Spanish Central System
 geology 282, *283*
 hydrochemistry 284, *285*, 286
 hydrogeology 282–283, *284*
 phreatophytes 281–292
spiny rush *see Juncus acutus*
spring flow, River Piddle 300–302
springs
 Alstonville Plateau basalt aquifer 193, *194*
 Pateira de Fermentelos lagoon 224, *225*, *226*, 227
 Spanish Central System **285**, 286
stable isotopes
 Barwon-Darling River 59–66
 Corangamite wetlands 73, *74*, 76–78
streamflow
 Alstonville Plateau 191
 Barwon-Darling River 61
 Gum Creek 191
 hyporheic zone 126–128
 Querença-Silves Aquifer system 308, 310–313
 River Piddle 300–302, 304–305
stygofauna, Yanchep Caves 34
sulphate
 Pateira de Fermentelos 228, 229, 230, *231*
 River Spree 30
sulphate reduction, Doñana aquifer system 177
sulphide, alkaline sulphide water 284, **285**, 286, 290–292
surface water *see* groundwater-surface water interaction; lagoons
Swan Coastal Plain
 Banksia woodland 197–205
  ecosystem dynamics 198–200
  response to drawdown 201–205
 rainfall 208, *209*
 wetland vegetation 207–218
  bushfires 213
  damplands 211, **212**, 213–214, 216–217
  lakes 211, **212**, 213–216
  monitoring 210–217
  sumplands 211, **212**, 213–214, 216–217
  water management implications 217–218
Szolnok aquifer 269

Tegelen Fault Zone 83, *84*
temperature profiling 81–91
 modelling 87–91
Tinto River 169, 170
toolbox *see* ecological water requirements, assessment toolbox
tourism, Doñana aquifer system 170
Transient Storage Model, hyporheic zone 126
transpiration 201, 203, *204*
 see also evapotranspiration
Tuart tree root mats, Yanchep Caves 34
Tweed Shield Volcano 188
Typha latifolia 243
Typha sp. 142, *143*

Ulva prolifera 153
Ulva rigida 153
Ulva rotundata 153
Upper Greensand **298**, 300, 302, **303**

vadophytes 199
vegetation
 Alstonville Plateau basalt aquifer 193, *194*
 Doñana aquifer system 169–170, 178–180
 dry beaches 242, *243*
 Manalana sub-catchment 12–13
 Pateira de Fermentelos lagoon 223–224
 Santa Fe province 142, *143*

vegetation (Continued)
Swan Coastal Plain
Banksia woodland 197–205
wetlands 210–217
wet beaches 243, 249
Venlo Block 83, *84*
viticulture, Pateira de Fermentelos lagoon 224
volcanism, Western Plains 69, 71

Wasaga Beach 238–239
groundwater 241, *242*, **243**
wet ecosystem 243
wastewater treatment 158–159, 222
water hyacinth 223–224
water milfoil 223
water table *see* phreatic surface
water violet 82
Weering Lake **72**, 78
wells
Doñana Coastal Aquifer 172
Pateira de Fermentelos lagoon 224, *225*, *226*
Spanish Central System 282
West Basin Lake **72**, 78
West Park Farm Formation 297, **298**, 301
Western Australia
mine dewatering 113–114
Swan Coastal Plain
Banksia woodland 197–205
wetland vegetation 207–218
Yanchep Caves 33–45
Western Plains, Corangamite wetlands 69–78, *70*
wetlands
Alstonville Plateau basalt aquifer 192
Corangamite, chemistry and stable isotopes 69–78
Doñana aquifer system *172*, 174, 175
role of groundwater 175–180
Sand River catchment, degradation 9–21
Swan Coastal Plain 207, *208*
response to groundwater decline 210–218
wijst fault-induced seepage 82, *84*
Wilgarup Lake *34*, 35
MODFLOW simulation 38
Wittenoom Formation 116
Woodland Beach 238
E. coli 245, 246, *247*, 248
groundwater *240*, 241, *242*, **243**
wet ecosystem 243, 244
wormwood 242

Yanchep Caves
artificial groundwater maintenance 33–45
groundwater flow 34, 35
modelling 37–44
groundwater levels 39
hydrogeology 35, *36*, 37
ion concentration 35, *37*
MODFLOW groundwater flow simulation 40–45
artificial recharge failure 43, **44**
long term water requirement 40–43
Yanchep National Park 33, *34*
groundwater flow modelling 37–45
recharge 39, 40–43, 45
Yonderup Lake *34*, 35
MODFLOW simulation 38, 45

Zig Zag Chalk **298**, 300
zinc
Pateira de Fermentelos lagoon 232
River Piddle *302*, 304
Zizaniopsis sp. 142, *143*

SERIES IAH-Selected Papers

Volume 1–4 Out of Print

5. Nitrates in Groundwater
   Edited by: Lidia Razowska-Jaworek and Andrzej Sadurski
   2005, ISBN Hb: 90-5809-664-5

6. Groundwater and Human Development
   Edited by: Emilia Bocanegra, Mario Hérnandez and Eduardo Usunoff
   2005, ISBN Hb: 0-415-36443-4

7. Groundwater Intensive Use
   Edited by: A. Sahuquillo, J. Capilla, L. Martinez-Cortina and X. Sánchez-Vila
   2005, ISBN Hb: 0-415-36444-2

8. Urban Groundwater – Meeting the Challenge
   Edited by: Ken F.W. Howard
   2007, ISBN Hb: 978-0-415-40745-8

9. Groundwater in Fractured Rocks
   Edited by: J. Krásný and John M. Sharp
   2007, ISBN Hb: 978-0-415-41442-5

10. Aquifer Systems Management: Darcy's Legacy in a World of impending Water Shortage
    Edited by: Laurence Chery and Ghislaine de Marsily
    2007, ISBN Hb: 978-0-415-44355-5

11. Groundwater Vulnerability Assessment and Mapping
    Edited by: Andrzej J. Witkowski, Andrzej Kowalczyk and Jaroslav Vrba
    2007, ISBN Hb: 978-0-415-44561-0

12. Groundwater Flow Understanding – From Local to Regional Scale
    Edited by: J. Joel Carrillo R. and M. Adrian Ortega G.
    2008, ISBN Hb: 978-0-415-43678-6

13. Applied Groundwater Studies in Africa
    Edited by: Segun Adelana, Alan MacDonald, Tamiru Alemayehu and Callist Tindimugaya
    2008, ISBN Hb: 978-0-415-45273-1

14. Advances in Subsurface Pollution of Porous Media: Indicators, Processes and Modelling
    Edited by: Lucila Candela, Iñaki Vadillo and Francisco Javier Elorza
    2008, ISBN Hb: 978-0-415-47690-4

15. Groundwater Governance in the Indo-Gangetic and Yellow River Basins – Realities and Challenges
Edited by: Aditi Mukherji, Karen G. Villholth, Bharat R. Sharma and Jinxia Wang
2009, ISBN Hb: 978-0-415-46580-9

16. Groundwater Response to Changing Climate
Edited by: Makoto Taniguchi and Ian P. Holman
2010, ISBN Hb: 978-0-415-54493-1

17. Groundwater Quality Sustainability
Edited by: Piotr Maloszewski, Stanisław Witczak and Grzegorz Malina
2013, ISBN Hb: 978-0-415-69841-2